Saint Martin and the Beggar by El Greco (1541–1614)

By permission of the National Gallery of Art, Washington, D. C., Mellon Collection. This picture embodies the legend from which the word "chaplain" is derived. See the account of the legend in the closing paragraphs of Chapter One of this volume.

THE HISTORY OF THE

CHAPLAIN CORPS, UNITED STATES NAVY

VOLUME ONE

1778 – 1939

Clifford Merrill Drury, 1897-
Captain, Chaplain Corps
United States Naval Reserve

NAVPERS 15807

TO
THE CHAPLAINS OF THE
UNITED STATES NAVY
WHO
HAVE BROUGHT THE BENEFITS OF RELIGION
TO NAVAL PERSONNEL ASHORE AND AFLOAT
IN ALL PARTS OF THE WORLD

★★ ★★ ★★

But he who fights against relentless foe
When silence reigns and cheer of eager van
Greets not his ear, but steadfast and alone
Drives back the host of sin, he is a man.

——Chaplain Henry Van Dyke, USNRF, 1918,
from a poem dedicated to chaplains.
(Chaplain Frazier's *Manual*)

ACKNOWLEDGMENTS

A work of this magnitude includes the contributions of many. Foremost among the Navy chaplains of yesterday who were diligent in gathering historical material pertaining to the history of the Corps was Roswell R. Hoes, who died 26 October 1921. His invaluable collection of pamphlets, old photographs, copies of official correspondence between chaplains and the Navy Department, and other material has been the primary source of information concerning the early history of the Navy chaplaincy. Chaplain David H. Tribou, a contemporary of Hoes, was also much interested in this history and carried on some research in that field. The personal recollections of Chaplains Hoes and Tribou provided additional valuable data for the history of the Chaplain Corps during the nineteenth century and the first part of the twentieth.

Most of the Hoes and Tribou material bearing on the history of the Navy chaplaincy passed into the possession of Chaplain W. W. Edel in 1921. Chaplain Edel has been keenly interested in the history of the Corps and continued the work of his predecessors. In 1944, while on a brief visit to the Naval Training Station, Sampson, New York, the author had access to the Hoes and Tribou collections. In 1946 Chaplain Edel turned these collections over, together with additions which he contributed, to the Chaplains Division[1] which in turn placed them on deposit in the Naval Records Collection, National Archives. Chaplain Clinton A. Neyman has also delved into the history of the Chaplain Corps and has freely made available the results of his study.

In the summer of 1944 Chaplain Edel, then at the Naval Training Station, Sampson, New York, initiated a project of having reproduced in oil the pictures of certain outstanding Navy chaplains. Clayton E. Braun, then seaman, second class, later photographer's mate, second class, painted fifteen portraits, each about 22″ by 27″, from pictures and photographs secured largely from the Hoes Collection. This splendid collection of paintings, beautifully framed, was presented to the office of Chief of Chaplains. Reproductions of fourteen of these paintings have been used in this volume and grateful acknowledgment is hereby given Photographer's Mate Braun and Chaplain Edel.

All pictures are from official Navy photographs unless otherwise noted.

A special word of appreciation is due Chaplain Robert D. Workman, who served as Chief of Chaplains when this project was initiated, and to Chaplain William N. Thomas who was Chief of Chaplains during its further progress and completion. Both chaplains in their official capacities rendered every assistance. Both have given the author the greatest possible freedom in deciding upon the contents and the manner of presentation. Any oversights, omissions, or misplaced emphases are his.

The author takes this opportunity to express his gratitude for the invaluable assistance rendered in the preparation of this volume by Commander A. L. Demaree, USNR, and Lieutenant Louise Kulka, W (C), USNR, whose services were made available to the Chaplains Division by the Office of Naval History for approximately eight months each beginning in October 1945.

Many others have contributed labor or talent to the preparation of this history including librarians, chaplains, line officers, several WAVE yeomen, civil service stenographers, photographers, Navy historians, and a number of civilians including wives of chaplains. The author hesitates to mention names lest some who assisted in gathering material, preparing copy, or correcting the manuscript or proof would be inadvertently overlooked.

To all who have been so generous in the contribution of material, of literary talent, or of advice, together with those who have rendered practical assistance with the pen and typewriter — a hearty "Thank You."

CLIFFORD M. DRURY

[1] From January 1946 to August 1948 the Chaplains Division was known as the Chaplains Activity.

TABLE OF CONTENTS

LIST OF ILLUSTRATIONS

ABBREVIATIONS USED

The following abbreviations have been used to indicate the location of source material consulted in the preparation of this work:

1. Nav. Rec. Coll.—Naval Records Collection, National Archives, Washington.

Most of the naval records prior to 1910 are in the Naval Records Collection and are serviced by the staff of the Naval Records and Library. This collection includes hundreds of original logs and muster rolls, maps, pictures, journals, files of classified documents, and several thousand volumes of original correspondence from and to the Navy Department dating back to 1790. These letters and other documents were classified into seven main groups, arranged chronologically, and bound at first by years and then by months. The assignment of Roman numerals to these groups is an arbitrary device of the author's to facilitate identification. The classification used is as follows:

- I. Letters from the Navy Department to Officers, Ships of War, (sometimes referred to as O. S. W.)
- II. Letters from Captains.
- III. Letters from Officers.
- IV. Miscellaneous Letters.
- V. Circulars and General Orders.
- VI. General Letter Book.
- VII. Commandant's Letters.

For example, the footnote: "Nav. Rec. Coll., IV: 2: 16" indicates letter number 16 of volume two of Miscellaneous Letters.

2. NRSO.—Naval Records Secretary's Office.

This is also a part of National Archives. Records go back to 1887 and in some cases earlier. This collection consists of items which were numbered and kept in files rather than being bound. The figures given in the footnotes, following the abbreviation NRSO, indicate the index number.

3. CoC—Chief of Chaplains' office.

Beginning with the appointment in 1917 of Chaplain J. B. Frazier, who acted as the first Chief of Navy Chaplains, a filing system was inaugurated which includes a "jacket" for each chaplain. This file often duplicates material carried in the official jacket kept by the Bureau of Naval Personnel. The file of the individual chaplain kept in the Chief's office contains correspondence and other material not found in the official jacket.

The Navy practice of referring to dates by day, month, and year (without commas) has been followed. The initial letters U. S. S. before the names of United States ships have been eliminated in the text.

4. R.S.—U. S. Revised Statutes.

INTRODUCTION

This is the first comprehensive and official history of the United States Navy chaplaincy ever prepared. Upon orders from the Secretary of the Navy, historians were selected to write the history of all phases of the Navy's activities during the recent war. A history of the Chaplain Corps was included in this overall program.

Chaplain Clifford M. Drury, who was on leave from San Francisco Theological Seminary where he occupies the chair of Church History, was called to Washington in March 1944 to work on this project. Chaplain R. D. Workman, then Chief of Chaplains, took a personal interest in the work and gave it much of his time. Chaplain Workman has continued his interest and assistance in the publication even though he is now retired. Until August 1945 Chaplain Drury served as District Chaplain of the Potomac River Naval Command and devoted only a portion of his time to the writing of this history. After VJ day he was able to give full time until his detachment from active duty on 1 July 1946. He returned to duty for three or four months during each of the summers of 1947 and 1948, and also in the early part of 1949, to complete the assignment.

While the immediate objective of the history-writing project of the Navy was focused on the recent war, the Chaplains Division felt that a study of the evolution of the Corps from its inception to World War II was also needed. Such a work was a logical and necessary prologue to the history of the Corps during the recent war.

This work, therefore, has been prepared with a three-fold purpose. In the first place, it presents a picture of the varied activities of chaplains from the days of the Continental Navy through World War II. Attention is given to the evolution of the Chaplain Corps itself. (The effectiveness of the chaplain's service was greatly increased when he no longer stood alone but became a part of a recognized Corps of the Navy, officially headed by a Chief of Chaplains.) Mention is made of the social and moral conditions under which naval personnel lived, because that which so vitally affects ethics and morality is logically a part of this study. Something of the religious life and activities of naval personnel on ships and stations where no chaplains have been on duty is included.

Secondly, this work has been written with the hope that it might be a guide and inspiration to the chaplains who compose the present Corps. The reasons for certain regulations, the significance of certain traditions, and the importance of the general policies that guide the activities of the individual chaplain as well as those of the Corps have their roots in the past. The lessons learned in the painful and costly experience of World War II must be recorded, while they are vivid in memory, for the benefit of the chaplains of tomorrow.

Finally, mindful of the necessity of having the support of the various denominations which have contributed members of their clergy to the naval chaplaincy, this history is presented to the public with the hope that it may be a factor in sustaining interest in the Navy Chaplain Corps. The Navy looks to the Naval Academy at Annapolis for many of its line officers. It looks to the churches of the United States for its chaplains. An intelligent and sympathetic appreciation of the opportunities and responsibilities of naval chaplains on the part of organized religion is necessary for the maintenance of a continuing supply of consecrated and well-qualified young men for this challenging field of religious endeavor.

The *History of the Chaplain Corps, United States Navy* will consist of two volumes of narrative and one volume containing the biographical and service-record sketches of 3,353 chaplains who served in the Continental and United States Navies down to the end of 1945. The latter volume was issued as a Government publication in 1948.[1]

This, the first volume of the narrative, presents the history of the Chaplain Corps from the days of the Continental Navy to 8 September 1939 when President Franklin D. Roosevelt declared the existence of a state of national emergency. Volume Two will continue the narrative from 8 September 1939 through World War II to 1949.

[1] For sale by Superintendent of Documents, U. S. Government Printing Office, Washington 25, D. C. Price $2.50.

Navy chaplains have long felt the need for a history of the Corps and it is with real joy and satisfaction that I now commend this work to the public. Nowhere outside of the armed services of the United States is it possible to find so many clergymen of so many different faiths working together in such close harmony for the spiritual welfare of so many. Here is a story that the church people of all denominations should read. And here is a guide and textbook for the chaplains still on active duty and for those who shall enter this service in the years to come.

William N. Thomas,
Rear Admiral, ChC., USN.

CHAPTER ONE

RELIGION AND THE SEA

From ancient times the sea has engendered in the hearts of those who have ventured forth upon the vastness of its waters feelings of fear, awe, and reverence. The mariner of yesterday who dared sail his frail craft out upon the uncharted wilderness of the sea had good reason to be afraid. Beyond the horizon was the mystery of the great unknown. Contrary winds, violent storms, the great loneliness of the sea, and the mythical leviathans of the deep, all contributed to his fear. Recognizing the risks he ran, the sailor turned in awe and sometimes in superstition to his deity for protection and guidance.

In these modern times the sailor has risen above many of his fears of the past. Most of the waters have been charted. With high-powered engines, he can run contrary to the wind and even defy the storm. Through the magic of radio and radar he can keep in touch with land. Weather reports forecast the wind, sea, and sky. Although science has largely replaced superstition, it has created new fears to take the place of the old. New and more terrible leviathans lurk in the deep. Death falls from the sky. War has always been an awful thing, and strong men still know what it is to be afraid.

An editorial in the *Army and Navy Chronicle* of 21 September 1837 developed the theme that "sea life has ever been thought congenial with religious feeling." The editor quoted an old Latin proverb from Horace, "Qui nescit orare, discat navigare," which, freely translated, is "Whoever would learn how to pray, let him become a navigator." The mariner, both ancient and modern, is by necessity an astronomer. In addition to serving as indispensable guides, the stars have from ancient times inspired man to meditate upon the wonders of creation. This is especially true of the sailor who walks the deck of a ship on a clear night. About him is the moving, pulsating surface of the ocean stretching out as far as the eye can reach. Above him is the infinity of heavenly spaces whose blackness is broken by the stars and the ever-changing moon. There is indeed an affinity between religion and the sea.

A Navy chaplain, Sydney Key Evans, comments on the inspiration the sea lends to religious thought:

> The Navy chaplain has particular helps in his work. The *very ocean itself* comes to his aid. In its calmer aspects it suggests the serenity, mystery, immensity and irresistible energy of God, and in its wilder moods His power, majesty, and awfulness. In all moods it is a constant reminder of man's littleness and dependence and has ever been one of his chief stimuli to resourcefulness, courage, heroism, and self-sacrifice.[1]

The sensitivity of the sailor to the material factors of his surroundings, especially in wartime, makes him one of the most religious of men. These religious impulses and serious meditations have not always been well directed. The religion of the seaman has often degenerated into a curious collection of salty superstitions. The vicissitudes of seafaring life have made it easy for the weak to indulge in vice. The Christian church has long recognized these facts and in various ways has sought to provide a spiritual leadership for the Navy's floating parishes. In modern days this endeavor has found expression in the chaplaincy.

The affinity between religion and the sea is older than Christianity. Fragmentary records of ancient days tell of religious ceremonies in connection with the laying of the keel, the placing of the mast, and the launching of the vessel. Even to this day a clergyman is usually called upon to take part in such ceremonies as the launching and commissioning of vessels. The early Latin mariners, when they ventured out upon the deep, carried with them images of their gods, called *pupi* (doll images). Sometimes an altar was built on the after-deck, and here their idols were placed. This deck, often elevated in the ancient craft, came to be known as the "poop-deck."[2] Carved images of patron saints were fastened to the prow, and these figureheads remained on our sailing vessels until the advent of steam.

[1] Evans, S. K., "Notes by a Navy Chaplain," *The Army Chaplain, April* 1931, p. 5.

[2] Skerrett, "The Baptism of Ships," *Naval Institute Proceedings,* August 1909, p. 545.

There has always been a close kinship between Christianity and the sea. Among the twelve disciples were several sailors from the Sea of Galilee. Jesus sometimes used a boat as a pulpit. The early church fathers referred to the church as the Ship of Salvation and compared it to Noah's Ark. The vaulted ceiling of a church was likened to an inverted keel of a ship and the main portion of the church, apart from the chancel, was called the nave from the Latin *navis,* ship. Among the most popular of the early Christian symbols was the fish, the Greek name for which formed an anagram whose letters summarized an early Christian creed.[3] Figures of speech from the sea occur frequently in Christian liturgy and hymnology. "Jesus Savior, Pilot Me," and "Jesus Calls Us O'er the Tumult" are but two examples.

Chaplains accompanied most of the exploring expeditions sent to the new world. When Francis Drake sailed around the world in his *Golden Hinde* in 1578-1580, Chaplain Francis Fletcher, a clergyman of the Church of England, was aboard. Chaplain Fletcher conducted the first Protestant service in the English language in what is now continental United States. The service was held in June 1579 on a beach located in what is known today as Drake's Bay, Marin County, California, about fifty miles north of San Francisco.

Other English sovereigns, following the precedent set by Queen Elizabeth, appointed chaplains to serve on the larger vessels of the British Navy. By the time of Charles I the custom seems to have been well established. Samuel Pepys, of diary fame, occupying a position in the British Government equivalent to the First Lord of the British Admiralty, was active in 1676 in urging captains of His Majesty's ships to take a "godly divine" with them. His diary shows his concern about the type of men selected. A recent authority on Pepys writes:

> It was laid down that henceforth the Secretary of the Admiralty Office was to notify the Church authorities of every ship ordered to sea that there might be a chaplain appointed to each, properly equipped with the King's warrant. . . .[4]

The origin of the title "chaplain" goes back to an old legend of Saint Martin of Tours (ca. 316-400). Martin, a soldier, with a group of well-dressed companions, met an importunate beggar shivering in his rags on a bitterly cold winter night at one of the city gates of Amiens. The others passed by heedless of the cries for alms. Martin, touched with compassion, opened his purse but found it empty. He drew his sword and with it divided his heavy cloak with its ample folds. One half he gave to the beggar and kept the other for himself. That night Martin in a dream saw Christ clad in that half cloak and was so moved by the vision that he sought baptism. Soon afterwards he abandoned his military career and devoted himself to the church. In time he became the patron saint of France, and his cloak was kept as a sacred relic by the French kings and was often carried with them into battle. The word "chaplain" is derived from the French word "Chapelain," originally the officer appointed to watch over the sacred cloak (medieval Latin, *cappa* or *capella).*[5]

While the skeptic may question the authenticity of the legend, yet one can accept the spirit of mercy and compassion exemplified by Saint Martin as being in full harmony with the spirit of Christ. Today it is assumed that this attitude of brotherly love is one of the inherent characteristics of all chaplains.

During the past century and a half chaplains of the United States Navy have shared the hardships and rewards that come to other naval personnel. They have carried on their ministry with these seagoing men, whose very sensitiveness to religion has been to the advantage of the chaplain, on sea, on shore, at home, in foreign lands, in peace, and in war. The chronicle of the activities of these padres of the sea begins with the days of the Continental Navy.

[3] The creed translated from the Greek reads: Jesus Christ, Son of God, Savior.

[4] Bryant, *Samuel Pepys,* II: 185.

[5] "Chaplain," *Encyclopaedia Britannica.*

CHAPTER TWO

CHAPLAINS IN THE CONTINENTAL NAVY

1775–1785

The history of the chaplaincy of the United States Navy necessarily parallels the history of the Navy itself. While this work is designed primarily to recount what Navy chaplains did, the conditions under which they lived and worked, and the results they accomplished, it will also be necessary to trace enough of the background of naval history to give perspective to the picture. Since ships and shore stations are the parishes of Navy chaplains, an elementary knowledge of the names and types of ships as well as some idea of the history of shore bases is essential to an understanding of the story of the Chaplain Corps.

The Continental Navy antedates the United States Navy. On 13 October 1775 a timid Congress by a margin of one vote authorized the building of two swift sailing vessels, one with fourteen guns and the other with ten. A Marine Committee of three members, later enlarged to seven, was appointed to conduct naval affairs.[1] After authorizing ships and an embryonic Navy Department, Congress directed its attention to the regulations which should govern the new Navy. Faced with the necessity of producing a set of rules immediately, it was natural that Congress should turn to the rules of the Mother Country for a model. Thus it happened that from the very beginning of the Continental Navy due consideration was given to Divine Services and to the place of chaplains on the larger vessels, since this was an inherent part of the British system.

The second article of the Navy regulations adopted 28 November 1775 read:

> The Commanders of the ships of the thirteen United Colonies, are to take care that divine service be performed twice a day on board, and a sermon preached on Sundays, unless bad weather or other extraordinary accidents prevent.

Although the chaplain is not mentioned in this article, the reference to a sermon implies that Congress intended that there should be an ordained clergyman on board.

The third article designed to protect the morals of those aboard ship was as follows:

> If any shall be heard to swear, curse, or blaspheme the name of God, the Commander is strictly enjoined to punish them for every offence, by causing them to wear a wooden collar, or some other shameful badge of distinction, for so long time as he shall judge proper. If he be a commissioned officer, he shall forfeit one shilling for each offence, and a warrant or inferior officer six pence. He who is guilty of drunkenness, if a seaman, shall be put in irons until he is sober, but if an officer, he shall forfeit two days pay.[2]

The first mention of a chaplain in the *Journals of the Continental Congress* refers to his share in the distribution of prize money. On 6 January 1776 Congress passed a resolution which included the provision:

> That the Commander in Chief have one twentieth of the said allotted prize money. . .
>
> That the lieutenants of marines, surgeons, chaplains, pursers, boatswains, gunners, carpenters, the masters' mates, and the secretary of the fleet, share together, and have two twentieth parts and one half of one twentieth part divided amongst them, equally of all prizes taken when they are in company.[3]

On 15 November 1776, more than a year after the first vessels had been authorized, Congress fixed the pay of officers in the Navy. The base pay of the chaplain was $20 a month, and, for purposes of comparison, the surgeon received $25. Officers in the pay grade which included the Navy chaplain, however, were to receive $4 a week subsistence when in domestic ports and when they were unable to live aboard their ships. This rate of pay compared favorably with that allowed Army officers. As early as 29 July 1775 Congress had fixed the pay of Army chaplains at $20 a month. This was raised to $33.33 on 16 January 1776.

[1] Alden and Westcott, *The United States Navy*, p. 10.

[2] *Journals of The Continental Congress*, III:378.

[3] *Ibid.*, IV:36.

832159—49——2

The article in Navy regulations calling for the performance of Divine Service twice daily and the delivery of a sermon on Sunday, together with the two previously mentioned references to a chaplain in the *Journals*, are the only known instances of official action by the Continental Congress regarding the Navy chaplaincy. These suffice, however, to show that Congress was aware of the spiritual needs of naval personnel and that it intended that chaplains be provided to ships whenever possible.

References to the presence and work of chaplains in the Continental Navy are few. The Library of Congress has a collection of John Paul Jones letters, two of which reveal Jones seeking chaplains in France for his ships *Ranger* and *Bon Homme Richard*. The first letter, written from Passy, France, on 12 July 1778, was directed to H[enry] Grand in Paris and reads:

In the selection of a Chaplain the following qualifications are deemed requisite "I could wish him to be a man of reading and of letters who understands, speaks and writes the french & english with elegance and propriety: For political reasons it would be well if he were a clergyman of the protestant profession whose sanctity of manners and happy natural principles would diffuse unanimity and cheerfulness thro' the ship. And if to these essentials are added the talent of writing fast and in fair characters, such a man would necessarily be worthy the highest confidence, & might, therefore, assure himself of my esteem and friendship; he should always have a place at my table, the regulation whereof would be entirely under his direction."[4]

It seems evident that Jones was looking for a man who could serve not only as the ship's chaplain but also as his private secretary. A few weeks later Jones was given command of the *Bon Homme Richard*. After some difficulty he assembled a crew of about 330, including only thirty Americans. Most of the crew were "raw French peasants."[5] Writing from L'Orient, on 30 April 1779 to his friend "Father John" Mehegan at Brest, the chaplain and secretary to Admiral Count d'Orvilliers, Jones said:

Having a number of French under my command, I am in want of a Chaplain. You know whom I would prefer if they are disengaged.[6]

Undoubtedly Jones was requesting a chaplain of the Catholic faith. There is no evidence of his securing either a Protestant chaplain for the *Ranger* or a Catholic for the *Bon Homme Richard*.

[4] Jones Collection, Letter No. 6783.
[5] Sands, *Life and Correspondence of John Paul Jones*, p. 156.
[6] Jones Colection, Letter No. 7073. See also *American Catholic Historical Researches*, p. 221.

Another reference to a chaplain is found in the following letter from the Marine Committee to Seth Harding, Captain of the Continental frigate *Confederacy*, regarding an unidentified chaplain who had gotten into trouble:

May 26th, 1779

Sir

I am ordered by the Marine Committee to desire that you will send up the Chaplin [sic] of your Ship to this place under a Guard so as to be brought before the Committee on friday evening next at six OClock, I am Sir

Yr. hble sert.
John Brown, Secy[7]

The nature of the offense or the outcome of the trial is not disclosed.

The first chaplain known to have served in the Continental Navy was the Reverend Benjamin Balch, a Harvard graduate and a Congregational minister, whose father had served in a similar capacity in His Majesty's expedition of 1745 in "King George's War" against Cape Breton. Benjamin Balch fought in the Battle of Lexington as one of the Minute Men. Later he served as an Army chaplain in the siege of Boston. On 28 October 1778 Balch reported aboard the frigate *Boston* under the command of Samuel Tucker. The muster roll, which carries the names of 287 men, indicates that the chaplain was paid at the rate of ninety shillings a month.[8] Some time after the capture of the *Boston* by the British in 1780 at Charleston, South Carolina, Balch began his service on the *Alliance*.

By 1781 the Continental Navy had only two frigates, the *Alliance* and the *Deane*, each of which carried thirty-two guns. When Balch served as chaplain on the *Alliance* from October 1780 to June 1781, the Anglophobic Irishman, John Barry, was her Captain.[9] During this time, the ship saw considerable action.

Chaplain Balch had with him two of his young sons, Thomas and Benjamin, who, because of their youth, were entered on the ship's payroll as drawing the pay of one man. The Chaplain's part in the *Alliance's* capture of two British vessels off Halifax follows:

The peril the ship was in brought out the desperate courage of every man on board the *Alliance*, the 'cloth'

[7] Paullin, (ed.), *Out-Letters of the Continental Marine Committee and Board of Admiralty August 1776–September 1780*, II:78.
[8] Sheppard, *The Life of Samuel Tucker*, p. 344. The original Journal of Samuel Tucker is in the Essex Institute, Salem, Mass.
[9] Clark, *Gallant John Barry*, pp. 193, 227, 239.

being no exception. Reverend Benjamin, armed cap-a-pie, was seen in the midst of the fray, and thereafter is said to have become known on the ship as the "fighting parson." His son, Thomas, was also in the fight, and when father and son met afterwards, it was with an embrace and with the words, "Thank God, my son."[10]

Benjamin Balch had twelve children, one of whom, William, was the first chaplain known to receive a commission in the United States Navy after the present Navy Department was established in 1798.

Soon after Chaplain Balch left the *Alliance* in October 1781, Captain Barry appointed James Geagan, a Navy surgeon, to serve as chaplain.[11] In those days many clergymen also practiced medicine and while it is possible that Geagan was ordained, the probability is that he was not. Since Barry was a devout Catholic, it has been assumed by some that Geagan was an Irish Catholic priest. Geagan served as chaplain or acting chaplain for about seven months. When the billet of ship's surgeon became vacant in July 1782, he was appointed to that position. In spite of his friendship for Captain Barry, he with a number of other officers left the *Alliance* at a French port sometime before 12 December 1782, because they had not been paid.[12]

Some of the colonies had naval forces of their own which may have included chaplains. A record of the Virginia Assembly for November 1781 shows that this body was then reducing its naval force and that "every person of the naval staff establishment, including the commissioner of the navy, chaplains" and others were being dismissed.[13]

Possibly chaplains served aboard some of the larger privateers, but of this there is no record. Congress encouraged these vessels to prey upon British shipping. Approximately 70,000 men embarked on such cruises and accounted for 600 British ships.[14]

The Continental Navy played a minor role in the Revolutionary War. Some ships authorized were never built or otherwise acquired. Others that were put into commission never saw action. However, the combined forces of the Continental Navy, the state navies, the privateers, and the invaluable assistance of the French Navy undoubtedly hastened the successful conclusion of the struggle. When the war ended the Continental Navy consisted of only four ships. By 1785 even these had been sold. Roughly speaking, the history of the Continental Navy spanned only ten years, from 1775 to 1785.

The records of the services rendered by Navy chaplains in the Revolutionary War are fragmentary. Benjamin Balch and James Geagen are the only chaplains who are known to have served in the Continental Navy. More important than the services of individual chaplains was the adoption by the Continental Congress of regulations which provided a place for religion and for chaplains in the Navy. These regulations were expanded in the rules which governed the new Navy of 1798.

[10] *Danvers Historical Collection,* VII:89-91.
[11] Clark, *op. cit.,* pp. 239, 247, 269.
[12] Griffin, *Commodore John Barry,* pp. 166, 188–203.
[13] Stewart, *The History of Virginia's Navy of the Revolution,* p. 120. A note in the Edel Collection gives the following information without stating source: "Land Grants to Chaplains in the Navy of Virginia in Revolution. John Braidfoote, Chaplain in State Navy of Virginia, received grant of 6,000 acres."
[14] Alden and Westcott, *The United States Navy,* p. 12; Sprout, H. and M., *The Rise of American Naval Power,* p. 11.

CHAPTER THREE

THE BEGINNINGS OF THE UNITED STATES NAVAL CHAPLAINCY

1789–1800

For thirteen years, 1785 to 1798, there was no American Navy and, therefore, no Navy chaplains. The Constitution adopted in 1789 invested Congress with the power "to provide and maintain a navy." The Congressional Act of 1789 establishing the War Department gave the Secretary of War jurisdiction over both the naval and military forces. Since the Navy Department was not established until 30 April 1798, both branches of our armed forces were under one administrative head for about nine years. Although several ships were authorized in the latter part of this period, no naval vessels actually put to sea under the authority of the Secretary of War.

The United States Navy was born out of a troubled international situation. As early as 1785 American ships had been seized by Algerian pirates and their crews held for ransom. Repeated outrages perpetrated by the Algerian corsairs on American ships and sailors finally aroused Congress to pass on 27 March 1794 "An Act to provide a Naval Armament." This Act called for the building of six frigates, four of forty-four guns and two of thirty-six. Joshua Humphreys of Philadelphia, who was selected to build one of the frigates, originated a new design for his vessel, the *United States,* which was so acceptable that it was used for the other frigates as well. In brief, his plan called for vessels which were longer, broader, and lower in the water than similar vessels of the British Navy. These fast-sailing fighting ships were the best of their time.[1]

There were then three categories of ships considered sufficiently large to warrant chaplains. First there were the three-masted square-rigged vessels which carried up to twenty-four guns, mounted on a single deck. These were called sloops. While they sometimes had a chaplain, this proved the exception rather than the rule. Then there were the frigates, which were larger than sloops, although similarly rigged. These carried from twenty-eight to forty-four guns on the spar deck above and the gun deck below. It was not until 1815 that the United States acquired its first ship of the line or line-of-battle ship. This type mounted seventy-four or more guns on two and even three decks.[2] These three types can be compared to the light cruisers, heavy cruisers, and battleships of today.

Although the Act of 1794 authorized the building of six frigates, the keels of only three were immediately laid. They were the *United States,* the *Constitution,* and the *Constellation,* honored names in the history of the United States Navy. Construction was halted in 1795 when a humiliating treaty was signed with Algiers. For about three years the bare skeleton, keel and ribs of ships, lay in their ways exposed to the elements.

Even though the United States had concluded treaties with Morocco and Algiers, none had been made with Tunis and Tripoli, which were also focal points of piracy and extortion for Mediterranean shipping. However, before anything serious happened in that area, American relations with France took an unexpected turn for the worse. The Jay Treaty signed with Great Britain in 1794 was viewed by the French as a violation of the Franco-American Treaty of 1778. In consequence the French began preying upon American shipping. On 21 June 1797 the Secretary of State reported that the French had seized thirty-two American ships in the previous nine months.[3] This situation was intolerable, and public sentiment rapidly crystallized in favor of aggressive defensive action.

Work on the unfinished frigates was resumed. On 7 September 1797 the *Constellation* was launched. On 21 October the *Constitution* slid down the ways. An ever darkening international horizon caused Congress to authorize the acquisition of twelve additional vessels, each of twenty-two guns, on 27 April 1798. Three days later Congress established the Navy Department and President Adams selected Benjamin

[1] Alden and Westcott, *The United States Navy,* p. 43.

[2] Sprout, H. and M., *The Rise of American Naval Power,* pp.42–3.

[3] Goldsborough, *Naval Chronicle,* I: 78.

U.S.S. CONSTITUTION

Stoddert to be its first secretary.[4] On 10 May 1798 the *United States* was launched. Continued French depredations on American shipping finally aroused Congress to vigorous action. On 28 May hostile action against armed French vessels hovering off the coast of the United States was authorized, and on 7 July Congress declared that the treaties with France were no longer in effect. Thus began the Quasi War with France which lasted until 30 Spetember 1800, a little more than two years. During this time, a fleet of about fifty vessels was sent to sea, only nine of which were heavy frigates.

The *Constellation*,[5] the first of the frigates authorized in 1794, sailed from Hampton Roads under the command of Captain Thomas Truxtun on 23 June 1798. On 13 July the *United States* under command of Captain John Barry, and on 20 July the *Constitution* under Captain Samuel Nicholson put to sea.

In 1799 the following additional frigates were added: *Chesapeake, Philadelphia, Congress, Insurgente, Essex, John Adams, Adams, George Washington, Boston,* and the *General Greene*.[6] The *President* and the *New York* were not added to the fleet until 1800. The *Insurgente* was captured from the French in 1799, but she foundered at sea the following year. As will be shown, chaplains were to serve on most of these vessels.

When peace was declared in the fall of 1800, public sentiment quickly demanded a curtailment of the naval establishment. A new administration under President Jefferson, bent on economizing, was elected in the fall of 1800. Secretary of Navy Stoddert, realizing that the new administration would be anti-navy, hastily acquired sites for Navy Yards at Portsmouth (New Hampshire), Boston, New York, Philadelphia, Washington, and Norfolk. These sites acquired during the closing weeks of a defeated administration laid the foundation for a permanent shore establishment of the Navy.

THE FIRST UNITED STATES NAVY CHAPLAINS

The second section of the Act of 1794 stated that there shall be employed on board each of the said ships of forty-four guns . . . one chaplain." His pay was to be $40 a month plus two rations a day.[7] When the President was authorized to equip and man the frigates *United States, Constellation,* and *Constitution* by the Act of 1 July 1797, a chaplain was detailed to each vessel with the same rate of pay as had been previously set.

According to the Secretary of the Navy's report of 24 December 1798 a chaplain was authorized on frigates of forty-four, thirty-six, and thirty-two guns, having a complement of 400, 340, and 260 men respectively.[8]

WILLIAM BALCH

First chaplain known to be commissioned, United States Navy, 1799–1801. From a painting by Clayton Braun.

Who had the honor of being the first chaplain in the newly established Navy in 1798? There is conclusive evidence that William Balch, a son of Benjamin Balch, was commissioned a chaplain in the United States Navy on 30 October 1799. On that date the Secretary of the Navy wrote to Captain James Sever of the *Congress*:

> Sir, I enclose you a commission for William Balch as Chaplain and John Mushainey as Boatswain for the Con-

[4] Technically, the first Secretary of the Navy was George Cabot of Massachusetts who declined the nomination after legally holding the office for fourteen days. Mayo, *Your Navy*, p. 17.

[5] Ships of the same class did not always carry the same number of guns.

[6] *Nav. Doc. Quasi War*, IV:484.

[7] *Gen. Reg. of Navy & Marine Corps*, p. 233.

[8] *Nav. Doc. Quasi War*, II:116–7.

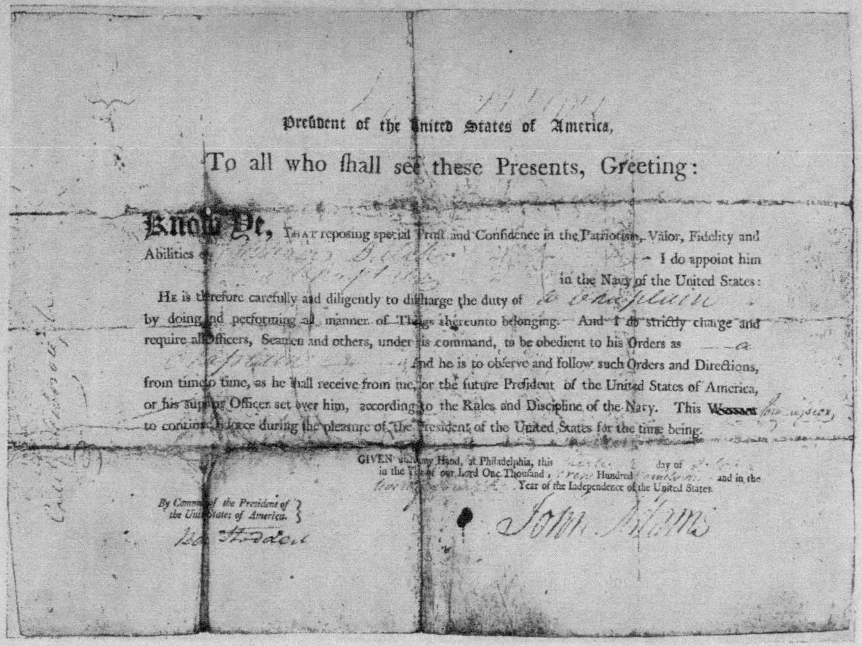

President of the United States of America,

To all who shall see these Presents, Greeting:

Know Ye, That reposing special Trust and Confidence in the Patriotism, Valor, Fidelity and Abilities of — I do appoint him in the Navy of the United States:

He is therefore carefully and diligently to discharge the duty of Chaplain by doing and performing all manner of Things thereunto belonging. And I do strictly charge and require all Officers, Seamen and others, under his command, to be obedient to his Orders as Chaplain And he is to observe and follow such Orders and Directions, from time to time, as he shall receive from me, or the future President of the United States of America, or his superior Officer set over him, according to the Rules and Discipline of the Navy. This Commission to continue in force during the pleasure of the President of the United States for the time being.

GIVEN under my Hand, at Philadelphia, this day of in the Year of our Lord One Thousand, Hundred and in the Year of the Independence of the United States.

By Command of the President of the United States of America.

John Adams

Courtesy Massachusetts Historical Society.

Reproduction of original commission granted Chaplain William Balch, 13 October 1799, and signed by President John Adams.

gress. You will require these Gentlemen to take the enclosed oath and return them to this office.[9]

This, the first commission known to have been given a United States naval chaplain, reads:

JOHN ADAMS President of the United States of America,

To all who shall see these Presents, Greeting:

KNOW YE, That reposing special Trust and Confidence in the Patriotism, Valor, Fidelity and Abilities of William Balch, I do appoint him a Chaplain in the Navy of the United States: He is therefore carefully and diligently to discharge the duty of a Chaplain by doing and performing all manner of Things thereunto belonging. And I do strictly charge and require all Officers, Seamen and others, under his command, to be obedient to his Orders as a Chaplain. And he is to observe and follow such Orders and Directions, from time to time, as he shall receive from me, or the future President of the United States of America, or his superior Officer set over him, according to the Rules and Discipline of the Navy. This Commission to continue in force during the pleasure of the President of the United States for the time being.

Given under my Hand, at Philadelphia, this Thirtieth day of October in the Year of our Lord One Thousand Seven Hundred Ninety nine and in the twenty fourth Year of the Independence of the United States.

John Adams

By Command of the President of
the United States of America
Be. Stoddert

A copy of the oath used by the Navy Department in 1805, thought to be the same as that used in 1799, follows:

I .... appointed ....
do solemnly swear to bear true allegiance to the United States of America, and to serve them honestly and faithfully against all their enemies or opposers whomsoever; and to observe and obey the orders of the President of the United States of America, and the orders of the officers appointed over me, and in all things to conform myself to the rules and regulations which now are or hereafter may be directed, and to the articles of war which may be enacted by Congress, for the better government of the navy of the United States, and that I will support the constitution of the United States.

Balch served on two ships. His duty aboard the *Congress* was cut short when that vessel was dismasted in a storm early in 1800. As Captain Barron of the *Chesapeake* needed a chaplain, Balch was transferred to that vessel on 1 May 1800 and served there a little more than a year.

While William Balch may have been the first commissioned chaplain to serve in the United States Navy, there is clear evidence that William Austin was serving as a chaplain, without a commission, aboard the *Constitution* nearly a year before Balch reported for duty on the *Congress*. The *Constitution* first sailed from Boston on 20 July 1798 under the command of Captain Nicholson who was relieved less than a year later by Captain Silas Talbot. The first name on the ship's muster roll which Captain Talbot signed in July 1799 is that of "William Austin—Chaplain."[10] According to this record, Austin had been paid to 31 May 1799, a fact that indicates he was serving on board prior to that date.

On 27 August 1800 Captain Talbot wrote the Secretary of the Navy stating:

I have always understood that Mr. Austin (a young Gentleman that has acted as Chaplain for about two years past) meant to leave the ship at her return from her late cruise, that his object in coming into the service was principally to acquire a little property to enable him to support himself a few years in the Commencement of the Study of the Law which he means to engage in; he is now absent and I cannot obtain immediately his answer on the Subject of Continuing in service. But I am almost Confident that he will decline. He has not had any warrant.[11]

Other sources indicate that Austin took passage from Cape Francois for New York aboard the American sloop *Chase* on 7 July 1800, which may have been the time he left the *Constitution*. Talbot intimated that Austin had been serving as chaplain for about two years and this places the date of his reporting sometime in the summer of 1798. However, Talbot himself did not take command of the *Constitution* until the summer of 1799.

On 17 May 1801 Austin wrote to the Secretary of the Navy on behalf of Captain Nicholson, who was then being criticized for his conduct when in command of the *Constitution*. Austin wrote:

I was a chaplain on board the frigate Constitution eighteen months, with Captain Nicholson, six, with Captain Talbot, twelve. My situation with both was similar. Captain Nicholson's politeness made me his companion; with him I lived, at his table I constantly sat, his cabin was my residence, and from it I seldom retired until his usual time of rest.[12]

Austin disproved the charge of intoxication which had been made against Captain Nicholson and signed his letter "Late Chaplain on board the frigate *Constitution*." This letter indicates that Austin was chaplain during the last six months of Captain Nicholson's command and therefore must have begun

[9] Nav. Rec. Coll.; I:3:10.

[10] Nav. Rec. Coll.
[11] *Ibid.*, Area 7.
[12] *Ibid.*, IV:2:89.

his duties about 1 January 1799, or ten months before William Balch was commissioned.

The first discovered reference to "Divine Service" on any United States ship-of-war is that found in the log of the *Constitution* for Sunday, 13 January 1799, which may have been Austin's first Sunday aboard.[13] An earlier reference to "prayers" is found in the Journal[14] kept by a James Pity (or Pettis) who was a member of the crew of the *Constitution.* On Sunday 12 August 1798 he noted: "a 11 oc A.M. call'd all hands aft to attend prayers. Doctr. Blake stood Chaplain." The muster roll lists a Charles Blake as surgeon's mate. Undoubtedly, for want of a regularly appointed chaplain, the ship's doctor read prayers. Pity referred to "prayers" on various Sundays but on 13 January changed his terminology from "prayers" to "Divine Service," thereby agreeing with a similar notation in the ship's log for Sunday, 13 January 1799 "At 10 call'd all hands aft and performed Divine Service."

After that date references to Divine Services appear with regularity both in the Journal kept by Pity and in the ship's log.

Since the practice of using unordained men as chaplains was common in the early days of the Navy, it cannot be assumed that Austin was a clergyman. There is no information regarding his education. Talbot stated that he served without "warrant" of commission. Thus Austin was appointed by the captain of the vessel on which he served and drew chaplain's pay. However, since he performed some if not all the duties of a chaplain, including conducting Divine Services, and was so listed in the official records of the ship, it appears that the distinction of being the first chaplain the United States Navy goes to him.

A section of the Act approved by Congress on 2 March 1799 contained the phrase: "Commanders of the ships of the United States, having on board chaplains." The plural use of the word "chaplain" suggests that Austin was not the only chaplain on duty at that date. No chaplain other than Austin, however, is known to have been in service before William Balch.

NAVY REGULATIONS GOVERNING CHAPLAINS

According to Captain Talbot's letter quoted previously, one of the primary reasons Austin entered the Navy as a chaplain was "to acquire a little property." This is reminiscent of the regulations of the Continental Navy which provided that chaplains receive their proportionate share of prize money. This section was included with but little change in the Act of 2 March 1799. Austin was not disappointed in his financial aspirations for he shared in the proceeds of the sale of those vessels captured by the *Constitution* during its cuise from July 1799 to August 1800.[15]

The Act of 1799 was susperseded by the Act of 23 April 1800, which contained many of the former provisions, including the ratio for the distribution of prize money. The possibility of gaining quick and easy wealth through the sale of captured ships was a powerful incentive in recruiting for both privateers and the Navy.

The provision governing Divine Service in the Act of 1799 was taken from the Navy regulations of 1775 with the addition of the words, "having on board chaplains." This implies that Divine Services were to be held only on ships which carried chaplains. Further revisions are found in the Act of 1800. The section, with the additions shown in italics, then read:

> The commanders of all ships and vessels in the navy, having chaplains on board, shall take care that divine services be performed *in a solemn, orderly, and reverent manner* twice a day, and a sermon preached on Sunday, unless bad weather, or other extraordinary accidents prevent it; *and that they cause all, or as many of the ship's company as can be spared from duty, to attend at every performance of the worship of Almighty God.*

The inclusion of the words: "in a solemn, orderly, and reverent manner" suggests that Divine Services had not always been so conducted. Judged by present day custom, compulsory attendance of all available hands at Divine Service twice a day as well as attendance at a sermon on Sunday seems rather stringent. No provision was made for members of religious minorities who might then have been in the Navy.

The newly-born Navy was without precedent or regulation regarding the selection and commissioning of chaplains. Sometimes captains chose their own chaplains as is indicated in the following letter from the Secretary of the Navy to Captain Samuel Barron dated 28 February 1800:

> To complete the officering of the Chesapeake, You require

[13] Log Book of the United States Frigate *Constitution*, 6 Dec. 1798–20 Oct. 1800. In the Naval History Collection, New York Historical Society.

[14] Nav. Rec. Coll.

[15] *Nav. Doc. Quasi War*, VI:283.

One Chaplain or School Master
Two Surgeons Mates
One Boatswain
One Gunner
One Sailmaker & one Carpenter—The first & four last of which you will be pleased to select proper characters for, and appoint. I will provide the Surgeons mates in due season perhaps the Gentlemen you mention will be one of them.[16]

Here the captain was given the authority to select his own "Chaplain or School Master," a right which was exercised by captains of naval vessels for more than twenty-five years.

The smaller frigates and ships-of-war were allowed schoolmasters instead of chaplains.[17] The schoolmaster received only $30 a month and two rations a day while the chaplain was allowed $40 and the rations. Many men who served as chaplains on the larger vessels might more appropriately have been listed schoolmasters, but because of the extra $10 a month were, by the indulgence of their captains, listed as chaplains. Occasionally even the lighter frigates carried chaplains when a strict interpretation of the law would have given them only schoolmasters.

This practice of permitting captains to choose chaplains from their ships' companies often resulted in unworthy men being given this high office. Most, if not all, of the men appointed in this way were not ordained. On the other hand, the few chaplains commissioned by the Navy Department during these early years were, with rare exceptions, either licensed or ordained clergymen.

The following letters, typical of others on file in the Naval Records Collection, reveal the mind of the general public regarding the chaplaincy:

Mr. Secretary,

I beg leave to recommend Mr. Luther Emerson of Maine as a suitable person for a purser or chaplain to some one of the Frigates or ships of war of the United States. He is a young man of liberal education & fair & unexceptional character.

I am, sir, . . .
Silas Lee[18]

In another letter the writer makes his own application.

Trenton N. Jersey July 9 1801

Honord Sir

From recent information I have understood their [sic] is a vessel belonging to the United States arived, which vessel is to be retained in the service of government. Thinking it probable their may be some vacancies in her or some of our vessels retained in service, I wish to get a berth as Purser, Clerk, or Chaplain. . . .

I am a young man who has been rather unfortunate, having had my foot lamed by Rheumatism, but not so as to be detrimental to me in the performance of this duty. I am a Orphan. I have nothing to depend on, for my father died while I was young. He had filled various offices in the State, he was a member of Congress in the year 1785. I had always the character of an Upright, honest man.

Geoe I. Houston.[19]

Nothing is said about the spiritual qualifications of these would-be chaplains. The applicants were often indifferent as to which one of several positions they obtained. If they could not secure a berth as a clerk or purser, then frequently a chaplaincy would do. This loose system of selecting chaplains led to many abuses before an attempt was made by the Navy Department about 1823 to correct it.

ADDITIONAL CHAPLAINS IN 1800

Even though the Secretary of the Navy on 24 December 1798 had recommended to Congress that a chaplain be placed on every frigate of thirty-two guns or more, yet there is record of but two men—William Austin and William Balch—who were made naval chaplains under this law before 1800.

Five more chaplains reported for duty in 1800. The first was Eli Vallette who began his fifteen months' tour of duty aboard the *Philadelphia* on 26 February 1800. Chaplain Vallette took with him on this cruise his ten-year-old son, Eli, who later became a rear admiral in the United States Navy. On 18 December 1850, this son, then Captain E. A. F. Lavallette, wrote to the Secretary of the Navy referring to his early experience. "I was rated Captain's clerk," he wrote, "my father performing both duties of Chaplain and clerk, and being large for my age, I performed occasionally midshipmen's duty."[20]

Matthew Flannery, who like Austin served without (a) commission, began his tour of duty on 12 May aboard the *Constellation*. Nothing more is known of the service rendered by Chaplains Vallette and Flannery.

On or about 12 July Robert Thompson (Thomson) reported for duty on the *President*. The *President's* log[21] bears such entries as these:

Sunday, 14 Sep. 1800 At 11 Called all hands to prayer.
Sunday 21 Sep. At 10 Called all hands to Prayers
Sunday 12 Oct. At 10 A.M. Called all hands to muster
½ past 11 all hands to Prayers

[16] Nav. Rec. Coll. I:3:218–9.
[17] *Nav. Doc. Quasi War*, VI:272.
[18] Nav. Rec. Coll., IV:2:16.
[19] *Ibid.*, IV:2:137.
[20] *Ibid.*, II:1850:302. The son changed his surname in 1840 by Act of the Pennsylvania legislature. A destroyer, the *Lavallette*, named in his honor, saw distinguished service in World War II in the South Pacific.
[21] Photostat in Naval Archives.

The entry for Sunday, 2 November 1800, contains the first reference to a chaplain discovered in a ship's log. It is as follows: "At 11 Divine Service was performed by the Chaplain."

Almost from the beginning of his career as a Navy chaplain, Thompson exhibited his proficiency in teaching navigation and mathematics. On 5 November 1800 Captain Thomas Truxtun of the *President* requested him to draw a large map of the *President's* "cruising track" for the benefit of the midshipmen.[22] On 14 November Captain Truxtun again addressed the Chaplain.

> Sir Since you have been on board the US ship *President* under my Command, every part of your conduct that has come within my immediate Notice has been modest, respectful, and decent, and your attention to the Young Gentlemen (Midshipmen) in teaching them Navigation, has met my most hearty approbation So much so, that I sincerely regret, particularly on their Account, your return home, but as a misunderstanding Subsists between you and the officers of the Gun, room, that I have not been able to reconcile, I could not do less than comply with your request to return to America, where I wish you Safe and a happy sight of your friends.[23]

The fourth chaplain to enter the service in 1800 was Noadiah Morris, a ship's clerk, who served as chaplain in the *Constitution* relieving Austin during the summer of 1800. The muster roll of this vessel listed him as "Acting Chaplain" on 5 November 1800.

The fifth and last was Samuel Chandler, who saw duty first on the sloop *Patapso,* which was launched in June 1799.[24] Aspiring to larger opportunities, he applied for the position of chaplain on the *United States* in a letter to Captain Barry dated 21 April 1800.

As has been stated, Captain Barry was a devout Catholic and it is possible that most aspirants to the naval chaplaincy preferred to serve under a Protestant skipper. Samuel Chandler was not one of these. In his letter of application he wrote:

> Brought up in the Church of England, I often visit the Catholic Church and am always pleased with the devout and becoming attention observed in them. I consider the different forms of religion only so many different roads to the same final happy home.[25]

[22] *Nav. Doc. Quasi War,* VI:521.

[23] *Ibid.,* VI:530–1.

[24] This vessel was first known as the *Chesapeake,* but when a frigate which was launched 2 Dec. 1799 was given that name, the sloop's name was changed. It is not probable that Chandler served as a chaplain on so small a vessel. He possibly served as a schoolmaster.

[25] Clark, *Gallant John Barry,* p. 466; Griffin, *Commodore John Barry,* p. 405.

Chandler also added, that if he could not be made chaplain, he would be glad to accept the position of schoolmaster. There is no evidence that he was ordained. He is reported to have been a graduate of Cambridge University and was probably well qualified to serve in the latter capacity. The muster roll of the *United States* listed him as chaplain on 31 December 1800, but it is possible that his service began before that date.[26]

Some midshipmen aboard the *United States,* one of them a son of Benjamin Stoddert, the Secretary of the Navy, "took up their studies under Chaplain Chandler in the schoolroom on the lower deck."[27] Chaplains Thompson and Chandler are the first of a long list of Navy chaplains to serve as schoolmasters for midshipmen.

In recapitulation, chaplains were assigned to the following seven frigates during all or part of 1799 and 1800, the years of the Quasi War with France.

Constitution
William Austin, 1 Jan. 1799 (?)–7 July 1800.
Noadiah Morris, 5 Nov. 1800–Mar. 1801.

Congress
William Balch, 30 Oct. 1799–1 May 1800.

Chesapeake
William Balch, 1 May 1800–1 May 1801.

Philadelphia
Eli Vallette, 26 Feb. 1800–15 June 1801.

Constellation
Matthew Flannery, 12 May 1800–15 June 1801.

President
Robert Thompson, 12 July 1800–30 Apr. 1801.

United States
Samuel Chandler, 31 Dec. 1800–.

Only two other heavy frigates, the *New York* and the *Insurgente,* were in service during any part of this period and these are not known to have had chaplains.

The number of United States Navy chaplains on duty up to 31 December 1800 totaled only seven and their combined known length of service was only about five years. Yet these were highly important years in the history of the naval chaplaincy, for during this time regulations were adopted, principles established, and traditions started which foreshadowed and moulded the subsequent history of the Chaplain Corps.

[26] Pay rolls and muster rolls cover certain stated periods only and cannot be accepted as showing the beginning and closing dates of a man's service unless so declared.

[27] Clark, *op. cit.,* p. 472.

Altar in the Commodore's main cabin, U.S.S. *Constitution*. Picture taken during cruise between July 1931 and June 1934.

CHAPTER FOUR

THE FIRST DECADE OF THE NINETEENTH CENTURY

1801-1810

The fortunes of the Chaplain Corps of the United States Navy were at a low ebb in the years 1801-1810. Indeed, at one time during the winter of 1806-1807, only one chaplain, Robert Thompson, was on duty. When the war with France ended in September 1800, the Government took steps at once to reduce its naval establishment to a peacetime basis. Naturally this reduction affected the chaplaincy.

Only two of the six chaplains on duty 1 January 1801 were retained. They were Chaplains Thompson and Chandler whose ability as schoolmasters may have accounted for their retention. Chaplain Balch resigned after receiving the following letter from the Secretary of the Navy dated 4 May 1801:

> The Rev'd William Balch
>
> The arrangements of Congress relative to the Navy have imposed on me the unpleasant task of permitting you to retire from the service. You will therefore consider your term of Service as expiring on the 10th instd.[1]

Chaplain Matthew Flannery was also released in May; Chaplain Eli Vallette in June. Chaplain Noadiah Morris served until February 1801, and while his connection with the Navy during the interval is not known, his name reappears in 1803 as a chaplain on the muster roll of the *Constitution.*

On 3 March 1801 Congress passed the Peace Establishment Act,[2] which suspended work on vessels being built, on dry-docks, and on shore establishments. All but eight heavy and six light frigates were to be sold, and only six of these were to be kept in commission. Although this Act authorized the retention of as many chaplains as there were ships in service which rated chaplains, yet, so far as is known, only two chaplains were on duty at any one time during the latter part of 1801 and the years 1802 and 1803.

[1] Nav. Rec. Coll., I:4:519.

[2] Another act of a similar nature passed 21 April 1806 is also known by the same title.

CHAPLAINS IN THE BARBARY WAR

In spite of the peaceful desires of the United States and the economizing policy of the Jefferson administration which called for a reduction of the naval establishment, continued depredations against American commerce by the Mediterranean Powers demanded that suitable action be taken. Consequently, President Jefferson dispatched to the Mediterranean in May 1801, a squadron of four vessels—the *President, Philadelphia, Essex,* and *Enterprise*—under the command of Commodore Richard Dale. Chaplain Thompson, aboard the *President,* was the only chaplain known to have sailed with the squadron. In a letter dated 11 May 1801, Captain Truxtun had recommended the Chaplain to Dale by writing: "Thompson you will find useful to the midshipmen as a tutor, in all the branches of Mathematical Science, connected with their profession."[3]

In February the Bashaw of Tripoli had repudiated his 1796 treaty with the United States, which he felt gave the rulers of Algiers and Tunis proportionately more tribute than he received. On 14 May he declared war by cutting down the pole bearing the American flag at the United States Consulate at Tripoli. The squadron on reaching Gibralter learned of the existence of a state of war.

When news of the crisis in the Mediterranean reached Washington, the administration was obliged to reverse its naval policy. All of the larger available war vessels were made ready for service and sent to the Mediterranean. The *Boston* sailed in October 1801, under command of the eccentric Captain Daniel McNeill.

On the evening of 19 January 1802 while the *Boston* lay at anchor at Toulon, Captain McNeill invited Chaplain Thompson of the *President* and three French officers aboard for dinner and then set sail for Tunis before his guests could depart. Commodore Dale, in command of the squadron, was in-

[3] *Nav. Doc. Barbary Wars,* I:449.

dignant. The incident came as a climax to a long series of grievances, and the Commodore laid them all before the Secretary of the Navy in a letter dated 24 January 1802. One of the charges was that McNeill had taken "Mr. Thompson the Parson belonging to the President, of[f] to sea with him . . ."[4] Dale requested an investigation.

The outcome of this incident, in which the only chaplain in the squadron was either deliberately or inadvertently "shanghaied," is not recorded. It is assumed that Thompson managed to get back to the *President* before she returned to the United States in May 1802. Thompson made a second trip to the Mediterranean during this war, while serving on the *President* from May 1804 to September 1805.

At the time war with Tripoli began, enlistment in the Navy was on a yearly basis. Therefore, the squadron which sailed from Hampton Roads in May 1801 was forced to return its crew within a year. The Government found it necessary to increase the enlistment period to two years. A second squadron consisting of the following vessels sailed for the Mediterranean in the spring of 1802: *Chesapeake, Adams, New York, John Adams, Constellation,* and *Enterprise*. The command was given to Commodore Richard V. Morris who after a year was relieved by Commodore Edward Preble.

On 15 February 1802 Chaplain Alexander McFarlan, an Episcopalian, the first from that denomination to serve as a naval chaplain, reported to the *Chesapeake*. At that time Samuel Chandler was chaplain of the vessel, having served in her since 1 June of the previous year. Chandler was notified by the Secretary of the Navy in a letter of 9 April that there was no longer any occasion for his services.[5] Several weeks earlier, the Secretary had written to the commodore in charge of the assembling squadron:

> This will be presented to you by Mr. McFarlane the Clergyman whom I mentioned to you as a proper person to be the Chaplain and Schoolmaster of the Chesapeak. He has been represented to me by Gentlemen of Respectability in the most favorable manner. His appointment in form will be sent to you for him.[6]

In those days the procurement procedure for naval officers was simple. Little emphasis was placed on physical requirements; if a man appeared healthy, he was accepted. Often the appointed officers were aboard their vessels weeks and sometimes months before being commissioned. On 2 March 1802 the Secretary of the Navy sent McFarlan's commission to the Commodore with the request that he administer the oath of allegiance and then deliver the commission. McFarlan saw service in the Mediterranean aboard the *Chesapeake,* and there is evidence that he was paid up to 7 June 1803. During the full period of his service, he and Thompson, as far as is known, were the only chaplains in the Navy.

Aboard the *Constitution* of the third Mediterranean Squadron was Chaplain Noadiah Morris, who had previously served on that vessel. It is evident that Commodore Preble had requested his services, for the Secretary of the Navy, Robert Smith, replying on 5 July 1803, stated: "I have no objection to you taking with you in the Constitution . . . Chaplain Morris."[7] Morris was listed until 10 December as "Actg Chaplain." He had started as a captain's clerk and had served at least twice as a chaplain before ending his career as a purser. The experience of Morris is a good example of the manner in which many shifted from one billet to another. There was no sanctity attached to the office of chaplain and special preparation for this work was not a requirement. Any one of the ship's company who could qualify as a schoolmaster might do. His interest in religion was incidental.

When Morris took over the duties of purser, Commodore Preble apointed Peter Leonard, a clerk, to succeed him as chaplain. Preble's letter to Leonard dated 1 December 1803 throws considerable light upon the way such appointments were then made:

> I do hereby appoint you Chaplain of the United States Ship Constitution under my command, willing and requiring you forthwith to take upon you the employment of Chaplain in her accordingly, and to be obedient to such commands as you shall from time to time receive from me or any other your superior officer. To hold the same employment until the pleasure of the President of the United States shall be known—Together with such allowance of wages, victuals and perquisites as are usual for the Chaplain of the said ship; and for so doing this shall be your warrant.[8]

The Congressional Resolution of 3 March 1805 thanking the officers of the United States Squadron under the command of Commodore Edward Preble for the attacks on Tripoli in July–September 1804 included the name "P. Leonard, Chaplain."

Chaplain Leonard resigned on 14 March 1805. The Secretary of the Navy replied on 15 March:

[4] *Ibid.*, II:28.
[5] Nav. Rec. Coll., I:5:384.
[6] *Ibid.*, I:5:335. Letter of 15 Feb. 1802.
[7] *Ibid.*, I:6:195.
[8] Edel Collection, from Hoes's notes which read: "Copy of original in Storeroom of Bureau of Construction and Repairs." See also *Nav. Doc. Barbary Wars*, III:251, 258.

I have this day received your Letter of yesterday's date tendering the resignation of your appointment as Chaplain. From the report made to me by Commodore Preble in relation to your Conduct, your resignation is a Subject of regret to me. I can not however refuse to accept it. I tender you my best wishes for your future prosperity.

Rt. Smith[9]

CHAPLAINS WITHOUT WARRANT OR COMMISSION

During 1804 and 1805, eight additional names appear in the Navy records as chaplains. Only one of these eight, Robert Dennison, who served for more than a year and a half aboard the *Constellation* and the *President*, is listed in any official register of Navy officers. The other seven were appointees from the ship's company. They were Schnetter, Budd, Cerdona, Cruize, Lyde, Ormsby, and Ravara, most of whom served for only a few months. At least two of the seven, Ormsby and Lyde, had the same experience as Noadiah Morris and were first captains' clerks, then chaplains, and finally pursers. Budd, after a few months as chaplain on the *Essex* and *Constellation*, became a midshipman and later a lieutenant in the Navy. These men saw duty on at least five of the frigates belonging to the Mediterranean Squadron.

The information regarding these men chosen from the ship's company to act as chaplains is scanty. The muster rolls do not give complete information relative to their length of service. There is no reason to believe that any one of this number was ordained.

On 25 January 1802 President Jefferson issued a new edition of *Naval Regulations* which contained the following outline of the duties of a chaplain:

1. He is to read prayers at stated periods; perform all funeral ceremonies over such persons as may die in the service, in the vessel to which he belongs; or, if directed by the commanding officer, over any person that may die in any other public vessel.
2. He shall perform the duty of a school-master; and to that end he shall instruct the midshipmen and volunteers, in writing, arithmetic and navigation, and in whatsoever may contribute to render them proficients. He is likewise to teach the other youths of the ship, according to such orders as he shall receive from the captain. He is to be diligent in his office, and such as are idle must be represented to the captain, who shall take due notice thereof.[10]

Section two reveals a surprising parallel with Articles II, III, and IV of the section governing the duties of the schoolmaster in the regulations of the British Navy. The British Articles read:

II.

He is to employ his Time on board in *instructing the Volunteers in Writing, Arithmetic,* and the Study of *Navigation, and in whatsoever may contribute to render them* Artists in that Science.

III.

He is likewise to teach the other Youths of the Ship according to such Orders as he shall receive from the Captain, and with Regard to their several Capacities, whether in Reading, Writing, or otherwise.

IV.

He is to be early every Morning at the Place of Teaching, and to represent the Names of *such as are idle,* or averse to Learning, to the Commander, in order to his taking Course for their Correction.[11]

The words in italics were incorporated in the revised *United States Naval Regulations* almost verbatim. Section two of the revised regulations concerning the chaplain contained three sentences, each of which summarized the thought of the corresponding section in the British provisions.

Conducting Divine Worship, preaching, and compulsory attendance are not mentioned in the revised regulations. The chaplain's duties were confined to the "reading of prayers at stated periods," the performing of funeral services, and teaching. The influence of these regulations dominated the selection of naval chaplains for at least fifteen years.

THE YEARS FOLLOWING THE BARBARY WAR

The war with Tripoli, which began in the spring of 1801, dragged on through four years. Peace was finally concluded on 10 June 1805. The terms of the treaty called for the discontinuance of all tribute payments on the part of the United States.

Another Peace Establishment Act was passed on 21 April 1806. The number of war vessels to be retained in service was to be determined by the President. "Ordinary seamen and boys" in the service were limited to 925. About a year later this number was increased by 500. Among the allotment of officers were thirteen captains and 150 midshipmen. As for chaplains, the "President may appoint, for the vessels in actual service, so many . . . chaplains . . . as may, in his opinion, be necessary and proper."[12]

In November 1807, the Secretary of the Navy stated that the United States had only two frigates

[9] Nav. Rec. Coll., I:6:534.

[10] *Op. cit.*, p. 18.

[11] *Regulations and Instructions relating to His Majesty's Service at Sea*, London, 1734, p. 136, quoted by Soley, *Historical Sketch of the United States Naval Academy*, p. 9.

[12] *Navy Register 1805–6*, p. 6; *Gen. Reg. of Navy and Marine Corps*, p. 280–1.

and four smaller cruising craft in service.[13] The actual personnel then totaled 191 officers and 1,954 enlisted men.[14]

As the number of naval personnel aproached zero, so likewise did the number of naval chaplains. At the end of 1806 there was but one commissioned chaplain in the service, Robert Thompson.

In March 1807, when relations with England were critical, William Petty was commissioned and ordered to the *Chesapeake*. In addition to his chaplain's duties, he acted as secretary to Captain Stephen Decatur until the Secretary of the Navy, having discovered Petty to be an English alien, advised Decatur in a letter of 16 July that:

> Since the appointment of Mr. Petty I have understood that he was not an American but an Englishman, and although I have no particular reason to distrust his integrity, yet under existing circumstances it would subject me to just censure if I were to continue him in the station he now holds on board of the Chesapeak. As your Secretary his duties must immediately cease, and you will be pleased to select and appoint a suitable person in his place, to whom you may allow the pay, & rations of a Chaplain in the Navy, and if he should wish it, and you will inform me of his name, I will send to you for him a commission as Chaplain.
>
> You may allow Mr. Petty to go on shore, with orders to report himself to me from time to time.[15]

So Petty was put on the beach. A note from Decatur to the Secretary of the Navy dated 12 February 1808 indicates that Petty had not always lived up to the standard of life expected of a chaplain. Decatur wrote:

> Mr. Petty the Gentleman who formerly acted as my Secretary, and Mr. Crump a Midshipman did a few days past disgrace themselves by boxing while on shore at Norfolk. . . . Mr. Petty has been frequently intoxicated.[16]

Perhaps as a result of this communication, Petty's commission was revoked on 18 February of that year. Decatur did not select a successor for nearly a year and then chose Edward W. Turner.[17]

Before the decade closed, four other men received commissions as chaplains. They were William Robinson, William Briscoe, Garret Barry, and Andrew Moore. Their religious affiliations were not recorded and little is known about these men.[18] It appears that Robinson, who was ordered in April 1809 to the *President*, then at the Washington Navy Yard, got into difficulties with the civil authorities. When deputies of the Circuit Court of the District of Columbia attempted to serve papers on him, Robinson resisted arrest and wounded one of the officers of the law. The Marshal of the District of Columbia called this matter to the attention of the Secretary of the Navy who promptly dismissed the offending and belligerent Chaplain.[19]

THE FIRST ACADEMY FOR MIDSHIPMEN

The outstanding feature of the history of Navy chaplains during the first decade of the last century is the story of Chaplain Robert Thompson's academy for midshipmen which he conducted at the Washington Navy Yard and aboard several frigates. Little evidence can be found to support the following sweeping criticism of chaplains as teachers which appears in a history of the United States Naval Academy.

> The system of giving instruction on board of cruising-ships, imperfect at the best, was barren of results where the chaplains were the teachers. The duty had no necessary connection with their profession, and the subjects they were compelled to teach were wholly foreign to their training and studies. They were appointed without any reference to their fitness for this work. . . .[20]

On the contrary, chaplains selected during the decade following the adoption of the regulations of 1802 were chosen more for their teaching ability than for their experience or training as ministers. Only one of the six chaplains in service on 1 January 1801 is known to have been ordained. Likewise only one chaplain among the twenty additional names which appear in the records in the decade following is known to have been ordained. These two were William Balch and Alexander McFarlan. The educational activities of chaplains were indeed not "barren of results." When the restrictions under which they labored, the inadequate teaching aids, the broken, irregular schedules while at sea, and even the illiteracy of some of the midshipmen are considered, there is reason to marvel that so much was accomplished.

After his first cruise to the Mediterranean, Chaplain Thompson was stationed at the Navy Yard in Washington where he taught midshipmen who were quartered aboard the *Congress*. This school was a

[13] Sprout, H. and M., *The Rise of American Naval Power*, p. 62.
[14] List in Nav. Rec. Coll.
[15] Nav. Rec. Coll., 1:7:445.
[16] *Ibid.*, II:1:50.
[17] The muster roll of the *Chesapeake* shows that Turner served from July 1808–14 May 1809 as chaplain. He may have been aboard before July 1808, however.
[18] The first *Navy Register* which indicates the religious affiliation of chaplains is a MS. Register for 1864.
[19] Nav. Rec. Coll., V:38, letter dated 30 May 1809.
[20] Soley, *Historical Sketch*, p. 9.

favorite project of Robert Smith, then Secretary of the Navy. The frigate remained "in ordinary," *i.e.*, in reduced commission, at the Yard from 1805 to 1812.

Indicative of Smith's interest in the school is his letter of 24 January 1804 to Chaplain Thompson:

> You will report to me the names of the Midshipmen who are at this time receiving instruction from you at this place, the time when they severally commenced their studies under you; the days they have each attended, and your opinion of the progress they have respectively made.[21]

Following his second cruise to the Mediterranean on the *President*, Thompson resumed his work as the one-man faculty for the training of midshipmen at Washington. On 23 July 1806 he received the following order from the Secretary of the Navy: "You will remain at the Seat of Government for the purpose of attending to the education of such officers of the Navy as may from time to time apply to you for instruction in Mathematics or in Navigation." [22]

That the Navy Department endeavored to provide suitable training for its midshipmen at this early date is evident from the following extracts from Thompson's correspondence.

On 13 January 1807 he wrote to the Secretary of the Navy:

> Sir, I take the liberty of awakening your recollection of my wish to have an appointment as Naval Mathematician and to superintend an Academy in this City which is to be entirely under your direction. There are some very useful officers on board the ships who have a number of children and find it difficult to have them educated. If you would use your influence to have the academy established upon principles most congenial to your own mind I would be willing to instruct any person belonging to the Navy that you might think proper to send and I have some reason to think it would meet the approbation of Congress. The reason the Bill which you presented Congress last session miscarried was because it was generally believed that there was an additional officer to be appointed instead of altering the title of one already appointed.[23]

Although this letter refers to the establishment of an academy for the "Children" of naval officers, it is evident from his other correspondence that he refers only to boys and young men.

A letter dated 9 March 1807 from the Secretary of the Navy to Captain Thomas Tingey, in charge of the Washington Navy Yard, reflects the Secretary's interest in the Chaplain's school. Smith wrote:

> Mr. Thompson, a Chaplain in the navy stationed here for the purpose of instructing the young gentlemen of the navy in the theory of navigation, states that he has been removed from the accommodations provided for him on board the Congress, and that he has now no publick place in which to perform the duties required of him. Whence has this arisen? I cannot consent that the object for which Mr. Thompson was stationed here shall be thus defeated. Adequate and comfortable accommodations must be provided for him on board one of the Ships in ordinary, immediately,—& I specially charge you with the execution of this order.[24]

On 26 March Thompson wrote the Secretary requesting books and instruments which he considered "absolutely necessary for the use of the department . . . at the navy-yard."[25]

On 27 April the Secretary sent the following directions to the Chaplain:

> You will consult with Capt. Gordon, and while the Chesapeak shall be fitting out, attend at your room of instruction, at such hours from sun rise to sun set and to 10 o'clock at night as Captain Gordon shall tell you the service will most conveniently admit.[26]

Thompson replied on 27 April giving details regarding his hours of instruction. He added: "I would wish also that you would mention to Captain Gordon that it is your intention to have them examined before they go to sea."[27] On 4 May Thompson was ordered to the frigate *Chesapeake* at Hampton Roads "for the purpose of instructing the young midshipmen attached to her in the Theory of Navigation."[28] On 27 July Thompson wrote to the Secretary urging the establishment of a naval academy for the "instructing of young officers of the Navy in the Theory of Navigation." He pointed out that should such an academy be founded at Washington midshipmen would be able to acquire there the theory of navigation "in less time and with less expense than (perhaps) they could in any other seminary on the continent, besides having the benefit of a most excellent mathematical apparatus, and seeing all the different parts of a ship, made by some of the best workmen in the United States."

Thompson's argument in favor of a three months' training course follows:

> My reason for saying they can acquire it in less time is, there will be no other branches taught in the seminary (unless ordered by yourself) they can also live cheaper, in each frigate there are eight state rooms, which accommodate the same number of officers; they could live there, and form a mess of their own, and draw their rations,

[21] Nav. Rec. Coll., I:6:284.
[22] *Ibid.*, I:7:193.
[23] *Ibid.*, III:1 of 1807:29.

[24] *Ibid.*, I:7:330.
[25] *Ibid.*, III:1 of 1807:149.
[26] *Ibid.*, I:7:373.
[27] *Ibid.*, III:1 of 1807:198.
[28] *Ibid.*, I:7:381.

832159—49——3

which would amount to six dollars per month, and this is less than they could procure their board in any respectable boarding house for one week, and their morals would be less subject to be corrupted, as they would be almost placed under your own eye. My experience, as a public teacher of navigation in the U.S. navy for upwards of seven years, under some of its most distinguished commanders, has enabled me to form some idea of the time sufficient for any young gentlemen of common abilities to acquire a sufficient knowledge of navigation to carry a ship to any known part of the globe, which I think they may do in three months, providing they are acquainted with arithmetic; the whole expense then would amount to only eighteen dollars. I am fully of opinion that the young gentleman who canont acquire a competent knowledge of navigation in that time is unworthy of a midshipman warrant in the U.S. navy.[29]

Thompson's classification displeased him since in reality he was a naval mathematician. He wrote the Secretary on 4 November.

Sir, I am once more under the necessity to solicit your influence of having my appointment as a Chaplain in the navy exchanged to that of naval mathematician; for nearly the space of eight years I have performed the duty of all the Chaplains that have been appointed since that time together with my own without receiving any additional recompense; the Chaplains (or rather Captains' clerks) who have been appointed since received the pay and emoluments of Chaplains whilst I have performed their duty; the name of Chaplain has become so unpleasant to me that I wish it exchanged, altho that exchange may not be attended with any additional salary.[30]

Since Thompson was still referred to as chaplain in 1810, it appears that his request was not granted. Chaplain Thompson's service extended throughout the first decade of the nineteenth century, during which time his principal duty was teaching. He was the first chaplain who looked upon the Navy as a career. He died in office in the late summer or early fall of 1810. The roots of the splendid Naval Academy at Annapolis are to be traced to the years 1804–1810 and to Thompson's classes aboard the *Congress* as it lay tied up to the dock at the Washington Navy Yard.

Since the Secretary was deeply interested in the academy for midshipmen, it was natural that he should have been concerned about Thompson's successor. On 26 October 1810 the Secretary wrote to Commodore John Rodgers:

Mr. Thompson, late Chaplain in the Navy, being dead, I wish to procure another Gentleman of suitable qualifications to supply his place in the Navy. Do you know of any person qualified for the station filled and the duties performed by Mr. Thompson? If you do, let me know his name and I will appoint him—if not, make enquiry for one.[31]

Andrew Hunter, 1811–1823

The one-man faculty of the "academy" for midshipmen at Washington Navy Yard. From a painting by Clayton Braun.

Among those who applied for this position was the Reverend Andrew Hunter, a Presbyterian clergyman, who had served as an Army chaplain in the Revolutionary War. Hunter directed his letter to William Helms, a member of Congress, who forwarded the application to the Secretary of the Navy. The Secretary replied on 21 February, saying in part:

If Doct. Hunter will engage to perform these duties, he shall receive the appointment, provided there should be no delay in informing me of his determination.—This is important—Several other applications have been made.[32]

Hunter evidently signified his willingness to undertake these duties even though he was fifty-nine years old. He continued his duties as the one-man faculty of the academy for midshipmen until his death in February 1823, a period of twelve years.

[29] *Ibid.*, III:2 of 1807:170. No professors of mathematics were listed in *Navy Registers* until 1835.
[30] *Ibid.*, III:3 of 1807:119.
[31] *Ibid.*, I:9:214.
[32] *Ibid.*, VI:10:301.

Mention should be made of the school for midshipmen which was conducted by Captain Bainbridge during the twenty months' imprisonment of the *Philadelphia's* crew at Tripoli. Bainbridge was able to regain from his captors by bribery some of the textbooks which had been aboard his ship. Lieutenant David Porter, who was among the prisoners, taught the midshipmen gunnery and navigation.[33]

SOCIAL AND MORAL CONDITIONS

Chaplains, by virtue of the implications inherent in their calling, are concerned with those social and moral conditions which affect spiritual values.

In the early nineteenth century, the Christian churches of America were gradually becoming articulate on social issues. Individual ministers were speaking out against such evils as dueling, slavery, and intemperance. Since certain customs and physical conditions under which Navy men of that day lived involved distinct danger to moral and spiritual values, it is altogether probable that some of the naval chaplains had stirrings of conscience on these same matters. However, no copies of their sermons or other pronouncements before 1810 have survived the vicissitudes of the years.

Among the practices observed in the early Navy was flogging. The apology for including in this history of chaplains' activities and of religion in the United States Navy a consideration of the barbaric custom of flogging is found in the conviction that any issue which vitally affects the spiritual values of individuals concerned is germane to the main subject. Chaplains, who on a Sunday morning preached the love of Christ and magnified the eternal worth of every individual, must have felt some pangs as they witnessed, perhaps immediately after Divine Service, the flogging of a sailor who may have broken, not a law of God, but some arbitrary law of man.

In the rules adopted for the Continental Navy is the following: "No Commander shall inflict any punishment upon a seaman beyond twelve lashes upon his bare back with a cat of nine tails. . . ."[34] This provision was carried over into the United States Navy. The "cat" was a short wooden stick about two feet long to which were attached nine heavy knotted cords, each about two feet long.

Flogging was administered by the boatswain's mates, of whom a contemporary wrote:

> Their duty is to enforce the orders of the officers, and, to enable them to do this, each is furnished with a hemp whip, consisting of only one lash, called a colt. They have to perform all the flogging, and the men hate them therefore as they would so many incarnate devils. In the ordinary flogging a colt is always used, but when all hands are called to witness punishment, another whip, composed of nine lashes and called the cats is employed.[35]

Often flogging took place on Sunday morning in the presence of all hands, who were assembled to witness the punishment. The following entry from the log of the *Constitution* is typical: "Sunday 6 Oct. 1799 Punished Lewis Amsden seaman with 12 lashes for riotous behavior." Flogging sometimes took place on week-days as this entry from the log of the *President* indicates:

> Wed 4 Oct 1809 At 9 punished David Kennedy with 100 lashes for desertion which is one half he has to receive, Conformable to Sentence of Court Martial.

Another social problem arose with the custom of permitting women to live aboard ship, not only in port but also at sea. Several commanding officers took their wives on cruises and occasionally permitted their officers to do likewise. Mrs. Isaac Hull frequently insisted on accompanying her husband and on several occasions took sisters or other women relatives along. These women occupied accommodations intended for the officers and this naturally created discontent.

Below the main or spar deck of a frigate was the gun deck. Beneath that was the berth deck where the living quarters of the crew were located. Aboard the *Chesapeake* when it sailed for the Mediterranean in the spring of 1802 were a number of wives of members of the crew. It was on this cruise that the berth deck became the birth deck. Midshipman Henry Wadsworth, kept a journal which gives the following quaint account of a baptismal service conducted by Chaplain McFarlan, together with a reference to the conduct of the "other Ladies of the Bay" who were not invited to the christening service:

> On the 22d Febry [1803] it being the day after we left Algiers: Mrs. Low (wife to James Low Captain of the Forecastle) bore a Son, in the Boatswain's Store Room: on the 31st inst [March].—the babe was baptiz'd in the Midshipmen's apartment: The Contriver of this business, was Melancthon Taylor Woolsey a Mid: who stood Godfather on the occasion & provided a handsome collation of Wine & Fruit: Mrs. Low being unwell Mrs Hays the Gunner's Lady officiated: Divine Service by Rev. Alex McFarlan. The Childs name Melancthon Woolsey Low:—All was conducted with due decorum & decency no doubt to the great satisfaction of the parents, as Mr. Woolsey's attention to them must in some measure have ameliorated the unhappy situation of the Lady who was so un-

[33] Benjamin, *The United States Naval Academy*, p. 27.
[34] *Journals of The Continental Congress*, III:378.

[35] Wines, *Two Years and a Half in the Navy*, I:39.

fortunate as to conceive & bare, on the Salt Sea. N. B. The other Ladies of the Bay*—viz Mrs Watson: the Boatswain's Wife, Mrs. Myres the Carpenter's Lady—with Mrs Crosby the corporal's Lady: got drunk in their own Quarters out of pure spite—not being invited to celebrate the Christening of Melancthon Woolsey Low.[36]

*The Forward Most part of the Birth [sic] Deck.

The practice of taking women to sea was forbidden by a Navy regulation in 1818 which read: "No captain shall carry any woman to sea without an order from the Secretary of the Navy, or from the commander of the fleet or squadron to which he belongs.[37]

Quite a different problem and a moral one was the presence of women of ill fame aboard our ships. Lovette in his *Naval Customs, Traditions, and Usage* states that an old definition of a man-o'-war's man was "Begotten in the galley and born under a gun." He adds that the term "Son of a Gun" was originally a term of reproach since it implied that the woman who gave birth to a child "under a gun" was of such loose morals that no one knew who the father might be.[38]

In the log of the *Essex* for the period 21 May 1801–9 August 1802 appears a set of General Orders taken from HMS *Superb* and copied by William Bainbridge with the following note of approval: "these being the best of the kind I have seen is the reason that I have given them in my Journal." Among the orders were the following:

10th In port women will be permitted to come on board but the indulgence is to be granted (as indeed all others) in proportion to the merits of the men who require them & upon their being accountable for the conduct of the women with them.

11th The Commanding Officer in Port will therefore permit such men to have women on board as he may choose & he will direct the Master at Arms to keep a list agreeable to the following form which he will carry to the Commanding Officer every morning for his inspection.

WOMEN'S LIST

Women's Names	*With Whom*	*Married or Single*	*When rec'd on Board*	*Conduct*

Article I of the "Act for the Better Government of the Navy" of 1800 placed squarely on the shoulders of the commanding officer the responsibility for the moral conditions aboard his ship. This article is as follows:

The commanders of all ships and vessels of war belonging to the navy are strictly enjoined and required to show in themselves a good example of virtue, honor, patriotism, and subordination, and be vigilant in inspecting the conduct of all such as are placed under their command; and to guard against, and suppress, all dissolute and immoral practices, and to correct all such as are guilty of them according to the usage of sea service.[39]

THE END OF THE DECADE

In spite of the haphazard way in which the twenty-two chaplains were appointed during the years 1801-1810, and in spite of the emphasis placed upon their duties as schoolmasters, the spiritual aspects of their work were not entirely neglected. With growing frequency Divine Services on the larger vessels in commission were conducted, if not by the chaplains then by the commanding officer or some other officer appointed by him.

The following extracts from the log of the *Constitution* are typical of the larger vessels during that period:

Sunday Oct. 8, 1809 At 11 mustered all Hands and read prayers.

Sunday Jan 14 1810 At 11 am, read Articles of War and performed Divine Service.

It is not known whether a chaplain was on board on the Sunday when Divine Service replaced the simpler reading of prayers. Henry Denison, who served as chaplain without a commission aboard the *Constitution* from 8 February to 17 June 1810, evidently conducted Divine Worship. The entry from the log for 23 March reads:

"11 am, mustered the ships company—read the Articles of War & performed Divine Service."

The first decade of the nineteenth century closed inauspiciously for the United States Navy and the chaplaincy. James Madison, who became President in March 1809, continued Jefferson's policy of a small Navy. Only three chaplains were on duty at the end of 1810. They were Garrett Barry on the *John Adams*, Henry Denison on the *President*, and Andrew L. Moore on the *Essex*. Two other chaplains were on the rolls, William Briscoe and Lemuel Morris, both of whom were called to active duty in the first part of the next decade.

---

[36] *Nav. Doc. Barbary Wars*, II:387.

[37] *Rules, Regulations, and Instructions for the Naval Service* (1818), Art. 24, p. 55.

[38] *Op. cit.*, p. 239.

[39] *Gen. Reg. of the Navy and Marine Corps* (1848), p. 261.
This article remains practically unchanged in present day regulations.

CHAPTER FIVE

A COLORFUL DECADE

1811-1820

The second decade of the nineteenth century, which included the War of 1812 is one of the most eventful and colorful in the history of the United States Navy. A few victorious and spectacular naval duels at sea together with the successes on the Lakes brought new prestige to the Navy. The stories of these thrilling combats, with the accounts of exploding shot, falling masts, boarding parties, marines firing from the tops, and the many instances of extraordinary heroism, fascinate even landlubbers who do not know the difference between port and starboard.

An appreciation of the need for the Navy which marked public opinion at the end of the decade under review was not evident at the beginning. The parsimonious policy of the administrations of both Jefferson and Madison had reduced the defensive power of our naval forces almost to impotency by 1810. This indifference to naval matters seems inexplicable when it is remembered that the British were blockading our ports as early as 1807. Working on the assumption—"Once an Englishman, always an Englishman"—they were boarding American vessels, seizing American sailors, and forcing them to serve in English ships. By the time war broke out, the United States Government had a list of approximately 6,500 alleged cases of impressment.[1] While there were other causes for the War of 1812, these repeated infringements of American sovereignty undoubtedly played an important part in the declaration of war.

Even with the shadow of war gradually deepening, sentiment in favor of a strong Navy was slow to crystallize. In December 1811, only seven months before the declaration of war, the House Naval Affairs Committee rejected a proposal from the Secretary of the Navy to acquire twelve 74-gun ships of the line and twenty additional frigates.

When war finally came, the Navy had but sixteen serviceable ships, and its personnel totaled in round numbers 4,000 but the men were well disciplined and trained. As was to be expected, the imperative need for naval strength stimulated naval construction and enlistments. Three 74-gun ships of the line were laid down in 1813, but were not completed in time to take part in the war. Naval personnel increased to approximately 15,000 by 1815 in addition to 2,700 marines. Theodore Roosevelt in his *Naval War of 1812* states that among the naval personnel in 1815 were thirty captains, 141 lieutenants, 510 midshipmen, fifty surgeons, twelve chaplains, fifty pursers, 5,000 able seamen, and 6,849 ordinary seamen and boys.[2]

When England's troubles with Napoleon were ending in 1814, she was able to concentrate a fleet of 616 seagoing vessels from powerful ships of the line to brigs and cutters. Of these, 124 were ships of the line and 126 frigates. The United States had none of the former ships and only seven frigates, of which the *United States, Constitution,* and *President* were the most powerful.[3] With her Lilliputian Navy, the United States dared challenge the might of the greatest sea power of the world!

In addition to the regular ships of the Navy, the United States had scores of privateers. Well over a thousand British merchant vessels were captured by Americans during the course of the war. These heavy losses together with their failure to invade from Canada contributed to the British Government's willingness to negotiate a peace in 1814.[4]

After the conclusion of the war, interest in the Navy continued. The Navy was occupied with various activities some of which were protecting commerce, suppressing piracy, and checking the slave trade.

NEW LEGISLATION AFFECTING CHAPLAINS

In answer to an inquiry made by a member of Congress regarding the duties and pay of a naval

[1] Alden and Westcott, *The United States Navy,* p. 64.

[2] *Op. cit.,* p. 47.

[3] Alden and Westcott, *op. cit.,* p. 67. Sprout, H. and M., *The Rise of American Naval Power,* p. 78, give much larger figures for England's Navy. They state that England had 219 ships of the line in 1814, and 296 frigates.

[4] Sprout, H. and M., *op. cit.,* p. 78.

chaplain, the Secretary of the Navy, Paul Hamilton, on 21 February 1811 wrote:

> The duties of a chaplain in the Navy are to read prayers at stated periods; to perform all funeral ceremonies; to lecture or preach to the crew on Sundays; to instruct the midshipmen & volunteers in writing, arithmetic, navigation, & lunar observations, & when required to teach other youths of the ship.

This roughly summarizes the regulations of 1802. Hamilton also commented on the pay of chaplains and pointed out the difficulties of enrolling and retaining the right type of men in the chaplaincy.

> The pay of a Chaplain is 40$ per month & 2 rations pr. day equal to 20 cents each making annual amount of pay & ration 626$. This is the pay of a chaplain attached to a vessel for a full term of service.
>
> The great difficulty which I have experienced in procuring gentlemen competent to these arduous & important duties induced me, some time before the death of Mr. Thompson, to order him from ship to ship, remaining on board each ship only a sufficient length of time to instruct the midshipmen in navigation & lunar observation, then proceeding to another & performing similar duties. For such Services Mr. Thompson was allowed 60$ pr. month & 5 rations pr. day equal to 1085$ pr. annum. These are the duties which at this time I should wish a chaplain to perform. His travelling expenses are allowed.[5]

Here the Secretary touched on a problem which was to become more acute—namely, the pay of chaplains. The pay of $40 a month first fixed by Congress in 1794 was not increased until 1835. With the gradual but steady rise in prices, it became increasingly difficult to retain well-trained men in the chaplaincy. Secretary Hamilton did not divulge his method of manipulating the records so that he could raise Chaplain Thompson's pay from the authorized $626 per annum to $1085.

By Act of Congress approved 2 January 1813, it was provided that each of the new 74-gun ships of the line should carry a chaplain and a schoolmaster, the latter to be paid $25 a month. Thus on the larger vessels chaplains would be relieved of a part, if not all, of their educational duties. The schoolmasters, however, were given a subordinate position aboard these vessels, sometimes being quartered with the midshipmen. This combination of low pay and little prestige did not attract or hold men of ability as schoolteachers.

On 16 April 1814 Congress authorized the President to increase by twenty-five percent the pay of naval personnel when engaged in any service "the hardships or disadvantages of which in his judgment rendered such an addition necessary." This Act was repealed 22 February 1817. It is not known whether any chaplains benefited by this provision. The Act of April 1814 repeated the stipulation that the pay of the chaplain be $40 a month plus two rations a day.

A revision of naval regulations, authorized by Congress in February 1815 appeared in April 1818. The section dealing with the duties of the chaplain amplified the 1802 regulations by assigning to him "when . . . required . . . the duties of secretary to a commodore." [6]

FORTY-ONE ADDITIONAL CHAPLAINS

In addition to the five chaplains previously mentioned whose period of services extended into the second decade of the nineteenth century, forty-one more were appointed or commissioned during the years 1811-1820. Of these forty-one, twenty-three served without any known warrant or commission. Thirteen of the twenty-three reported for duty during the years 1812 and 1813. There was a change in policy after the end of the War of 1812 when appointments from the ship's company became increasingly rare.

Eighteen chaplains were commissioned by the Navy Department and of this number eleven appear to have been ordained either prior to their selection or sometime thereafter. Such was the case of John Cook who entered the service in 1812 but was not ordained until 1824. However, the general rule that a chaplain be ordained before appointment was increasingly observed after 1818. Of the eleven ordained clergymen who served as chaplains in the second decade, one was a Unitarian, one a Presbyterian, and nine were Episcopalians.

The men who were ordained were increasingly inclined to look upon the chaplaincy as a life work. Of the eleven ordained men, all but two or three served ten years or longer. Moses Chase was a chaplain from 1819 to 1822. He was reappointed in 1841 and served until 1861, a total of about twenty-three years. Addison Searle was appointed in 1820 when he was twenty-eight years old and remained in the service until his death at sea in 1850. The real strength of the chaplaincy rested upon these ordained men who entered the Navy with the intention of making it a career.

Philander Chase, Jr., was only eighteen or nineteen when he entered the service in 1818. He was

[5] Nav. Rec. Coll., VI:10:301.

[6] *Rules, Regulations, and Instructions for the Naval Service of the United States*, p. 136.

Philander Chase, Jr., 1818–1820
Youngest chaplain to serve in the Navy, appointed when 18 or 19 years old. From picture, Edel Collection.

the son of a pioneer Episcopal bishop of Ohio. In the bishop's autobiography, the following statement appears concerning his brilliant son:

Such was his moral and religious deportment that he was admitted a lay reader, and a candidate for holy orders, under the supervision of Bishop Griswold. This was done at the instance of Commodore McDonough, who had for some time past known his pious and manly character, and being well assured of his competent learning, had made application to him to become a teacher on board the Guerriere,—of which vessel he had the command,—and go with him to Russia, and thence to the Mediterranean Sea, in the place and with the pay of chaplain.[7]

In all probability Philander Chase, Jr., was the youngest man ever to receive a chaplain's commission in the United States Navy.

The imprudence of accepting candidates solely on the basis of letters of recommendation furnished by them was evident in the case of Nathaniel R. Smith who received his commission from Secretary Paul Hamilton on 12 September 1811. A letter from Purser Samuel Hambleton to Captain David Porter dated 17 October of that year reveals the story.

[7] Chase, *Reminiscences*, p. 168.

A forgery was lately practised on the Navy department also. A Mr. Smith from Boston, a *clergyman*, presented himself with such strong Recommendations that the Secretary appointed him off hand Chaplain of the frigate Congress and attended him to the Navy Yard and introduced him in a particular manner. No wonder his letters were strong for he wrote them *himself*. This was soon after discovered & the reverend Mr. Smith was dismissed.[8]

The fraud angered Secretary Hamilton and he wrote the following order to the Captain in charge of the Navy Yard on 9 October:

I have discovered, not without great concern, that Mr. N R Smith, lately appd Chaplain to the Congress, has imposed false papers upon me. You will therefore, Sir, in presence of the officers and crew of the Congress, require him to deliver up to you, his commission, & other papers he has received at this Department, & inform him, that in consequence of the discovery made upon him, he is dismissed from the Service.[9]

There was a growing tendency in the Navy Department and on the part of ranking officers to favor the appointment of ordained men. An incident in point concerned the assignment of a chaplain to Commodore Charles Stewart's ship, the *Franklin*. It appears that after the Commodore had selected unordained Thomas Burrows on 4 June 1817 for the position, the Secretary of the Navy expressed a wish that the Reverend Nathaniel Andrews, an Episcopal clergyman who had been appointed to the *Congress* on 11 September, serve on the *Franklin* instead. Stewart replied on 3 November to the Secretary's proposal:

I have recd your letter by the Reverd Mr. Andrews, who is desirous of being appointed to the Franklin. As I have selected and brought on with me a young Gentlm as sectary, it is too late to make the arrangement for Mr. Burrows, otherwise it should have been done; when Mr. Burrows solicited my consent to his being appointed to act as Chaplain, I expressly stated to him that I should prefer a gentleman in orders, but as the compensation was insufficient it was very difficult to procure one of that description. Mr. Andrews being a regularly appointed Chaplain, and fully competent to the duties required by the act of Congress, has an undoubted right to the situation onbd. the Franklin, and it was understood by me of Mr. Burrows that his only object in requesting the appointment to act in that situation was a convenience to himself and would enable him to draw the higher emoluments of the office.

If the situation is occupied by a regular Chaplain the duties of the office can be performed agreeably to the meaning and intent of the Law of Congress, but if irregularly occupied the more important duties canont be permitted without degrading religion and bringing it into

[8] Hambleton Correspondence, New York Hist. Soc.
[9] Nav. Rec. Coll., I:9:485.

disrepute. Perhaps the department could dispose of Mr. Burrows in such manner as would prevent his feeling any disappointment or could permit him to proceed as supernumerary in the Franklin & be transferred to one of the vessels in the Mediterranean. The department will please to do in the case what may seem proper to them.[10]

The exchange was effected. It is interesting to note that the Commodore preferred "a gentleman in orders" and felt that unordained men could not perform the important duties of a chaplain's office "without degrading religion and bringing it into disrepute." This recognition of the importance of appointing ordained men marks a significant milestone in the history of the Navy chaplaincy. On 15 January 1819 the Secretary of the Navy sent the following letter to three newly appointed chaplains, all addressed as "Rev." even though Philander Chase was then only a candidate for the ministry and was not ordained until 1820.

Rev. Philander Chase
" James Everett
" James Brooks

The President of the U. S. having appointed you a Chaplain, I have the pleasure herewith, to hand you your Commission dated 28th Dec. 1818.[11]

The simultaneous commissioning of three Navy chaplains, two of whom were ordained and one soon to be, was a real step forward. Gone were the days, with but rare exceptions, when clerks, pursers, or others from the ship's company could be appointed to that sacred office. More was required from the chaplain than simply the ability to teach the three "R's" to the midshipmen.

OUTSTANDING CHAPLAINS OF THE SECOND DECADE

One of the chief characteristics of the War of 1812 was the number of brilliant ship duels. Several United States naval chaplains were on board American ships in these encounters, and at least three have attracted the interest of historians by their conduct in combat.

On 29 December 1812 the *Constitution* emerged victorious from her battle with HMS *Java*. Congress on 3 March 1813 voted its official thanks to those aboard the *Constitution*, and among the officers was "John Carlton, Chaplain." Nothing more is known of him except that he was commissioned a sailing master in September 1816. He was doubtless one of the temporary acting chaplains chosen from the ship's company.

Chaplains Samuel Livermore and Thomas Breese were also acclaimed for feats of valor. Both had been appointed from the ship's company to the chaplain's office and both afterwards became pursers. Under the present Geneva Convention, both would have violated the provision which classifies chaplains as noncombatants, for each, in the excitement of battle, took part by firing guns or wielding the sword. Free from inhibitions which would have restricted the activities of an ordained clergyman, they felt free to take part in actual combat.

Samuel Livermore, a Kentucky gentleman, happened to be in Boston at the time Captain James Lawrence was hastily assembling a crew for the *Chesapeake*. Just outside Boston, HMS *Shannon*, under command of Captain Philip B. V. Broke, was eagerly awaiting combat. Livermore, a personal friend of Lawrence, applied for a berth aboard the *Chesapeake* and Lawrence appointed him chaplain.

On 1 June 1813 the *Chesapeake* sailed out. Although the ships were evenly matched, the advantage lay with the British who had a better disciplined and trained crew. In the bitter struggle that followed, Lawrence fell mortally wounded. The British under the personal leadership of Captain Broke boarded the *Chesapeake*, and it was then that Chaplain Livermore stepped into the center of the picture. According to one account, he seized a pistol and fired at Captain Broke. The bullet missed its mark but struck a seaman.[12] According to another account, the Chaplain, with a cutlass, inflicted a serious head wound on the Captain which incapacitated him for the remainder of his life.[13] Both accounts agree that the Captain fought back and with his "mighty Toledo blade" knocked the Chaplain to the deck with a severe wound on the arm.[14]

The British were victorious and took the *Chesapeake* and the survivors to Halifax. Captain Lawrence, in the delirium that preceded his death four days later, imagined that he was still in battle and kept repeating the words: "Don't give up the ship." Among the prisoners was Chaplain Livermore. Although the cartel signed by the British and

[10] *Ibid.*, II:5 of 1817:9.
[11] *Ibid.*, "Appointments and Resignations," II:295.

[12] Maclay, *History of the United States Navy from 1775 to 1893*, I:456–7.
[13] Parker, *Recollections of a Naval Officer*, p. 25.
[14] It is thought that Captain Marryat in his *Mr. Midshipmen Easy* used Chaplain Livermore as the model for the character of Chaplain Hawkins. See Parker, *op. cit.*, p. 25.

Americans governing the exchange of prisoners classified chaplains as non-combatants, it is not likely that Captain Broke was in any mood to consider Chaplain Livermore as such. In all probability Chaplain Livermore was not only the first United States naval chaplain to be wounded in combat, he was also the first to be captured.[15]

Speedy arrangements were made at Halifax for the exchange of prisoners. Chaplain Livermore arrived at Newport, Rhode Island, on 24 June, and was soon back in the service as a purser. The destroyer *Livermore,* launched at Bath, Maine, 3 August 1940 was named in honor of this intrepid chaplain and purser.

The second chaplain to take up arms in the heat of battle was Thomas Breese, who was with Commodore O. H. Perry on the *Lawrence* in the battle of Lake Erie on 10 September 1813. In the account of that struggle, after all but

> one of the guns [on the Lawrence] are now dismounted, this one still keeps up its fire; for the commodore, with the brave purser Hambleton, and Chaplain Breese, aided by two or three men, are working it themselves. At last the purser falls, his shoulder shattered by a grape-shot. Presently this gun, too,—the last one,—is disabled, and the "Lawrence" cannot fire a shot.[16]

Perry, realizing that if his Flagship surrendered the whole fleet would do likewise, got into a boat with a few companions and was rowed to the *Niagara* and continued the fight.[17] It is assumed that Chaplain Breese accompanied his Commanding Officer. The Resolution of Congress dated 6 January 1814 thanking the officers and men who took part in the battle of Lake Erie included the name of "Thomas Breese, Chaplain."

Both Livermore and Breese served as chaplains on other vessels. Livermore's name appears on the muster roll of the *Guerriere* from October 1813 to March 1814, while Breese was chaplain on the *Java* from July to October 1815. It was fitting that both of the "fighting parsons" should have served on vessels captured from the British. The memory of Thomas Breese lingered long in the Navy. His name was sung in an old Navy song: "Here's a health to you, Tom Breese," and in the *Life of Rear-Admiral Charles Henry Davis,* it is stated that Thomas Breese "was perhaps more universally beloved, than any man in the service."[18]

At the famous battle of Lake Champlain Commodore Thomas Macdonough's acting chaplain was Francis H. Penny.[19] While Penny's contribution as a chaplain fails to appear on the official report of the battle, there is evidence that the spiritual as well as the military leadership came from the Commodore himself. Few officers in the Navy were more religious or better able to perform the functions of a chaplain than the thirty-one year old Commodore. Ranked by one authority as the "greatest figure in our naval history" down to the Civil War, it was said that he "feared his foes not at all, but his God a great deal."[20] Long a deeply religious man, he early accepted the teachings of the Episcopal Church. He had an intimate knowledge of the Bible and his favorite book was the Epistle of James, which he regarded as peculiarly adapted to the sailor's mind. To a visitor aboard his Flagship, the *Saratoga,* a few days before the battle, he quoted illustrations drawn from sea life, as "For he that wavereth is like a wave of the sea driven with the wind and tossed," and "Behold also the ships, which though they be so great, and are driven of fierce winds, yet are they turned about with a very small helm, withersoever the governor listeth."[21]

It is not surprising that Macdonough called his officers on the *Saratoga* to the quarter-deck for a few minutes prayer as the stronger English fleet stood in for the battle. He repeated the prayer appointed by the Episcopal Church to be said before a fight at sea ending in these words:

> Stir up thy strength, O Lord, and come and help us; for thou givest not always the battle to the strong, but canst save by many or by few. O let not our sins now cry against us for vengeance; but hear us thy poor servants, begging mercy and imploring thy help, and that thou wouldst be a defence unto us against the face of the enemy. Make it appear that thou art our Saviour and mighty Deliverer, through Jesus Christ our Lord. Amen[22]

---

[15] Chaplain Hoes in his manuscript notes mentions a Chaplain Henry Robinson as another casualty in the War of 1812. Robinson was aboard the *President* in her losing fight with the *Endymion* 15 January 1815. Official records, however, do not list Robinson as a chaplain.

[16] Soley, *Boys of 1812,* p. 258. Purser Samuel Hambleton, a brother of Chaplain John Needles Hambleton, once made claim for extra compensation for services he stated he rendered as acting chaplain.

[17] Perry took with him his fighting flag with the famous words "DON'T GIVE UP THE SHIP." This flag now hangs in Bancroft Hall at Annapolis and under it each incoming midshipman takes his oath.

[18] Davis, *op. cit.,* p. 65.

[19] Penny's name appears on the rolls of "Lake Champlain and White Hall Stations" (1813–4) as Acting Chaplain. According to this record he received a total of $1,163.62 prize money.

[20] Roosevelt, *Naval War of 1812, pp.* 389, 399.

[21] Macdonough, *Life of Commodore Thomas Macdonough,* p. 155.

[22] Hunt, "A Forgotten Commodore—Thomas MacDonough," *Naval Institute Proceedings,* Nov. 1931, p. 1486.

The Commodore's report of the victory to the Secretary of the Navy, too, was in harmony with his religious conviction.

Sir; The Almighty has been pleased to grant us a signal victory on Lake Champlain in the capture of one frigate, one brig, and two sloops of war of the enemy.

His prayer to God before battle and his thanks to God after the victory were as consistent with his character as was his generous and humane treatment of the vanquished foe.

The services of another chaplain, David Phineas Adams, were more spectacular and of greater significance than those of either Livermore or Breese. Adams was a member of the class of 1801 at Harvard. He became a member of the faculty of Columbia College, later Columbia University, in New York City, as professor of mathematics and astronomy.

Captain James Lawrence may have been responsible for Adams' interest in the Navy. When the latter applied for the position of "teacher of mathematics and navigation," Lawrence himself submitted the application to the Secretary of the Navy, writing that he took such liberty out of "a wish to benefit the service." He added:

Should you be pleased to give him an appointment similar to that of Chaplain Thompson deceased, I . . . request as a particular favour that you will order him to the Argus for a few months, as I have a number of Midshipmen aboard, all promising young men who would be much gratified with so good an opportunity of making themselves perfect in Navigation, particularly as Mr. Adams is considered one of the best Lunarians in the United States.[23]

Lawrence was not alone in his evaluation of Adams' abilities, for the chaplain came highly recommended. Commodore John Rogers, too, although he already had a chaplain on the *President*, desired that Adams receive orders to his frigate.[24]

Adams was appointed on 10 May 1811, the commission being mailed to the "Revd David P. Adams." He was not ordained and the Secretary, learning of his error is using the title "Revd," hastened to correct it. Adams, like his predecessor, Chaplain Thompson, would have preferred to serve under the title of naval mathematician but found it necessary to accept the title of chaplain. He was first assigned to the *President* but on 15 July was transferred to the *Essex*, under command of Captain David Porter.

Adams was aboard the *Essex* when she made her famous cruise to the Pacific in what proved to be one of the most memorable voyages made by an American war vessel. The *Essex* rounded Cape Horn in February 1813, and was the first naval vessel to carry the United States flag into the Pacific. After provisioning his ship at Valparaiso, Chile, Porter sailed forth on a raiding expedition on British whalers in the Pacific. By September he had successfully captured twelve of the twenty whalers and spread havoc in that important British industry.[25]

Porter captured so many prizes that he found it necessary to destroy some of them because he had no crews to man them. Every available officer was used and even twelve-year-old Midshipman David Farragut was given command of one prize.[26] The legend is that young Farragut was accompanied about deck by a burly quartermaster who saw to it that proper respect was paid to the midshipman's orders.

The need for capable officers who knew navigation was so urgent that even the chaplain was placed successively aboard several prizes. Chaplain Adams commanded three vessels for short periods. They were the twenty-gun auxiliary vessel *Atlantic*, renamed the *Essex Junior;* the brig *Georgiana;* and the whaler *Sir Andrew Hammond.* When the last named vessel was taken, several whales were tied alongside and a quantity of blubber was waiting to be dried and casked. Chaplain Adams supervised the stowing of some $80,000 worth of whale oil.[27] He is the only naval chaplain known to have been in command of a war vessel flying the United States flag.

The British finally caught up with Porter at Valparaiso and on 28 March 1814 gave battle and forced the surrender of the *Essex*. During the engagement, Chaplain Adams was found in the time-honored battle station for chaplains, the sick bay. In his official account of the fight Captain Porter commended the chaplain (among others) and stated that the professional knowledge of the two acting surgeons and "the benevolent attentions and assistance of Mr. D P Adams the Chaplain, saved the lives of many of the wounded."[28]

The disarmed *Essex Junior* was used as a cartel vessel to carry the paroled American prisoners back to New York. Chaplain Adams and several others were exchanged. Adams was one of the three whom Porter and the victorious British Captain dispatched to England to present affidavits "respecting the

[23] Nav. Rec. Coll., IV:1:57. Letter of 26 Jan. 1811.
[24] *Ibid.*, III:1 of Jan.-Mar. 1811:282.

[25] Alden and Westcott, *The United States Navy*, p. 81.
[26] David Farragut (1801-1870) entered the Navy when he was only nine years old.
[27] Edel, "Navy Chaplains from 1775 to 1917," p. 5.
[28] Nav. Rec. Coll., II:4 of 1814:144.

Prize." On 24 February 1815, he reported his return at Norfolk to the Secretary of the Navy.[29]

Later Adams was attached to the *Franklin* but was given additional duty surveying Hampton Roads and vicinity. While in the Pacific, he had surveyed one of the islands of the Galapagos group. The original holographic map of Hampton Roads, which Adams drew in 1815, now hangs in the Adams Memorial Chapel at Norfolk with a letter to Stephen Decatur written on the map.[30] On 9 July of the following year the Secretary of the Navy ordered Adams to make surveys on Chesapeake Bay.[31] Writing to the Secretary on 15 November, Adams apologized for the slowness of his work and added that he had already acquired "an abundance of materials for constructing an exact chart of the coasts of Chesapeake Bay to the Southward of the Parallel of New Point Comfort Light House."[32] This is the first time that such surveys had been made. Following the signing of the Treaty of Ghent, Adams was called upon to survey the boundary of the northern states. In the latter part of the decade he was appointed chaplain of the receiving ship *Columbus* where he had opportunity to continue the teaching of boys and midshipmen.

As Chaplain Thompson was the outstanding Navy chaplain of the first decade of the nineteenth century, so the versatile Chaplain Adams rose above his associates in the second. His unusual abilities and high character brought added respect to the office of chaplain.

Another chaplain who served during the years 1811 to 1820 played an outstanding role in naval history. This chaplain did not take part in any thrilling battle where he might have distinguished himself beyond the call of duty. He did not command an American vessel of war. Instead he is remembered principally because he won command of the mind and heart of a young midshipman and from his chaplain's bridge set the course for a great life. The chaplain was Charles Folsom, a Unitarian minister, and the lad was David Farragut, who later became the first admiral of the United States Navy.

Twenty-one-year-old Folsom began his duty as chaplain aborad the *Washington* on 2 May 1816, a ship to which Midshipman Farragut, then fifteen, was attached. The two were still in the ship's company in late 1816 when she joined the Mediterranean Squadron. The Squadron had been sent to settle accounts with the Barbary States, who had taken advantage of the War of 1812 to renew their attacks on American shipping.

A cordial feeling soon existed between Farragut and Folsom, a relationship that is traceable perhaps to Farragut's happy associations earlier with Chaplain Adams, when he and other midshipmen studied under the Chaplain on the whaler, *Sir Andrew Hammond.* It is known that young Folsom took a special interest in the lad, and Farragut returned the friendship with a loyal devotion which lasted for more than fifty years.

When Folsom was appointed Consul at Tunis in the fall of 1817, he wrote to Commodore Chauncey requesting permission to take young Farragut ashore with him.

> The grounds of my request are the following: Mr. Farragut has been, almost from infancy, in the naval service, with exceedingly limited opportunities of improving his mind. His prospects in life depend on his merits and abilities in a peculiar manner, as he is entirely destitute of the aids of fortune or the influence of friends, other than those whom his character may attach to him.
>
> During his connection with this ship a favorable change has been observed in him. He has acquired a sense of character and a manly tone of thinking from which the best results are to be expected. His desire of cultivating his mind, which at first was feeble, has grown into an ardent zeal. His attention to his studies of late, the manner in which he has repaid my endeavors to advance his knowledge, his improving character, and his peculiar situation, have conspired to excite in me a strong interest in his welfare, and a wish to do all in my power to promote his education. At this critical period of his life, the opportunity he may enjoy with me may prove of incalculable advantage to him, while his conduct for some time past affords a pledge that he will not be disposed to abuse it.[33]

The request was granted and Folsom and Farragut left the *Washington* at Tunis on 25 October. Looking back on that experience Farragut wrote: "I remained at Tunis nine months, pursuing my studies under Mr. Folsom, for whom I had formed a sincere attachment. I studied French, Italian, English literature and mathematics."[34] In later years Farragut was always quick to acknowledge the debt of gratitude he owed the "young Yankee parson," as he called his friend and tutor. Farragut remained throughout his life a deeply religious man.[35] Even

[29] *Ibid.*, III:1 of Jan.–Feb. 1815:137.

[30] The map has been loaned for exhibition purposes by Chaplain W. W. Edel. An account of this chapel will appear in a subsequent volume.

[31] Nav. Rec. Coll., IV:1 of 1815:207.

[32] *Ibid.*

[33] Farragut, *Life of David Glasgow Farragut*, p. 61–2.

[34] *Ibid.*, p. 63.

[35] A beautiful stained glass window dedicated to Admiral Farragut in the naval chapel at Annapolis shows him lashed to the rigging of his ship, the *Hartford*, as she moved into Mobile Bay. Farragut's Bible is on display in the Chapel at Annapolis.

though Adams and Folsom were primarily teachers, both must have had great spiritual strength of character to have so influenced young Farragut. To Folsom must go to the major credit for he enjoyed the closer associations. Had Folsom done nothing more than mold the character of this midshipman, who became one of America's greatest heroes, he would have merited the highest praise of his countrymen.

CHAPLAINS AS SCHOOLMASTERS

Until the Government established the Naval Academy at Annapolis in 1845, the main burden of preparing junior officiers for their future duties rested upon chaplains. The Secretary of the Navy expected them to be schoolteachers. On 4 May 1818 he wrote to the commanding officer of the *Guerriere* saying: "You will be pleased to select a Gentleman possessing the necessary qualifications to officiate as chaplain & instruct the junior officers in the theory of their profession."[36] The commanding officer preferred a schoolteacher to a chaplain and requested permission to appoint a Mr. Bonfils. On 24 June the Secretary replied: ". . . the Department declines sanctioning such an appointment. It is the duty of the Chaplains on board our public Vessels, under the direction of their respective commanders, to take charge of the education of the Junior officiers."[37]

Several schools for the training of midshipmen came into existence during the years prior to 1845. As has been noted, there was such a school under Chaplains Thompson and Hunter at the Washington Navy Yard during the first and second decades. Commodore Isaac Chauncey, writing to the Secretary of the Navy on 30 November 1814, described a school established at Sackett's Harbor on Lake Erie in that year:

> Sir. I have the pleasure to inform you that I have established a Mathematical School under the direction of my Chaplain the Revd. Mr. Felch who is fully competent to the duties of such a School. More than One hundred Officers attend this School, as they can be spared from duty and about Sixty Lieutenants and Midshipmen attend daily who make great progress in the various branches of Mathematics Navigation etc.[38]

Chaplain Felch wrote to the Secretary on 17 February 1815 requesting compensation for extra duties.

> Sir: When fleet came in last fall, Commodore Chauncey ordered me on shore to open a Naval School, for the officers and boys on this station—There are attending ninety-five officers studying Navigation, Mathematics, Astronomy, Philosophy Geography &c and twenty one boys in various studies. For the officers, I have no assistant; and you will readily perceive, Sir, that my duties must be of the most arduous nature. So unremitting an attention is necessary that may health is somewhat impaired by it. . . .
>
> Perhaps some compensation may be due me, for this extra Service, and I beg leave respectfully to ask your attention to the subject.
>
> Dr. Hunter, at Washington, received an additional pay from the contingent fund, although his School was nothing in comparison with this. . . . And it would seem a hardship, that I should receive no more pay, than those who do no duty either as Chaplains or School-Masters; who hold the station as a sinecure.[39]

Felch summarized his activities during the years 1816 to 1822 in a letter to President John Quincy Adams on 24 October 1825. He mentioned having "opened a Naval School at the Navy Yard at Charlestown, [Boston."] and revealed that he

> was from January 1st 1816 to May 26th 1822 continually employed alternating in surveying . . ., instructing the officers . . ., and attending to . . . [his] duties in the Navy.[40]

Felch has the dubious distinction of having been one of the first Navy chaplains, if not the first, to be "permitted to resign" from the service for conduct unbecoming an officer and gentleman. When faced with accusations involving dishonesty, intemperance, and failure to perform his duties, Felch voluntarily submitted his resignation on 29 August 1825 rather than face a court-martial.[41] A few other chaplains in the course of the years have brought disgrace to their office and to the Chaplain Corps but the number has been small.

Writing to the Secretary of the Navy from the Navy Yard at New York on 20 June 1823, Chaplain Cave Jones stated:

> In your communication of the 3d inst. you were pleased to direct me to organize a school on board of the Washington, for the instruction of the young officers. As I find there is a school-master on board of that ship for the same purpose, I beg for your further instructions as to the plan which you wish to be pursued.

[36] Nav. Rec. Coll., I:13:179.
[37] *Ibid.*, I:13:206.
[38] Chauncey Letter Book, 1814–1815, New York Hist. Soc.

[39] Nav. Rec. Coll., III:1 of 1815:113. Felch received additional pay totaling $614.35 for extra services drafting charts and "preparing system of studies for Midshipmen." (See volume marked "Congress, No. 3, 1818–1822," pp. 182–4.)
[40] Nav. Rec. Coll., Felch File.
[41] The Protestant Episcopal Convention of 1822 found Felch guilty of four charges and so certified to his Bishop. Senate Bill No. 544 was introduced 15 January 1855 calling for the granting of certain public lands to Mary Felch, the widow of Chaplain Cheever Felch. Mrs. Felch stated that she was left a widow with six small children in indigent circumstances when her husband died shortly after his release from the Navy. The Senate Committee on Public Lands recommended passage.

Jones then referred to the Secretary's report to Congress "at their last session" recommending "the establishment of a Seminary for the Navy, to embrace a general course of instruction, similar to that for the Army, at West Point." The Chaplain added that it would afford him "great pleasure to aid in the promotion of such an establishment."[42]

CERTAIN MORAL CONDITIONS DURING THE SECOND DECADE

The assembling of men anywhere away from the restraining influences of home, church, and community invariably gives rise to certain evils which are intensified by such factors as the length of separation involved, the physical conditions under which men live, and lack of spiritual leadership. Through the years the United States Government has sought to alleviate conditions unfavorable to morale and morality through the improvement of living quarters, food, and medical care; in the introduction of libraries, and recreational facilities; in the revision of certain naval regulations; and in the improvement of financial compensation for Navy personnel and their dependents. Certain chaplains took an early stand on some of these problems. The main constructive influence of the Corps, however, was felt at a much later date.

Living conditions aboard the old sailing ships were hard. The unbalanced diet often gave rise to scurvy. The public conscience of the first decades of the nineteenth century permitted such practices as flogging, now condemned as brutal and inhuman. Other customs then permitted which are now forbidden were dueling, the daily issuance of grog, and the bringing of prostitutes aboard ship when in port.

Dueling was a serious problem. According to one report, the United States Navy lost two-thirds as many officers from dueling during the first fifty years of its history as from the wars during the period. Stephen Decatur took a stand against the practice and required his officers to pledge themselves to refrain from dueling until they had referred the dispute to their commanding officer. Yet this same Stephen Decatur felt it necessary to accept a challenge issued by Captain James Barron. The two met at Bladensburg, near Washington, D. C., on 22 March 1820 where Decatur was mortally wounded. The death of this popular idol was one of the incidents that finally aroused public sentiment and resulted in the insertion in *Naval Regulations* of an article forbidding naval personnel to take part in a duel.[43]

The "Medical and Surgical Journal" of the frigate *Java* for 1816, which may be taken as typical, throws considerable light upon the prevalence of venereal disease aboard naval vessels at that time. The report for 1 July, when the ship was at sea, indicates that sixteen of the thirty-one on sick report were incapacitated because of syphilis. The writer of this Journal observed that "of the different diseases with which men were attacked, those of a syphilitic nature was most numerous."[44]

The history of flogging in the American Navy has some gruesome chapters. While the naval Act of 1799 forbade a captain to inflict more than twelve lashes for any one offense, a tyrannical officer could easily circumvent the law by charging a man with several offenses. This same Act forbade the inflicting of more than one hundred lashes except for a capital crime. Men are known to have died under the lash. Even though a man committed no offense for which he could be tried, he might still feel the sting of the lash. The boatswain's mate of those days often carried a short whip in his cap called a "colt" or "starter" which was wielded simply on an order from an officer.

The logs of various United States war vessels of the second decade indicate that it was customary to muster the crew on a Sunday morning, read the Articles of War, perform Divine Service, and then detain the crew to witness any flogging that might be prescribed. The following entries from the log of the *Congress* are typical:

1 Dec. 1816 Divine Service Performed. Mustered the Crew.

13 Apr 1817 Mustered the Crew & performed Divine Service.

6 June 1819 at 11 read the Articles of War and mustered the crew.

20 June 1819 at 11 called all hands to witness the punishment of Willard Scott and Isaac Waldroon, Marines. The former with 12 lashes for quarrelling and the latter with 6 lashes for Drunkenness.

13 Feb 1820 Mustered the Crew and read the Articles of War. Punished John Black with 12 lashes.

There were two periods during the years 1799–1861 when the number of floggings were proportionately high—the first, 1808–1814, and the second, 1840–

[42] Nav. Rec. Coll., III:4 of 1823:124.

[43] Bell, *Room to Swing a Cat*, pp. 142, 155. See also *Nav. Reg.*, 1920, art. 8, (5).

[44] Original Journal in New York Hist. Soc. It was not until 1838 that gonorrhea and syphilis were clearly differentiated. See Louis H. Roddis, *A Short History of Nautical Medicine*, p. 56.

1849. In the earlier period there were forty-six instances in which one hundred or more lashes were authorized, including the following six cases of three hundred lashes each:

Abraham Bristol, Marine, 10 Oct 1807, Desertion Embezzlement and Theft

William Brown, Bugleman, 15 April 1814, Cowardice and neglect of duty[45]

Edward Jones, Sea. 2 Nov. 1811, Mutiny and seditious conduct

John Perry, Common Sea. 2 Feb. 1813, Desertion and Mutinous Language

Patrick Smith, Marine, 12 Dec. 1809, Wounding a Seaman

John Wells, Marine, 11 Feb. 1813, Desertion[46]

During the War of 1812, the most common cause for flogging appears to have been desertion. Other causes were mutiny, seditious language, disobedience, assault, theft, and drunkenness.

It is well to note, however, that no sermons of chaplains who served during this second decade are extant. Hence, whatever they may have written to mitigate the horrors of this form of punishment has been lost.

IN CONCLUSION

The second decade of the nineteenth century ended with about twelve chaplains on the rolls. The exact number is difficult to ascertain because some were on extended furlough. Of these twelve, eight were Episcopal clergymen and one was a Presbyterian. Adams, the mathematician, was one of the unordained men still on duty. On the whole, the outlook for the naval chaplaincy at the close of 1820 was considerably brighter than a decade earlier.

[45] Punishment mitigated to 100 lashes by President James Madison.

[46] "Index for General Courts Martial and Courts of Inquiry, 1799–1861." See Manuscript Summary, Nav. Rec. Coll.

CHAPTER SIX

"LIMITED TO NINE"

1821–1840

A writer in the *Army and Navy Chronicle* for 22 November 1838 raised the question: "Why is it that so few chaplains are appointed in the Navy." He added:

> For many years the number was limited to nine; and that appeared to be a charmed number, for it was vain to expect that any one, no matter what his qualifications for that situation were, could be appointed, unless a death or resignation should reduce the number.[1]

An examination of the annual *Navy Registers* for the years 1821 to 1840 shows that an average of nine was consistently maintained. If a chaplain died in office, a new appointment was promptly made. When Chaplain Andrew Hunter died at the Washington Navy Yard on 24 February 1823, his successor, Burgess Allison, was commissioned within two weeks. Cave Jones succeeded John Ireland at the New York Navy Yard within two months after the latter's death in March 1823. There seems to have been an attempt to choose successors of the same denomination.

This arbitrary quota of nine chaplains, several of whom might be on extended furlough, was a serious handicap to those on duty who were faced with the growing needs of a slowly expanding Navy. When the third decade opened the American Navy had the following frigates in commission: *Constitution, Constellation, Congress, Columbia, Java, Guerriere,* and *Macedonian.* The *Congress* was used as a receiving ship at Norfolk. Both the *Macedonian* and the *Congress* were broken up in the fourth decade. There were also the three line-of-battle ships, *Independence, Franklin,* and *Washington.* In addition to the ten frigates and line-of-battle ships were several shore establishments which also required chaplains.

To aggravate the problem further, other large ships were being commissioned during the years 1821–1840. During the decade 1821–1830, the *North Carolina, Delaware, Ohio, Potomac, Brandywine,* and *Hudson* were added, and from 1831 to 1840, the *Macedonian* (to replace the first vessel of that name), *Pennsylvania,* and *Columbia.* All of these nine ships rated chaplains. Since the shore establishments at Boston, New York, Philadelphia, Washington, Norfolk, and Sackett's Harbor had chaplains on duty for varying lengths of time during the years under review, only a very few were available for sea duty. Instead of nine chaplains, there should have been at least twice as many. This fact was recognized by the Navy Department in 1841 when eleven additional chaplains were appointed.

During these twenty years of peace, several squadrons patterned after the Mediterranean Squadron were established. A Pacific Squadron was organized in 1821. Other squadrons were the West India in 1822, the South Atlantic in 1826, the East India in 1835, and the Home Squadron in 1841. The African Squadron was organized to suppress the slave trade in 1843.[2] In addition, the Navy sponsored a number of exploring expeditions. When possible, at least one chaplain was assigned to each squadron and each expedition.

THE OBSERVATIONS OF GEORGE JONES

From these world-wide cruises came a number of volumes on travel, some of them written by chaplains. These books not only describe strange scenes in foreign countries but reveal the daily activities of those who lived within the "wooden walls" and throw considerable light on the life and activities of chaplains aboard men-of-war.

The first work to be considered is a two-volume set entitled *Sketches of Naval Life* by "A Civilian" which appeared in 1929. The author was George Jones who was on board the *Brandywine* as a schoolmaster in August 1825. His first letter written from the ship is dated 19 August and in it he describes his position:

[1] *Op. cit.*, p. 330.

[2] Sprout, H. and M., *The Rise of American Naval Power*, pp. 94–5.

My office is that of schoolmaster, but what this means, or what are its duties, I cannot exactly tell: I only know, that those who understand these things, say it is a better situation than that of clerk for which I thought of applying, and recommended it accordingly. The chaplaincy is now accessible only to clergymen, and if the case were different, I should not have applied for it. The midshipmen are to be my pupils, and a *fine, industrious*, set of pupils they will be, to judge from present appearances.[3]

The next day Jones wrote: "A man had died and I was called upon, for the second time to perform the service, as our chaplain has not arrived." The *Brandywine* sailed on its maiden voyage from the Potomac for France on 9 September with General Lafayette as a passenger. The distinguished General was escorted aboard the *Brandywine* by the Secretary of the Navy on 8 September, at which time he was welcomed with a cheer from the crew and a salute of seventeen guns. Jones, in describing the scene, writes that "two bands of musicians" were on hand "and all was life and confusion." In the midst of this excitement the newly appointed chaplain, the Reverend James G. Ogilvie, a Presbyterian clergyman from New York, arrived. Jones, in describing Ogilvie's arrival and ignominious retreat wrote:

Among those who came and went to-day, was our chaplain: he had his trunks brought on board, but I believe was frightened off by the uproar and riot: at all events, he tendered his commission to the Commodore, and throwing himself into the next boat, left us to ourselves. If he is so easily frightened as that, why, it is well that he is gone: a chaplain in the navy must "be made of sterner stuff." I regret exceedingly that we go off without one.[4]

Ogilvie, who had been appointed for the *Brandywine* on 5 September wrote a humble letter of explanation to the Secretary of the Navy on the 9th.

I was unable to realize the situation which I had anticipated to fill until I had arrived on board of the Brandywine & I must appeal, Sir, to your own sensibilities for any apology that when about to leave my family and home and to become initiated in those scenes of life to which I had been little accustomed and when I had found that there was no possibility of ending any voyage without arrival at France, the principal object of my appointment and its acceptance, and that during so long a course of absence my wife might be taken from me by death and my litle boy be left without a protector, I should waver in my determination and yield to the impulse of nature by hastening to meet that family and to be again connected with those associations which I had too precipitately relinquished.[5]

Ogilvie claims that he acquainted Captain Morris and General Lafayette with the state of his feelings and that he "left the Brandywine under the sanction of their advice." They may have felt that a faint-hearted chaplain was worse than none. Ogilvie's commission was never confirmed. Several months later he wrote the Secretary expressing his regret that he had not gone with the *Brandywine*. In April 1826, Edward McLaughlin, another Presbyterian clergyman, was commissioned in his stead.

Since there was no chaplain on the *Brandywine*, Jones was asked to officiate as such from time to time. In making some interesting observations on the characteristics of the American sailor, he wrote:

The sailor's character is a strange compound. He is kind, because he seldom witnesses distress; generous, because his pay is competent and sure; fearless, because familiar with danger; thoughtless, because under no responsibility; changeful in his feelings; because there is little occasion for control; a drunkard, often from fashion; and dissipated, because from under the restraints of society. From all these proceeds his most striking characteristic:—there is no dependence to be placed on his resolutions. . . . His vices too are glaring: he stands convicted, weeps, resolves, and sins again. . . . There was considerable seriousness in our ship, at one time last summer: no one knew exactly how it originated; and it progressed silently, yet was extensive enough to attract attention from all on board. About thirty, I believe, discontinued drinking grog, and near twelve were under strong religious feeling. I noticed it first in an unusual application for religious tracts, and a more general reading of the Bible: they formed a kind of association or agreement to encourage and sustain one another, and met in the evening in small groups on the upper deck, to sing hymns and converse.[6]

Chaplains in the service contemporary with Jones came to similar conclusions. Chaplain Colton thought that sailors were the "most thoughtless, improvident beings in the world," but that they underwent "hardship, privation, and suffering with unclouded cheerfulness."[7] Chaplain Stewart said that they were warm-hearted and generous, had a disregard of self and a recklessness of danger, had humor, a gay spirit, and great credulity, and, while generally victims of vice, were not invariably so.[8] Chaplain Rockwell noted that some sailors were "frugal honest, and trustworthy," but quickly added by "far the greater number, are reckless, profligate, intemperate, and profane."[9] It was his opinion that most "common sailors are inveterate liars."[10]

Another characteristic of sailors which was frequently mentioned by chaplains in their writings was

[3] *Op. cit.*, I:2.
[4] *Ibid.*, I:15–6.
[5] Nav. Rec. Coll., III:6 of 1825:132.

[6] Jones, *Sketches of Naval Life*, II:236–7.
[7] Colton, *Deck and Port*, p. 35; Colton, *Ship and Shore*, p. 26.
[8] Stewart, *A Visit to the South Seas*, I:5.
[9] Rockwell, *Sketches of Foreign Travel*, II:389.
[10] *Ibid.*, p. 397.

their extreme superstition. As a protection against the devil they nailed a horseshoe on the foremast; they considered Friday an unlucky day to sail and Sunday a lucky one; to kill an albatross they thought would bring disastrous consequences; and the presence of a shark about a ship, they felt, was a fatal omen for the sick on board.[11] It was in this mental atmosphere that the chaplain worked. Chaplain Rockwell referred to the sailors' "weak and childish superstition."[12]

Jones felt that a brighter day had dawned for the naval chaplaincy because of new policies adopted by Secretary of the Navy Samuel L. Southard, who took office on 16 September 1823. According to Jones:

> We are going to have Chaplains of a different character from the former ones, and this is a subject that interests me greatly. Until the present Secretary came into office, little discrimination was used in admitting persons to that birth [sic]; and while it had a few of good scientific acquirements, the moral character of most was indifferent. Mr. Southard resolved to admit none but clergymen, and these of good standing in their several denominations. It is getting to be a respectable station, and our young men of piety and talents are beginning to look to it.[13]

Jones had his ideal of the perfect chaplain. He must be gentlemanly, prudent, fearless, and firm.

> He must not encourage vice even by a smile; but must not be quick at noticing faults: he must feel constantly, that he is there, not to condemn others, but to reform them. He should check more by example than speech; and the quieter he is about it the better. If conversation takes an improper turn among the company he is in, the best reproof is to leave them; and this will apply to most other things: it is a reproof no can can quarrel with, while it will be felt by all.
>
> He should be well educated, and above all, though I bring it last, he must be pious; and his piety must be of that noble, and elevating, and purifying kind, that comes not from men and systems, but from the Bible. There is no place so trying to a religious person, all through, as a man of war. He is in close contact with his fellows, men on whom his life should be an constant reproof, and who therefore scan his life, and feelings, and words, and his very thoughts, with a searching eye. He may have Christian society; perhaps he will not. And no one, but he who has felt it, knows what it is to live months and years, where there is none to whom you can throw open all your feelings, and look for sympathy; none to say to you "God speed you, my brother;" to live among those who talk familiarly of things you have scarcely heard named before; where oaths are common language, and sabbaths regularly profaned; where matters you have loved as holy and sacred, are unknown, or spoken of with a smile: and all this, till your own heart begins to catch the infection, and you start with alarm to find the disease you came to cure fastening on yourself. This is a Chaplain's birth, and you ask, who is sufficient for these things? I answer, He is, who is sufficient for all things, and the man who does not depend on God for sufficiency, has no business there.[14]

Jones also wrote about the opportunities of a chaplain aboard ship and pointed out that he was the one individual who could associate freely with both officers and men. He felt that the chaplain, with a knowledge of the needs and preferences of the personnel gained through this contact, should "supply himself with books adapted to their capacities." Here is the first reference to a chaplain's interest in a library. Jones added:

> Small histories, and story books will be most acceptable, and these he may blend with those of a better character. I have never been able to keep a supply of tracts, though I have procured them as often as I could. Their size, and their language generally, are well adapted to a ship: after a distribution, I have always seen the men reading them all around the decks, where I believe they met with more attention and respect than they would have done ashore.[15]

Jones wrote frankly about the regulation which required prayers every morning and evening aboard ship. He felt that compliance was difficult and perhaps even impossible. "The law," he stated, "as it will always be a dead letter, had better be repealed: but the services of the Sabbath ought to be a matter of more attention than they are." Jones thought that some provision should be made for men to be seated throughout the service rather than forcing them to stand, sometimes in the hot sun. He touched briefly on the problem of discipline aboard ship. "There is no place where the searching motives of religion are more needed than in a ship. Our gangway has often been witness to this," he wrote. Floggings took

---

[11] Colton, *Deck and Port*, p. 165; Colton, *Sea and Sailor*, p. 34; Colton, *Ship and Shore*, p. 190–1; Rockwell, *op. cit.*, II:392.

[12] Rockwell, *op. cit.*, II:392.

[13] Jones, *Sketches of Naval Life*, II:237–8.

[14] *Ibid.*, II:239–40.

[15] *Ibid.*, p. 242.

The first ship's library was placed aboard a United States naval vessel in 1821. The initiative in this venture was furnished by Mr. William Wood, a New York philanthropist. Just before the *Franklin*, under the command of Commodore Charles Stewart, sailed for a three years' cruise of the Pacific, Mr. Wood, with the permission of the Commodore, addressed the crew on the subject of a "Seaman's Library." His remarks were greeted with cheers and the officers and men immediately subscribed approximately $800. With this money 1,500 books were chosen and procured by Mr. Wood who later performed a similar service for the *United States* and the *Erie*. The Commodore promptly set aside an apartment aboard the *Franklin* as a library and appointed a librarian. Upon the return of the ship, what was left of the collection became the nucleus of the Seaman's Library at the Brooklyn Navy Yard. (Nav. Rec. Coll., IV:6 of 1824:31; I:15:358; II:4 of 1824:12; *Southern Literary Messenger*, Dec. 1840, pp. 799 ff.)

832159—49——4

place at the gangway. "Give them religious principle," he urged, ". . . and then you may throw iron gags and manacles into the sea."[16]

Jones remained on the *Brandywine* as schoolmaster until March 1826, when he was transferred to the *Constitution*. He returned to the United States in 1828, completed his theological education, was ordained in 1833, and was appointed a Navy chaplain on 20 April of that year. One of the truly great chaplains of the United States Navy, he gave his life to this work. His part in the establishment of the Naval Academy at Annapolis and his experiences with Commodore Perry's expedition to Japan will be discussed in a later chapter. Jones died at Philadelphia in 1870 in his seventy-first year, still an active member of the naval chaplaincy.

THE CRUISE OF THE *VINCENNES*

The first United States Navy vessel to sail around the world was the sloop *Vincennes*, which made the voyage in 1829–1830. Chaplain Charles Samuel Stewart, who boarded the *Vincennes* at Callao, Peru, has left an interesting account of his experiences in the two-volume work, *A Visit to the South Seas in the U. S. Ship Vincennes, during the years 1829 and 1830*. Stewart's knowledge of the Hawaiian language made his appointment as chaplain of the *Vincennes* a logical selection. Prior to his appointment to the naval chaplaincy in November 1828, he had served as a missionary under the American Board of Commissioners for Foreign Missions in the Sandwich Islands, now the Hawaiian Islands, during the years 1822 to 1825, and had written books on the Hawaiian and other South Pacific Islands.

Stewart sailed for the Pacific from Hampton Roads in the *Guerriere* on 14 February 1829 to join his ship. In his book the Chaplain gives intimate details regarding a chaplain's life aboard a frigate in the early nineteenth century. It is a pleasant, restful, and interesting life he pictures. He said that usually the chaplain's accommodations on a frigate were in the wardroom adjoining those of his fellow "non-combatants," the purser and surgeon. Stewart, however, much to his delight, was assigned a stateroom in the cabin. While his room was not on the same deck with other officers of his grade, he could and did associate freely with them.

[16] *Ibid.*, II:244–5.

His duties do not appear to have been pressing. An hour or two was spent each morning on deck, "for air and exercise." In the afternoon he visited the sick, a call which seems to have been awaited eagerly by those confined to that dingy portion of the ship called sick bay.

Most of the chaplains' day was spent in professional and official duties in his own room where he studied and wrote. Here in an apartment "such as to afford almost the retirement of a study at home," with good light and fresh air and relatively little noise and confusion, he worked.

Nor could the chaplain have received warmer cooperation from the Commodore in the discharge of his duties. Soon after Stewart reported on the *Guerriere* he was informed by the Commodore that he desired daily public prayers on board when the weather permitted. While the regulations of 1800 had called for daily prayers, the rule had, in general, been disregarded in the naval service. It was finally decided aboard the *Guerriere* that the hour of sunset was the most convenient one for all concerned. This ten minute service consisted of reading a hymn or a few verses from the Bible, or perhaps a half-dozen appropriate remarks which were "to prepare the thoughts and feelings for the more hallowed exercise of a short prayer." It was soon the opinion of many officers aboard that "a more powerful auxiliary in the discipline of a ship could not be adopted."

The Commodore approved of the Chaplain's religious services, his private appeals to the men to lead better lives, the distribution of religious tracts, and other activities which he felt enlisted the "feelings of the seamen in favor of virtue and piety." Stewart was given special permission to have members of the crew visit him in his stateroom at times which did not interfere with their duties. Since the Chaplain's stateroom adjoined the Commodore's apartment this was indeed a concession. Stewart said he received encouragement from his seniors "beyond the most sanguine anticipations."

When the day's work was completed Stewart could meet socially with other officers in the forward or dining cabin. It was octagon in shape and was about twenty feet in diameter. Conspicuously displayed was an array of battle axes, cutlasses, and pistols. On the upper bulkheads were glazed sashes which could be raised or lowered at pleasure to admit light and air from the ports. In winter months this room was appropriately and rather elegantly furnished with a warm carpet on the decks and hangings of moreen.

In this cozy room officers encircled the center table to read and converse. Here in this delightful spot when the "winter storm whistles boisterously through the masts and rigging," the officers could move nearer the "cheerful fire of a bright stove."

The following revealing extracts taken from the Chaplain's writings reveal much about the religious life of the ship's company aboard the *Guerriere:*

A more interesting and attentive audience than that formed by five hundred of our crew at worship on the Sabbath, I have seldom addressed; and every look, and the whole appearance of the men, after the first sermon I preached, as I passed among them while at dinner to distribute a set of tracts, plainly told they were far from being indifferent to the services of my office, and regard me personally with feelings of kindness and good will.[17]

Stewart continues with his description of evening prayers aboard the *Guerriere*:

It is no common spectacle thus presented by our ship, when, as the curtains of the night begin to drop around us, the busy and varied occupations of so large a company is seen to cease, and, at the appointed signal, all, from the highest to the lowest, quietly gather to the altar we have erected, to offer to heaven an evening sacrifice of thanksgiving and prayer.

. . . .

To believe it an unwelcome and irksome duty to the crew, is a mistake. There may be individuals who regard it as such, but they are few indeed in comparison to the many, who give the most evident proofs of the interest and satisfaction with which they engage in it. . . . All give the most serious and respectful attention; while a youthful company of some fifty or sixty, the flower of our crew, usually press closely to me with more than ordinary interest.

. . . .

From the observations already made on the effect of this regulation, I am fully persuaded that . . . this single service, properly performed, would soon be found to do more in promoting the good order of a crew, than all the hardness of the rope's end, backed by the terrors of the cat-o'-nine-tails. This is far from being my own solitary opinion —it is that of many of the officers on board. Prayers had scarce been established a week, before one of the most skillful and popular, but at the same time, one of the most gay and thoughtless of their number, in expressing his sentiments on this subject, closed with the following remark:—"Whatever may be said to the contrary, Mr. Stewart, there is nothing like a service of religion in elevating the character of a crew—it makes different men of them, and it is the only thing that will do it. . . ."[18]

On her way to the West Coast, the *Guerriere* remained for some days at Rio de Janeiro. While Chaplain Stewart enjoyed his visits ashore at this beautiful southern city, he was relieved when the ship headed again to sea. He had seen little of the crew in port except on the Sabbath and now he was happy again amidst his charge. He wrote that "no sounds I had for many days heard came more cheerfully and welcome to my heart than those of the boatswain's pipe, calling all hands to prayer, the first evening we were at sea." It was evident too that the retirement and quiet of his little room, with opportunities for reading and reflection, afforded a grateful change from the excitement of days ashore.

Chaplain Stewart joined the *Vincennes* when the *Guerriere* reached Calleo, Peru, in July 1829. His new ship was engaged not only in exploration but also in creating goodwill for the United States among the inhabitants of the Pacific islands. Too frequently American citizens' chief concern had been to exploit the natives or to satisfy lustful desires. When the *Vincennes* visited Hawaii, her Commanding Officer, Captain Finch, presented to the King a letter dated 29 January 1829 from the Secretary of the Navy in which the Secretary complimented the Hawaiians on their progress in the Christian religion, and reassured the King regarding the conduct of American citizens.

Our citizens who violate your laws, or interfere with your regulations, violate at the same time their duty to their own government and country, and merit censure and punishment. We have heard with pain that this has sometimes been the case; and we have sought to know and to punish those who are guilty. Captain Finch is commanded to inquire into the conduct of our citizens, whom he may find at the islands; and as far as he has the authority, to insure proper conduct and deportment from them. . . .[19]

The Chaplain was introduced to the King in this letter as follows: ". . . the Rev. C. S. Stewart, who resided for some time with you, has received the favor of his government in an appointment to an office of religion in our Navy, and will visit you in company with Captain Finch."[20]

Finch and Stewart were active in attempting to restrain any exploitation on the part of American citizens. They also took advantage of every opportunity to speak against certain pagan practices of some natives, as cannibalism and human sacrifices, rumors of which occasionally reached their ears.

One native chief told Captain Finch that his was the first ship in which they were ever told it was wrong to fight. The heavily armed sloop with its guns, muskets, and other weapons seemed to contradict the peaceful words of the captain, but it was explained

[17] Stewart, *op. cit.*, I:32-3.
[18] *Ibid.*, II:33-5.

[19] *Ibid.*, II:128.
[20] *Ibid.*, II:129. Introduction used as foreword in Vol. III of this series.

to the chief that such an array was for the purpose of securing peace both at home and abroad.[21]

The Navy chaplain had opportunity to prepare the way for Christian missions in places where the missionaries had not yet come. Of this Stewart wrote:

> I explained to them some of the leading principles of the Christian religion, the nature of missions, and the character and object of missionaries: that they were men and women of enlightened and powerful nations, who at a sacrifice of many advantages and enjoyments in their native countries—left their fathers, and mothers, and sisters, and brothers, behind them and went voluntarily to live with people such as themselves; to introduce among them the arts of civilized life; to give them books and writing; and above all, to communicate to them the knowledge of the true God, and the salvation of the soul. . . .
>
> I canont but hope that the presentation of the subject will not be utterly forgotten; and that it may have some influence . . . in preparing the way for the welcome reception and kind treatment of any who may happily be sent by those who look for the salvation of the world to lead them in the paths of truth and righteousness.[22]

The *Vincennes* returned to the United States in June 1830. Chaplain Stewart had the distinction of being the first United States Navy chaplain to encircle the globe.

OTHERS LOOK AT THE CHAPLAIN

During the third and fourth decades of the nineteenth century, several books appeared which contained descriptions of the work of Navy chaplains. In addition to the writings of Jones and Stewart, other volumes deserve attention. Among these is *Two Years and a Half in the Navy* by E. C. Wines, which was published in 1832.

Wines joined the *Constellation* as schoolmaster on 14 July 1829 when the ship was about to sail for the Mediterranean. The chaplain aboard was James Everett, who was not in good health. Wines left his impressions of the religious life aboard the frigate in the following extracts from his book:

> The laws of the United States make it the duty of the chaplain to perform divine service twice a day and preach a sermon on the Sabbath, unless bad weather or other extraordinary occurrences prevent it. The former of these duties is never fulfilled, and the latter rarely. On board the Constellation prayers were generally read of a Sunday morning, but only one sermon was preached during the whole of the cruise. I mean not to cast any reflections on our chaplain. He is a man of genuine piety and sterling worth, but he was a settled invalid, and unable to perform the active duties of his office. The moral and religious culture of the seamen on board of our public vessels is too much neglected. Sailors, though generally rough, profane and fond of grog, are not the worst of men. I see no good reason why our sloops, as well as frigates, and line of battle ships, should not be supplied with chaplains. The labours of an efficient, pious and intelligent body of chaplains in our navy, I canont but believe, would be attended with the happiest results.[23]

The first Sunday at sea fell on 16 August when, to the disappointment of Wines, there was no religious service. He wrote in his journal: "The fourth commandment is generally but little regarded on board of a man of war. This surely ought not so to be."[24]

On Sunday, 13 September, after the *Constellation* had been at sea for about a month, Wines observed:

> Mr. E—read such portions of the beautiful service of the Episcopal church, as could be read without responses. The sky, the air and the surrounding waters, in their purity and stillness, harmonized beautifully with the sacred character of the day, and combined with other things to render the scene one of the most interesting and gratifying I have ever witnessed. I have never seen a more decorous or attentive audience on land; and as His servant, in an humble but simple tone, offered up our thanks for His mercies, acknowledged our dependence and guilt, and invoked His clemency and protection. . . .[25]

On Sunday, 15 November, Divine Service was again performed on board, and Wines recorded his observations of the awe-inspiring surroundings of a worship service at sea:

> I have often wished my friends in America with me on these occasions. How grand, how awful to worship the Deity, encircled by the noblest image of himself, and canopied by heaven's unclouded azure! What can produce sentiments of devotion so strong, so pure, so sublime as the service of the church, pronounced in the ocean-temple? In addition to the solemn, tranquil, devotional character of the service itself, the matchless sublimities by which you are surrounded, are fitted to remind you, with no ordinary force, of the perfections of that mysterious Being, who stills the raging of the tempest, and gives its rainbow-tints to the frailest flower that blooms on the mountain cliff.[26]

In the summer of 1834 the Reverend Charles Rockwell, who had received an appointment as chaplain of the Seamen's Friend Society for Marseilles, France, applied to the Navy Department for passage on the *Potomac* to his new post. His request was granted and Rockwell was aboard when she sailed for the Mediterranean in October of that year. Since there was no chaplain on the frigate, Rockwell served as such and was so acceptable that the Captain urged him

[21] *Ibid.*, II:128.
[22] *Ibid.*, I:305, 307.
[23] *Op. cit.*, 1:34.
[24] *Ibid.*, I:58.
[25] *Ibid.*, I:72.
[26] *Ibid.*, I:130–1.

William Ryland, 1829-1846

Walter Colton, 1830-1851

George Jones, 1833-1870

Charles Rockwell, 1834-1837

FOUR EARLY NINETEENTH CENTURY CHAPLAINS

to continue in this capacity after he had reached his destination. Rockwell consented, serving as chaplain and receiving a chaplain's pay until October 1835. Chaplain James Everett then reported on board for duty and Rockwell accepted the position of clerk. Everett left the ship on 7 April 1836 and Rockwell again served as chaplain, but received only clerk's pay, until 9 March 1837.[27]

Rockwell was a writer of considerable merit and from his facile pen flowed some of the most fascinating travel descriptions published in the United States in his generation. His two volume work, *Sketches of Foreign Travel and Life at Sea,* includes interesting descriptions of experiences aboard the *Potomac.* The following is an account of a Divine Service held in the open on the spar-deck after a violent storm:

And here let us turn to another scene. On the Sabbath which succeeded the events described above, our ship, free from danger, was ploughing her way through the high-rolling waves; and though the motion of the ship, the loud surging of the billows, and the piping of the winds, seemed to forbid our assembling for worship, still, even those from whom it might have been least expected, were most anxious that public and united thanks should be given to God, for our signal deliverance from danger and from death. All, therefore, collected together, in one dark, dense mass, upon the deck. Their heads were uncovered, and, in the centre of the group, was one who was to be the organ of their devotions. It was an hour of tender, yet sublime, emotion. The sea rolled high around, while the heavens were our only canopy, and the lofty masts, like so many towering spires, pointed to the rest above. The voice of the speaker was raised to its loudest tones, that thus it might be heard above the noise of the elements. A portion of that sacred ode was read, which, with so much power and beauty, describes the peculiar perils of those "who go down to the sea in ships, and do business in the great waters,—who see the works of the Lord, and his wonders in the deep. For he commandeth and raiseth the stormy wind, which lifteth up the waves thereof. They mount up to the heaven, they go down again to the depths; their soul is melted because of trouble." And when, in view of deliverance from such peril, the writer exclaims, "O, that men would praise the Lord for

[27] Rockwell petitioned Congress for the difference in pay but there is no indication that his request was granted. See 25th Cong., 2nd sess., *Report of House Committee,* No. 996.

his goodness, and for his wonderful works to the children of men;" these words had to us a meaning, such as they had never had before.[28]

While some chaplains were not always faithful in conducting services every Sunday, Rockwell stated that this was done regularly during the time he was aboard the *Potomac*. He described the service on the first Sunday after the ship left Gibraltar.

> It was the Sabbath, and we moved quietly on in our floating chapel, with nothing to lessen the peculiar interest which one can hardly help feeling in religious worship at sea. All are present at such times but the officer of the deck, together with four or five at the wheel by which the ship is steered, and a few in the tops, and most of these can hear the voice of the preacher. . . . A ship's company on board a man-of-war are commonly a very attentive congregation, where the services are adapted to their circumstances and feelings. . . . Choirs of singers are sometimes organized on board our men-of-war, but we rely wholly upon a fine band of music to give impressiveness and power to those tunes which, from our earliest years, had been associated in our minds with the words of sacred song.[29]

This is one of the first known references to a choir aboard a United States naval vessel and to a ship's band assisting in Divine Worship. Rockwell envied the fine choir on board the *Delaware*, where Chaplain George Jones was serving. In bemoaning the *Potomac's* deficiency, he said that she "might have had one, had it not been that those, who would have been leaders in the business, were such utter reprobates, that they would have brought public worship into disrepute."[30] Chaplain Jones was fortunate in having men in whom musical ability blended with piety.

Rockwell described a portion of his parish:

> In speaking of those who man our ships of war, I shall begin with such as are rated as boys. Of these, we had nearly thirty on board our ship, many of whom were taken from the House of Refuge, in New York, or were the sweepings of the streets of our large cities. Some were children of poor parents, who had been placed under the care of some sailor of their acquaintance, to take their first lesson in ship craft, and, I may add, in devil-craft, too, on board a man-of-war; for surely, a boy must be a dull scholar, who, in such a place, would not learn far more evil than good. These boys were from ten to sixteen or seventeen years of age, and some of them, from having been familiar, from their earliest years, with vice and crime, in almost every form, were among the most hardened, hopeless vagabonds in the world; and yet, they had so much shrewdness and intelligence, and such perfect self-possession in all circumstances, that one could not but feel a peculiar interest in them.[31]

One of these lads remarked to the chaplain: "I came to sea on purpose to get rid of religion, but I have more of it here than I did on shore." Chaplain Rockwell conducted a Sunday School in addition to supervising the week-day school taught by one of the better educated men of the crew. Rockwell felt that the immoral influences aboard ship were so injurious to the boys that he wrote: "There are few situations in which I would not place a boy, sooner than on board a man-of-war."

Among the projects sponsored by this chaplain to improve the moral and social conditions was the establishment of a library, even though he found forty men aboard who could read but poorly or not at all. Of this library Rockwell wrote:

> To remedy in some degree this gross neglect, several of us obtained such books as we were able to get, and, at Gibraltar, we purchased for the crew, at their request and at their own expense, school-books, such as geographies, grammars, arithmetics, etc., to the value of $100. Thus, were many of them furnished with the means of improving their minds; and, so important is the influence of books in making seamen peaceful, contented, and happy, that it would be good policy, on the part of our government, to furnish every ship of war with a well selected and appropriate library for the use of the crew. Some of our larger ships have had libraries of several hundreds of volumes, purchased by the men on board, and great good has resulted from them; but, from having no system on the subject, the books have been disposed of at auction, or by lot, at the end of a cruise, or left to mould and waste away at some naval depot, instead of being carefully preserved and transferred to some other ship, where they might be useful.[32]

These words are reminiscent of statements on the same subject by Chaplain Jones. When the *Delaware* and the *Potomac* were in the Mediterranean Squadron, Jones and Rockwell talked over their problems. "It was very gratifying to me," wrote Rockwell, "to meet one for whom I had so high an esteem, and who, from the soundness of his judgment, his accurate knowledge of men, and his experience in the Navy, was so well qualified to advise me as to the duties of my office."[33] This is the first mention of an older chaplain passing on to the initiate the benefit of several years of experience in the Navy. This was the approved method of indoctrinating new chaplains into the duties of their office and continued until the establishment of a school of naval chaplains at Norfolk during the Second World War.

[28] *Op. cit.*, I:13–4.
[29] *Ibid.*, I:27.
[30] *Ibid.*, II:395.
[31] *Ibid.*, II:384.
[32] *Ibid.*, II:396.
[33] *Ibid.*, I:31–2.

Since Chaplain Everett was in the Mediterranean Squadron for a part of the time that Rockwell served aboard the Potomac, it is possible that the three chaplains met to discuss matters pertaining to their work. If this surmise be true, it was the first time that a fleet conference of Navy chaplains was held.

Reforms, as the placing of libraries aboard Navy ships, advocated by such chaplains as Jones and Rockwell, were finally adopted generally. Naval personnel and chaplains in particular have since been blessed by these early pioneer chaplains whose labors and agitations brought about benefits now accepted as a matter of course.

On 6 May 1838 a United States naval squadron consisting of the *Columbia* and the *John Adams* left Hampton Roads for a cruise around the world. Two accounts of this voyage appeared in book form, both in 1840. The one, *The Flag Ship, or a Voyage Around the World,* was by Fitch W. Taylor, the Commodore's secretary who also served as chaplain. The other, *Cruise of the Frigate Columbia Around the World,* was by William Meacham Murrell, a member of the crew.[34] Both authors served aboard the *Columbia* and in many instances described identical experiences.

Murrell describes a Divine Service held on the spar-deck of his frigate:

> On Sunday mornings, immediately after quarters, should the weather permit, all hands are called to muster. The summons is instantly obeyed, by every one proceeding to the quarter-deck, (the sick alone exempted,) where the minister stands in readiness, arrayed in his clerical robes, and the capstern covered with the national flag, to answer the purpose of a pulpit. The commodore takes his station on the weather side of the chaplain; the lieutenants, and all other commissioned and warrant officers on the weather side of the deck; the forward officers at the fife-rail, and petty officers at the fore part of the main-mast. The blue-jackets take up their position abaft the mizzen-mast, clad in white frocks, with blue collars, white trousers, and straw hats, looking the picture of cleanliness; whilst the marines are stationed and drawn up in rank, on the lee side of the deck, headed by their commanding officer, all in full uniform. The whole *en masse* forms a very picturesque group.[35]

Since Chaplain Taylor who officiated was an Episcopalian, it was natural that he should wear the vestments of his church. Telling of another service, Taylor describes the capstan which served as a reading desk and pulpit. "Over the top of the capstan," he wrote, "which is some four feet in diameter, a flag is thrown, in preparation for the expected service."[36] The service was comparatively short since the ship's

[34] The *Columbia* sailed around the world in 1838–1840 under the command of Commodore George C. Read. Murrell's book is one of the earliest written by an enlisted man in the United States Navy.

According to Murrell, the *Columbia* spent 459 days at sea and 313 in the eighteen ports visited. The total distance covered was "54,796 knots," making an average of about "121 knots" for every day at sea (p. 229).

[35] Murrell, *op. cit.*, pp. 132–3.

[36] Taylor, *A Voyage Around the World,* I:18.

Divine Services on a frigate. The chaplain uses the capstan for a pulpit.
From Colton's *Sea and a Sailor.* (Published 1851.)

company had to stand uncovered. In tropical climates it was customary to spread an awning over the quarter-deck during the service. Taylor further described such a service.

> Surely no one can, for the first time, contemplate such a scene on the deck of a man-of-war, without interest—nearly five hundred souls, their persons attired in their neatest dresses, often deemed a rough people, but now exhibiting a beautiful aspect of propriety and neatness, and profoundest stillness, gathered for solemn worship on the decks of a majestic frigate, bounding yet fleetly on her way of ocean. . . .[37]

BURIAL AT SEA

Sickness and accident took their toll of the ship's company. Murrell describes how dysentery afflicted the officers and men when the *Columbia* was off the coast of China. Two, three, and sometimes even four died in one day when the scourge was at its worse stages. Taylor, in the following passage, describes a burial service at sea:

> After quarters, the succeeding morning, "all hands" were piped "to bury the dead." The sailor, who fell last night from the main-top-gallant yard, was to be given to the deep. He had been laid out, as usual in such cases, by his messmates, on the half-deck, with the flag of his nation thrown over him. His messmates were his watchers during the night, and now, at the hour of his burial, they bore him to the leeward gangway of the frigate.
>
> . . . . In full view of these, nearly five hundred gazers, rested the plank upon the upper step of the gangway, on which the unconscious sleeper, sewed in his hammock, with thirty-two pound shots at his feet, was reposing, with the stars and stripes wreathing his rough bier, as his honorable pall. Six of his messmates, as his bearers, held the plank in its horizontal position, ready to launch their brother of the ocean into the blue sea. And nearest them, stood the Commodore. The chaplain advanced to his side, commencing the services, as all, uncovered and with the silence of the dead, listened to the affecting ritual. . . . "Forasmuch as it has pleased Almighty God, in his wise providence, to take out of this world the soul of our deceased brother, we therefore commit his body to the deep!" And in the breathless stillness of the momentary pause, the solemn plunge was heard, which spoke louder than the thunder of ordnance to the heart, as the dead man was sinking to deeper and yet deeper fathoms, until the eloquent silence was again broken by the chaplain's voice as he added, "Earth to earth, ashes to ashes, dust to dust; looking for the general resurrection in the last day, and the life of the world to come through our Lord Jesus Christ; at whose second coming in glorious majesty to judge the world, the earth and the sea shall give up their dead. . . ."

[37] *Ibid.*

> The services ended—the crew were again piped to their places—and then we were on our course again, to other lands.[38]

The chaplain had charge of the burial ceremonies at sea and the call "All hands to bury the dead" was a frequent one. Typhus, small-pox, typhoid fever, pneumonia, dysentery, and other diseases took a large toll. One chaplain noted three deaths within less than thirty days, while another commented on nine burial services in a less number of weeks and finally the loss of more than seventy men out of a crew of five hundred on their trip around the world.[39] Another observed, "You seldom meet with a gray headed sailor. . . . He dies in the midst of his days and often in his full strength," and again, his "occupation and habits shake his life-glass and hurry out its sands."[40]

There was something peculiarly melancholy and impressive about burial at sea. Naturally, there was no coffin, no hearse, no procession, and no tolling of the bell—no similarity to such a service ashore. After the body was laid out in a suitable part of the deck, it was covered with a flag. Finally the corpse was brought up, sewed in its hammock with a cannon ball attached to the feet. It was then laid on a board by the open gangway, and all hands uncovering, the chaplain began the ceremony. When he read "we therefore commit his body to the deep", one end of the board was raised and the body plunged into the sea.[41]

SELECTION AND APPOINTMENT OF CHAPLAINS

Certain men, who had served as chaplains for the sake of higher pay, had brought the chaplaincy into disrepute since they lacked training and, sometimes, religious conviction. It seems to be human nature for the average person to remember for an inordinately long time the breaches of the moral law committed by clergymen or acting clergymen. So it was with the anecdotes of the misdeeds of certain chaplains who served in the early days without warrant or commission. These tales evidently became luscious morsels of gossip and were passed from tongue to tongue among naval personnel for several decades. Chaplain Rockwell related some of the stories he heard:

> I have it from those who were themselves eye and ear witnesses of what they state, that one of these chaplains,

[38] *Ibid.*, I:27–9.
[39] Colton, *Deck and Port*, pp. 60–1; Taylor, *A Voyage Round the World*, II:66, 327.
[40] Colton, *Ship and Shore*, p. 26.
[41] *Ibid.*, p. 31; Colton, *Deck and Port*, p. 60; Jones, *Sketches of Naval Life*, I:36.

so called, the moment he closed public worship would turn to his servant, and in the hearing of all, with a profane oath or curse, order him to bring him a glass of grog,—that another used to lie drunk on the deck every night, and that another, who went on shore with the officer who told me the fact, to bury a man, was so drunk when they landed, that the officer himself was compelled to read the funeral service, and leave the chaplain lying on the bottom of the boat.[42]

When Samuel Southard became Secretary of the Navy on 16 September 1823, however, he instigated a reform. Southard issued a directive forbidding the appointment of anyone to the chaplaincy who was not an ordained clergyman.[43]

Though the Secretary discouraged appointments of any but professionally qualified men, it appears from contemporary testimony that commanding officers sometimes preferred good men chosen from the ship's company to ordained chaplains. A case in point is that of John A. Bates. On 27 December 1823 Captain Isaac Hull of the frigate *United States* explained to the Secretary of the Navy the occasion for appointing Bates.

Commodore Chauncy (then performing the duties of the Secretary of the Navy) and myself looked over the list of Chaplains and found that there was not one unemployed that possessed either Religion or good Morals, and not feeling himself authorized to make an appointment, he said he did not see any impropriety in taking Mr. Bates, particularly so, as himself and other Commissioners knew him to be a correct young man and every way respectable and beloved by every one, that has been associated with him.

. . . .

. . . I do not hesitate to say that he has more religion and good morals, than two thirds of the Commissioned Chaplains in the Navy; indeed, I do not know two that I should be willing to have in the Ship.[44]

On 18 June 1827 Captain Jacob Jones of the *Brandywine* notified the Secretary that he had made his schoolmaster, Mr. Soter, "a gentleman of education and worth," chaplain so that he might obtain the increased pay that went with such an appointment. The Secretary replied that the appointment was to cease when the ship returned to the United States.[45]

When Chaplain Stewart left the *Guerriere* for the *Vincennes* in November 1828, the Captain appointed his schoolmaster, Charles May, to the vacated position. After the *Vincennes* had returned from its voyage around the world, the Secretary of the Navy on 6 December 1831 wrote to Commodore Thompson: "I will thank you to inform me by what authority Ch. May, who is reported as Chaplain on the Guerriere, was impowered to act in that capacity."[46] The Commodore replied on 10 December saying in part:

I beg leave to represent that the appointment was my act, and the authority for it the power delegated by the Navy Department, since the creation of the navy, to persons in my station to make acting appointments, and always exercised as far as I am informed, without reprehension or interdict till the present time; - further, that this power was conferred on me, personally, by the tenor of my instructions, both written and oral, upon leaving the United States. That this discretionary power has been soundly exercised in the instance which forms the subject of this enquiry, I have no reason to doubt, as well because I considered that great advantages would result, and now believe that they did result, from the orderly performance of religious services in the squadron, as that no censure was passed on the act of the late Secretary of the Navy, to whom it was duly reported, and no revocation ever made by him of the appointment.[47]

On 16 December Secretary Southard notified May that his apointment would cease on the arrival of the *Guerriere* at Norfolk.

The principles Southard established regarding the selection and apointment of chaplains were carried on with increasing strictness by his successors. So far as has been discovered, the last chaplain to be appointed from the ship's company was John F. Girard, who served on the *Potomac* from 4 July 1840 to 14 September 1841. Thus a practice which existed in the United States Navy for more than thirty years gradually came to an end. As the older commissioned chaplains died or were otherwise separated from the service, more care was taken in the selection of their successors.

THE PAY OF CHAPLAINS AND RELATED PROBLEMS

Adequate financial compensation was necessary, however, before the work of chaplains in the United States Navy was actually placed on a sound basis. The chaplain's stipend of $40 a month together with two daily rations, first fixed by Congress in 1794, was not changed until 1835. During the latter part of this period, chaplains suffered from the penurious policy of the Government. The reforms regarding the selection and appointment of chaplains were of no value if qualified men felt they could not maintain a

[42] Rockwell, *Sketches of Foreign Travel*, II:427.
[43] Contemporary writers refer to this ruling but no record has been found of the directive itself.
[44] Original Hull Letter Book, Naval History Coll., New York Hist. Soc.
[45] Nav. Rec. Coll., II:May-June 1827:92 and I:17:355.
[46] *Ibid.*, I:20:169.
[47] *Ibid.*, II:Dec. 1831:35.

decent living for themselves and their families on the salary provided.

In 1822 the Secretary of the Navy estimated chaplains' pay for the following year as follows: those on sea duty and most shore stations $480 and two rations a day, totaling $662.50. For duty at Washington, New York, and Norfolk Navy Yards, an additional $250 a year was allowed for house rent which brought the total to $912.50. In a few instances chaplains on shore duty received special allowances for candles, servants (one ration per diem for each servant), wood, and occasionally additional pay for extra duty. The Secretary of the Navy in 1829 estimated that chaplains on duty at the Navy Yards received, in cash and kind, a total of $1,161.75 a year. The extra allowances which amounted to a considerable sum were denied chaplains on sea duty.

As early as 1823 a chaplain protested against certain discriminations which existed in the treatment of members of his corps. In a letter to the Secretary of the Navy dated 22 October 1823, Chaplain Cave Jones at the Brooklyn Navy Yard stated:

> You are doubtless, Hond. Sir, well aware that the contingent emoluments form a very important part of the provision for those who hold offices under Government. This is peculiarly the case with regard to the Chaplains; the pay provided for whom is so utterly inadequate as effectually to prevent the acceptance of such appointment by gentlemen of respectability & standing in society. It is to this point, as it concerns my own situations, that I beg leave respectfully to solicit your attention.
>
> By reference to the standing regulations of the Yard, it appears that every Officer, of every grade connected with the establishment, with the single exception of the Chaplain, is allowed, besides house-rent, fuel, candles, & a servant. My predecessor had nearly the same allowance, until the month of January 1821, at which time, for causes unknown to me, the emoluments, except house rent, were withdrawn.[48]

Jones felt doubly grieved about the situation when he recalled the fact that he had been given additional duty from which his predecessor had been exempted. Nothing came of his protest.

During the first years of the fourth decade a concerted effort was made by chaplains to obtain an increase of pay. On 14 November 1831 Chaplain Hervey H. Hayes wrote a pungent letter to the Secretary of the Navy. Even though Hayes had been a Navy chaplain for less than four years, that experience was sufficient to inspire him to describe with paint-brush phrases the discouragements and difficulties under which he labored. The letter is given in full because of the light it throws upon the problem.

[48] *Ibid.*, III:8 of 1823:124.

> Allow me respectfully to submit to your consideration the depressed state of the Chaplaincy of the Navy, arising from the degraded character into which it has been sunk by former incumbents, & from the embarrassments with which we now contend, both from this cause, & the inadequacy of our pay.
>
> As, Sir, the whole machinery of the Navy moved by your hand, & your eye has no doubt looked through its past operations, you must have seen how badly the wheel, in which I am a *spoke*, has worked. To my mind the reason is clear. It has for the most part been made of bad materials. So notorious is this, all over this land, & in some foreign ports, that to be known as Chaplain of the U.S.N. is far from being a passport to confidence or favour. The cheap rate at which we are obtained indicates that we cannot be very sound. It seems to me bad policy to place bad timber in any part of a ship, though it might be obtained for nothing, & that stick should be 1st rate which is set as a safeguard to all the rest, whatever it may cost. It would be no more *absurd* to set *a feather in the tail of a Frigate* or to make a rudder of *punk* to conflict with the billows of the ocean and bear her safely onward, than to place a weak, ignorant or inefficient Chaplain on board to contend with the swelling of human corruption, & guide a ship's company upward in moral glory. Of all places, a man of strong intellectual & moral power is demanded for this. I humbly conceive, Sir, the most important if not the principal field of a Chaplain's useful influence is to be with the officers. At any rate, things are so constituted in a ship that, if the Chaplain is not esteemed & respected by the officers, his influence is at an end with the men. Now, a higher degree of literary attainment & moral behaviour is expected in a Chaplain than in any other officer, & if he be deficient he fails to bear a salutary moral influence into the bosoms of the officers—fails to draw forth respect from them—fails to move upon the morals of the men—utterly fails of adding any honour to the service—utterly fails of answering all the ends of a Chaplaincy, and becomes a blot and dead weight upon the Navy. A single strong man well armed & provided with intellectual, literary & moral ammunition is worth hosts of barren dwarfs. A weak Chaplain, so far from being a useful officer, is an absolute injury—fit not even for ballast in a ship. The same weight in pot-metal would be worth more & might be had at a cheaper rate.
>
> But, Sir, such men as are demanded must be supported, or not secured. He who raises the standard of morals in the Navy is engaged in shedding glory on his country & his country ought not to leave him or his family to famish. I feel, Sir, morally certain that an efficient & respectable chaplaincy—a chaplaincy that will carry such utility or reflect honour upon the service, can never be attained & preserved on the present footing. *A race of invalids* who can run into it & at any moment fly under the wing of an indulgent Surgeon & escape the calls to duty, or a band who have been cut loose from employment for want of proper qualifications, may no doubt be retained as a sinking weight on the honor & morals of the service. The deep interest, Sir, you have shown in advancing the present interest & future prospects of the Navy, inspires me with confidence to invite your attention to this subject, &

respectfully to pray for your favour in removing the embarrassments that press upon the chaplaincy from the smallness of their pay.

Chaplain Hayes then proceeded to give five reasons why the chaplain's stipend should be increased.

1st, it is the common sense of all officers in the Navy, so far as I know, that a Chaplain's pay, especially at sea, is incompetent to his necessary expenses abroad & a decent support of a family at home, & that it is disproportioned to the pay of other officers. While the pay of other officers has been raised, that of Chaplains remains the same as when the comparative poverty of this country forbade more pay to her officers than a mere subsistence.

2nd. The laws of nature-the Laws of God-& the Laws of these U. States allow every man a family, & I cannot believe that this *proud nation* wishes any servant of hers to go abroad on the business of his Country, straitened at every point & even driven to meanness by the scantiness of his pay. Nothing can be plainer than that a Chaplain cannot live, in a style corresponding to that of his associates & in a style which his respectability & consequently his usefulness demands on $480 a year and two rations per day, & out of this also support a family at home.

From the fact his mess-mates are allowed 4 rations per day & a Chaplain but 2, I fancy that it cannot be well known that the expenses of a Chaplain's living on board, are necessarily the same as those of a Surgeon or Lieutenant. All these mess together & must pay their equal proportions to the aggregate expense unless, indeed, he voluntarily absent himself & degrade his influence, as some have done, by a course of unworthy parsimony.

3rd. I cannot conceive of a single reason, in the nature of the case, why the pay of a Chaplain should be less than that of a surgeon. If he be properly & regularly qualified for his office, the Chaplain has passed through *nine* years laborious, classical preparatory study, which is more than is demanded by the Surgeon, & who will say that the task of taking care of the souls of men is less arduous, difficult or responsible than the care of their bodies? The Lieutenant has perhaps spent an equal length of time preparatory to his rank, but not more laboriously, & the whole course at the expense of the Government, whereas the Chaplain's preparatory expense has been wholly from his own pocket.

4th. While other officers are rising in pay as time advances, their dependents' increase & experience raises the value of their services, the *poor Chaplain* is doomed to remain as stationary as the *intellect of the brute creation* & his energies are paralyzed by the darkness of his prospects.

It is an entirely mistaken notion which some have taken up that any *minister of the Gospel* is at once fitted to enter successfully upon the duties of the chaplaincy. An acquaintance with this peculiar class of men is no less important to a Chaplain's success than experience is important to the surgeon. So long as the chaplaincy is a perpetual flux no great good can reasonably be expected.

5th. As a Chaplain has no rank, except what his pay gives him, & as rank in war is scarcely less influential than caste in India, a Chaplain's usefulness is materially enfeebled by the smallness of his pay.

Allow me, Sir, respectfully to submit to your attention & favourable consideration these views of this subject.

I have written as though I were among the more favoured ones who are blessed with families, but my pay is too threatening for adventure. I am, however, deeply interested in the condition of my more highly favoured brethren. It pains me to the very heart to hear them reproached for their penuriousness. I know it has been hard to get bread for their children that some have been driven into narrow paths. I feel so identified with the chaplaincy that reproach brought upon it by any of its members is a reproach upon my own reputation & a bar to my future usefulness. "If one member suffers the whole body suffers."[49]

The Secretary acknowledged the letter on 19 November and stated that the communication would "receive due attention."

On 10 January 1832 Chaplain Hayes addressed another letter to the Secretary requesting a change of orders. "It is well known to the Department," wrote Hayes, "that I have seen more sea-service, in proportion to the time I have been in the Chaplaincy, than any other Chaplain in the Navy, and I have also done more duty." Hayes had served aboard the *Java* for about three and a half years and when ordered to sea again had protested. "It really seems to me," he wrote, "a hard case to be compelled to perform so much of the uninviting service, on pay of $660 per year while other Chaplains are permitted to remain constantly on pleasant stations with nearly twice the compensation."[50]

Hayes's remarks concerning favored chaplains who remained on shore duty with its attendant financial benefits were pointed. Several chaplains on the rolls had had little or no sea duty. Addison Searle had none from January 1821 to March 1835. For several years Searle was on leave. James Everett was at the Charlestown Navy Yard from June 1822 to June 1829. Edward McLaughlin, who was commissioned in 1826, had his first sea duty in 1830 and then for only five months. There is no record that Chaplains Addison, Brooks, Cook, Ireland, Fenner, or Ryland were ever sent to sea. When it is recalled that the total number of chaplains in service was more or less limited to nine it is clear that those available for sea duty could be counted on the fingers of one hand.

The case of Chaplain Ryland is of interest. While Andrew Jackson was a United States Senator from Tennessee, he attended services in Washington at the Foundry Methodist Church where the eloquent Wil-

[49] *Ibid.*, III: 10 of 1831:71.

[50] *Ibid.*, III: Jan. 1832:62. See also *The Sailor's Magazine*, Jan. 1832, p. 152 and June 1832, p. 299, for references to the reluctance of Navy chaplains to go to sea.

liam Ryland was pastor. Even though such orators as Webster, Clay, and Calhoun were then members of the Senate, Ryland had the reputation of being one of the most forceful speakers in the Capital City. Four times he was elected chaplain of the United States Senate. When Jackson became President in 1829, he appointed Ryland, then fifty-nine years old and in poor health, chaplain of the Washington Navy Yard. In spite of his poor physical condition, Ryland threw himself into his work and for the next thirteen years rendered excellent service to all attached to the Yard. He was greatly beloved, yet the Administration was criticized for

> undue favoritism toward Mr. Ryland. It was alleged that while other chaplains had to take their turn at sea for 2 or more years, Mr. Ryland was allowed to remain on shore, for an indefinite time, and it was urged that such a practice was contrary to all right, and an infringment of the regulations of the Navy. Mr. Dickinson [Secretary of the Navy] it seems sympathized with these views, and notified Mr. Ryland that he must hold himself in readiness for a three years' voyage. The old chaplain waited upon the President, and told him frankly that his age, and the state of Mrs. Ryland's health made it impractical for him to go to sea. The President told him to go home and make himself easy; that Mr. Dickinson did not know what he was about; and that neither he nor any other man should disturb him.
>
> General Jackson sent for the Secretary, and asked him how he could have the heart to send an old man to sea, leaving a sick wife behind him; and he told him moreover that he could only excuse him on the ground of his being an old bachelor, and therefore not competent to judge of family affairs. So the subject was dropped and the venerable chaplain was permitted to hold on to the sceptre of his office on land. . . .[51]

The attrition rate among naval chaplains was very high in the twenties. The following chaplains died during the years 1823 to 1829: Adams, Hunter, Ireland, Cave Jones, Allison, and Andrews. During the years 1825 to 1830, the following, several of whom had less than three years' service, resigned: Felch, McCarty, Addison, Ridgley, Brooks, Ogilvie, and McLaughlin. By 1831, when Chaplain Hayes was writing his protestations to the Navy Department, only two of the older chaplains were on duty. They were Searle and Everett and neither was in good health. Only three or four, including Colton and Stewart, were available and fit for sea duty. Hayes was undoubtedly right when he claimed that the main difficulties were the lack of adequate financial support and the presence of too many old and infirm chaplains on shore duty.

[51] Sprague, *Annals of the American Pulpit,* VII:393–4.

Hayes gave vent to his indignation about existing conditions in a third letter to the Secretary of the Navy dated 27 January 1832.

> No grade of officers is required to spend 6 years out of 7 at sea. There are officers enough in every grade except ours to give each as much time on shore as they spend at sea, & every Chaplain in the service except myself has been allowed more than this. If, Sir, I am doomed to carry half a dozen sick Chaplains on my back through life, it is a clear case that I shall be greatly fatigued before the journey is half ended, especially as each of them will be doubly fed, the whole time to myself.
>
> Where is the man in all the earth that can take such a load without a murmur? Were the *Apostle Paul* here with all his meekness good nature, I have no doubt he would put up his back to shake them off. To keep me constantly at sea on $660, & others on land upon $1140, is like keeping one horse perpetually on the jump over rough roads, on *bran,* while others are at their ease in comfortable stalls, on *oats.* I would commend the spirit of such a horse were he to throw himself out of the traces. I have often admired the spirit of Baalam's jackass in opening his mouth in remonstrance when severly beaten. . . ."[52]

The Secretary cancelled the orders which would have sent Hayes to sea and assigned him to the Navy Yard, Pensacola, Florida.

Another chaplain to protest against the inadequate pay was Walter Colton who repeated in a more subdued but none the less forceful tone some of the arguments of Chaplain Hayes. Chaplain Colton made the statement that certain chaplains were obliged for financial reasons to leave the wardroom mess of their brother officers and eat with the enlisted men in the cockpit. Colton wrote under date of 16 November 1831:

> Sir,
>
> Allow me respectfully to solicit your attention to a subject involving the respectability, influence and happiness of an important, though unimposing branch of the Naval Service. I take it for granted that the country is willing to afford a competent support to those who surrender their time, talents and lives to her service. A Chaplain in the Navy frequently finds himself compelled to solicit an exemption from foreign Service in order to escape from the most perplexing pecuniary embarrassment. His pay is $660. a year: this barely adequate to his personal necessities. But if he be a married man without other resources —and this is frequently the case—he is obliged when ordered out of the country to leave his family to the precarious provisions of charity or to starve. It is no wonder, under these circumstances, that he importunes the Department for a release from foreign duty. He would evince a gloomy exemption from the sympathies of our common nature were he not to do it.
>
> If it be asked why he cannot practice a more rigid economy in consideration of those whom he has left at

[52] Nav. Rec. Coll., III:Jan. 1832:161.

home—the true reply is that this extreme frugality is in most cases out of his power. He messes with the officers of the Wardroom, and must sustain his proportion of the expenses. Their salaries are greatly in the advance of his, and they can, without any embarrassment, endulge in a liberality at the table which is ruinous to him. This has compelled a Chaplain in some instances, for the sake of saving a few dollars for his family, to desert his rank and mess in the cockpit—an expedient that would render him liable to the charge of parsimonious meanness, were it not forced upon him by the inadequacy of his pay. The Government is therefore under a sort of humane necessity to excuse from foreign-duty those Chaplains who have families, and consequently the perils, privations and hardships of sea service must devolve upon those who are without these domesticities and responsibilities. There is no course left to the Department that will not involve one portion of us in the extremities of want, or the other in the oppressiveness of unrelieved duty.

If there is a want of efficiency in the Chaplains of the Navy, it can be traced to other causes than a moral apathy. The want of a comfortable subsistence must be a weight upon the most generous devotion of which the human heart is capable. It can hardly be expected that a man will be warm and enthusiastic in a pursuit where every impulse of fresh zeal only plunges him the deeper into poverty and distress. He must be an angel that, in such circumstance, would hold on unwearied and undismayed. . . .[53]

On 20 January 1832 a memorial was submitted to the Senate by Navy chaplains appealing for an increase in compensation. It was written by Chaplain Colton who incorporated in it many of the statements he had written to the Secretary. The memorial concluded with these words:

It will occur to your honorable body, that if there are to be chaplains in the Navy, they should be men of decided talent, learning and piety. No others can command the respect of the officers, or accomplish the good intended. Men of this stamp may enter the service for a few years, from motives of health, or foreign travel, but they cannot be retained in it, upon the present footing of chaplains.

Your memorialists, therefore, respectfully pray for the passage of a law, in which the compensation of chaplains in the Navy, may be increased to a proper and suitable amount.[54]

A pay reform for all naval personnel, including chaplains, was long overdue. Congress on 3 March 1835 passed the law entitled: "An Act to regulate the pay of the Navy of the United States." By this Act chaplains were to receive $1,200 per annum when on active duty, and $800 when on leave of absence or while "awaiting orders." They received, as did all other officers, one ration a day when attached to seagoing vessels, and an allowance of ten cents a mile when traveling. All other allowances were discontinued.[55]

This Act of 1835 appears decidedly unfair to chaplains. Under it surgeons received $1,000 per annum for the first five years and a $200 increase for each subsequent five years until $1,800 was reached at the end of twenty years. Chaplains, on the other hand, had no longevity increases. Lieutenants of the line were granted a stipend of $1,800 when serving as commanding officers, $1,500 when on duty in other capacities, and $1,200 when awaiting orders.

Widows of deceased chaplains were granted pensions on the same basis as widows of other officers, namely, one-half the officer's base pay. Mary Hunter, the widow of Chaplain Andrew Hunter, received a monthly pension of $20 beginning February 1823. Such a pension had to be renewed by act of Congress from time to time. During these years under review, 1821-1840, Mrs. Cave Jones and Mrs. Hannah Everett were also receiving widows' pensions following the deaths of their husbands.

THE DUTIES AND EXPERIENCES OF CHAPLAINS

A new rule, appearing under the date of 28 November 1833, made attendance at prayers, following Sunday morning muster, compulsory for all naval personnel attached to Navy Yards.

The hour of muster was to be set sufficiently early to permit those desiring to attend Sunday morning service to do so in their accustomed places of worship. The new order read:

When Chaplains shall be attached to Navy Yards, they shall read prayers directly after muster on Sundays, and the Officers and men belonging to the Navy Yard and ordinary and the recruits shall attend, with any other persons who may think proper, and who may be permitted by the Comdn. of the Yard.[56]

Out of one of these morning services held in the New York Navy Yard by an unidentified Episcopal chaplain sprang an anecdote involving the legendary Commodore Chauncey who was for so long Commandant of that Yard. According to the story, the chaplain read a church notice which, he innocently added, was by order of the Bishop of the Diocese.

"By *whose* order, did you say?" suddenly interrupted the Commodore springing up.

"By order of the bishop of the diocese," meekly replied the chaplain.

[53] *Ibid.*, III:Nov. 1831:82.

[54] 22nd Cong., 1st sess., *Sen. Doc.*, No. 30.

[55] Callan and Russell, *Laws of the United States relating to the Navy and Marine Corps*, pp. 296–7.

[56] Nav. Rec. Coll., V:1:310.

"Well," thundered the Commodore, "the notice will not be obeyed. I'll let you know that *I am Bishop of this diocese.*"[57]

The effectiveness of a chaplain's work then, as now, was largely influenced by the attitude of the chaplain's commanding officer. Where no naval regulation existed to cover a situation, the chaplain was often at the mercy of the arbitrary orders of an unsympathetic officer.

The manner in which chaplains' duties and prerogatives were gradually being formulated in these early days may be illustrated by the experience of Chaplain Edward McLaughlin stationed at Portsmouth, New Hampshire. In a letter to the Secretary of the Navy on 17 February 1827, the Chaplain contended that he should be allowed free visitation to the patients of the hospital and that this was being denied him by the doctor in charge. He wrote:

> There have been lately two deaths, of the sickness of which cases I was never apprised until notified to attend the funerals. One of these men I was told died in great distress of mind and was very desirous to see me during his short illness.

McLaughlin requested that he be notified whenever there were patients in the hospital who were seriously ill. The doctor replied that "a compliance was impossible." After the Secretary read the Chaplain's letter, he made the following notation on the back:

> Write to Dr. Wmson [Williamson] that it is desirable where there is an attending Chaplain he should be admitted unless the sick requires another.[58]

In this way another policy regarding the duties of chaplains was established. Today it is expected that chaplains visit the sick bay daily and be ready at any time to give spiritual aid to the sick and the dying.

Chaplain McLaughlin, who reported aboard the *Brandywine* on 8 February 1830, sent the following account of his religious activities aboard his ship to the editor of *Sailor's Magazine:*

> Divine service was generally attended on the Lord's day morning, when the weather permitted, and at these seasons a remarkable sobriety and attention were manifest. . . . Shortly after the Brandywine went to sea from the port of New-York, we established two weekly prayer-meetings which were regularly and well attended, and with increasing interest. . . .[59]

Writing to the Secretary of the Navy on 28 June 1831, Chaplain McLaughlin included the following report of his activities relative to the distribution of religious literature on the ships of the West Indies Squadron.

> During the voyage, the following distribution was made of the tracts and other religious publication . . . to the frigate Brandywine, twenty thousand pages; the sloop-of-war Falmouth, thirty thousand; the sloop Eric, twenty-eight thousand; the sloop Peacock, twenty thousand; the United States schooner Shark, fourteen thousand two hundred. . . . I found all the ships plentifully supplied with bibles.[60]

A description of evening prayers aboard the *United States* while the ship was lying at anchor with some English men-of-war at Mytelene appeared in a letter dated 14 August 1834, which was published in the *Sailor's Magazine* for February 1835. The chaplain on board was C. S. Stewart.

> [Evening prayer] is performed immediately after lowering the colours at sunset, and when all hands are "piped to hammocks," an hour at which the commanders and superior officers of the English ships were usually lounging after dinner, in the stern galleries of the stately castles in which they float; and we have been told again and again by the most distinguished of them, from Sir Josias down, that they knew not when they have been more happily impressed, or more delighted, than when listening to the devotional strains, breathed over the anchorage by our band, and looking down from their more lofty vessels on the groupings of our ship's company assembled for prayers.[61]

Chaplain Addison Searle summarized his duties at the Navy Yard, Boston, in a letter to the Secretary of the Navy dated 31 August 1841.

> Since my orders to this Station, on the first of November 1837, I have done the duty of Chaplain on board the receiving ship Columbus, as well as at the Yard, with the only exception of the month of August 1840. I officiate on Sundays on board the Columbus, attend the Sunday School & the Apprentice boys & such of the crew as are disposed to attend, and preach at the Yard in the afternoon. Besides attending the sick & the burials, I teach the Apprentices the church service & sacred music every week. The Sunday School, commenced nearly two years ago, is directed & superintended by myself.[62]

That additional activities, beyond formal naval duties, were occasionally being assumed by chaplains may be seen from the unique experience of Chaplain Timothy J. Harrison when his ship, the *Java,* put in at Monrovia, the capital of Liberia. Here, on Sunday, 20 February 1831, he had an opportunity to speak at the Methodist Church before a congregation of colored people who had been sent there by the American Colonization Society. The Chaplain was accom-

[57] Parker, *Recollections of a Naval Officer,* p. 122.
[58] Nav. Rec. Coll., III:Jan.-Feb. 1827:94.
[59] *Op. cit.,* Dec. 1830, p. 114.
[60] Nav. Rec. Coll., III:5 of 1831:223.
[61] *Op. cit.,* p. 178.
[62] Nav. Rec. Coll., III:Aug. 1841:282.

panied by several officers of the *Java*. They found "a large, well-dressed, respectable looking congregation of worshippers" who manifested keen interest in his remarks.[63]

The fundamental duties of the chaplain have not changed with the passing of the decades. Officiating at Divine Services, visiting the sick and the dying, counseling those with problems, and distributing religious literature still occupy the major part of a chaplain's working hours.

SCHOOLS FOR MIDSHIPMEN

Since chaplains had long been identified with education aboard ship, it is not surprising that Chaplain David P. Adams was placed in charge of a school for midshipmen in August 1821 on board the *Guerriere* at Norfolk (then called Gosport). For a time Chaplain Moses B. Chase assisted. Captain A. Sinclair reported the condition of the school to the Secretary of the Navy on 19 June 1822.

> Mr. Chase, the late assistant of Mr. Adams, in the Nautical School on this Station, being about to receive orders as a Minister of the Gospel, has exprest a wish to resign his situation as School Master; and as the present reduced state of the institution justfies our dispensing with his services, I have, in concurrence with Mr. Adams, consented that he shall do so.
>
> . . . .
>
> In the original plan, it was my decided opinion that it should be somewhat upon that of the Military Academy at West Point. . . . Mr. Adams has now requested me to solicit you to authorize it to be put on that footing. . . .[64]

Secretary of the Navy Branch in his annual report for 1829 referred to the schools for midshipmen then being conducted at Norfolk, New York, and Boston. The *Navy Registers* for 1827, 1828, and 1829 indicate that Chaplain Cave Jones was on duty at the "Naval School, New York." The three schools were still in existence in 1833, at which time Norfolk had the largest enrollment—thirty-one. Following the death of Chaplain Hunter in February 1823, little is known of the school which once flourished at Washington. In 1838 a naval school was established at the Naval Asylum in Philadelphia, but chaplains are not known to have been associated with this endeavor. Attendance on the part of midshipmen at these schools was entirely voluntary.

Beginning with the appointment of professors of mathematics in 1831, the chaplain's responsibilities for the education of midshipmen gradually decreased. By 1835 it appears that the major educational duties at the naval schools had been transferred from chaplains to the professors of mathematics.

An editorial in the New York *Observer* offered the novel suggestion of combining the duties of chaplain with that of professor of mathematics so that clergymen would be available for more ships.

> While the character and deeds of the navy are fully appreciated by the religious portion of the community, the limited provision made by the government for its moral and intellectual improvement induces many to withhold that cordial support which otherwise might be accorded. The most feasible plan of correcting this evil, would seem to be, by uniting the office of chaplain and professor of mathematics in the same individual, thus securing to every ship in commission, the invaluable services of a clergyman; as all vessels, except brigs and schooners, are entitled to professors. Those duties may easily be performed by one competent person.[65]

Chaplain George Jones was greatly interested in the education of midshipmen and on 30 January 1839 wrote a long letter to the Secretary of the Navy revealing the defects of the existing system and outlining in detail his proposal for the establishment of a naval academy.

> I have, during nearly seven years, been occupied more or less in teaching Midshipmen & during a yet longer period, I have been an attentive observer of their studies as conducted by others. As the result of all this I have no hesitation in saying not only that the old system of instruction on ship-board is almost entirely inefficacious, but also from the necessities of the case it never can do much good. An improvement has recently been made by appointing Professors of Mathematics instead of schoolmasters, & the severe examination which candidates for this situation have to undergo must secure to the young officers in every case competent instructors.[66]

For more than forty years, or until the establishment of a naval academy at Annapolis in 1845, chaplains took an active part in the education of the midshipmen. Even after the Academy was founded, certain chaplains continued their collateral duties as teachers.[67]

SOCIAL AND MORAL CONDITIONS

During the third and fourth decades of the past century, the nine chaplains on duty were becoming increasingly sensitive to moral issues and were more articulate.

Rockwell was one of the first to condemn excessive drinking. He stated that aboard his vessel a daily

[63] *Sailor's Magazine*, Aug. 1831, pp. 387-8.
[64] Nav. Rec. Coll., II:2 of 1822:31.
[65] Reprinted in *Sailor's Magazine*, May 1838, p. 381.
[66] Nav. Rec. Coll., III:Jan. 1839:169.
[67] See Drury, "Famous Chaplain Teachers of Midshipmen," *Naval Institute Proceedings*, May 1946.

allowance of half a pint of whiskey was given each man. This whiskey was mixed with water in a large tub and the diluted liquor was called grog. This was served three times a day. "Intemperance in the use of ardent spirits," wrote Rockwell, "is to the seaman literally the mother of abominations, and the prolific source of most of his degradation and deep and bitter woe."[68]

In the Diary of Chaplain Thomas Lambert the following entry occurs under date of 16 April 1837:

> Today I preached on board the Frigate Constitution, by request of Com. Elliott, upon the subject of temperance. The discourse was intended for last Sabbath, but it was deferred in consequence of the weather. All the officers of the squadron that could be spared, were present.[69]

This may be considered typical of the interest that other chaplains took in the subject of temperance.

Those especially addicted to drink would sometimes smuggle liquor aboard to supplement their grog, and when on shore would spend all their money for drink. An increasing number of merchant ships of this period discontinued serving grog. This influenced sailors who wanted grog to enlist in the Navy.

One of the lighter punishments aboard a man-of-war was the temporary denial of a man's allowance of grog. Some sailors even preferred the cat-o'-nine-tails to such a restriction. Certain officers felt that since the denial of grog was such an effective punishment for light offenses, the legal abolishment of the grog ration would be detrimental to discipline.

Drunkenness was a major cause for flogging. An editorial in *Sailor's Magazine* for May 1831 under the heading "Whipping in the Navy" quoted the Honorable Mr. Ellsworth as saying:

> A very intelligent lieutenant of the navy, recently informed me that a great proportion, I think he said eight-tenths of the whipping on board of our national ships, was made necessary by intemperance. He said it was a monstrous evil; and who would doubt it, if it caused nothing else but the brutal and debasing practice of whipping?[70]

Again in June 1840 the editor returned to the subject of intemperance in the Navy. He gave the following quotation from a letter written by a naval officer:

> Intoxicating drinks are the cause of all the trouble we have on board our ships. Men are frequently confined in double irons, and for what? Beastly intoxication, insubordinate and mutinous conduct, insolence, or desertion, while under the influence of liquor. Nine men out of ten, yes ninety-nine out of a hundred, that are brought to the gang-way to receive a dozen lashes with a cat-o-nine tails on the naked back are flogged for offences committed while under the influence of ardent spirits.[71]

Some chaplains were active in promoting temperance by persuading sailors to sign a total abstinence pledge. Chaplain McLaughlin comments on temperance activities on the ships of the West Indies Squadron in 1830:

> On board the United States sloop-of-war Falmouth, seventy seamen were reported to me as having totally abstained from the use of ardent spirits and between forty and fifty on board the Brandywine; and these were among the most healthy, cheerful, and orderly in the ship. During the cruise I never knew a complaint against one of them. What was the total number in the squadron, I could not ascertain, but have reason to believe it was considerable, and that the total disuse of spiritous liquors is increasing in our navy generally.[72]

On 13 January 1830 three Navy surgeons unanimously condemned the use of spirits by midshipmen in a report submitted to Secretary Branch. The growth of the temperance movement in the Navy induced the Department to issue an order dated 15 June 1831 which permitted seamen to draw a commutation of pay, at the rate of six cents per ration, in lieu of grog.[73] Within a year, this order was having a marked effect. "By accounts from several of our vessels of war, on foreign stations," one contemporary noted, "it appears that more than half their crews have voluntarily relinquished ardent spirits, and accepted the cost of their customary allowance in the small sum substituted by order of Mr. Woodbury."[74]

About a year later, there was still further evidence that the "cause of Temperance" in the Navy had been

> steadily advancing. Many, probably, still recollect the letter of Commodore Biddle from the Mediterranean the last summer, stating, that "of the 1107 men in his squadron, 819 had stopped their ration of spirits, and that on board the John Adams not a man drew his grog."[75]

Of the progress of temperance aboard the *Potomac* alone, Rockwell was able to state at the end of 1835:

> About one hundred of our men draw their grog, of whom one third are petty officers. The character of those who have given it up, and the results of the whole experiment, fully prove that there is no need of having a drop of the poison on board the ship.[76]

[68] Rockwell, *Sketches of Foreign Travel*, II:411.
[69] Diary in possession of Capt. Harold E. Cook, USN (Ret.), who has granted permission to use quotations.
[70] *Op. cit.*, p. 281.
[71] *Op. cit., pp.* 308–9.
[72] Nav. Rec. Coll., III:5 of 1821: 223.
[73] Nav. Rec. Coll., V:1:285.
[74] *Sailor's Magazine*, May 1832, p. 281.
[75] *Ibid.*, June 1833, p. 293.
[76] *Ibid.*, Dec. 1835, p. 132.

The success or failure of the temperance movement on individual ships was determined largely by the attitude of the commanding officer. The records indicate that while the movement was strong on certain vessels, the opposite situation prevailed on others.

Other signs of improved moral conditions aboard United States naval vessels during this period were not wanting. An observer in the *Constitution* in 1842, at the request of the editor of a prominent periodical, gave an illuminating report on "moral and religious improvement in our Navy" during the previous decade.[77] Profanity, while far from disappearing completely, was much less heard among the men than formerly and fewer officers were found who felt that "deck-duty could not go on without a volley of oaths now and then."

A decided change in the attitude of naval personnel on matters of religion was also noted. Some years earlier, the observer said, "religion was looked upon as quite out of place in a man-of-war: a religious officer was a rare individual; and when any person gave serious and practical attention to these important matters, he was considered as, in a measure, disqualifying himself for the service; and by dropping the high and chivalrous parts of an officer's character, as no longer fitted to do the Navy credit." He noted that formerly "when an officer became religious, he was looked upon somewhat in the light of an insane person, with a kind of feeling towards him of avoidance," but this was now past. The seamen too were experiencing the same change, although "among them it is more recent, and marked by greater fluctuations."

EVIL INFLUENCES ON YOUTH ABOARD SHIP

Closely linked with the prevalence of intemperance in the Navy was the influence of the environment aboard ship upon the young boys and midshipmen. According to a circular order of 1 April 1826 ships-of-the-line were allowed forty boys; ships-of-the-line, second class, thirty; frigates, first class, fifteen; and frigates, second class, ten. This meant that one boy served two guns, or one gun on each side of the ship. It was assumed that only the guns on one side would ever be in use at the same time. The boys' ages usually ranged from fourteen to eighteen.[78] The *Brandywine* in 1825 carried twenty-five boys. A Navy officer, writing to the editor of *Sailor's Magazine*, bemoaned the fact that so many boys learned to drink after entering the naval service. It was not until 29 August 1842 that a law forbade the issuance of spirits to those under twenty-one and allowed a monetary equivalent.[79]

In addition to the boys were the midshipmen, often in their early teens. Midshipmen, by virtue of their office, were given considerable authority over enlisted men, some of whom were old enough to be their grandfathers. W. M. Murrell, a member of the crew, reflected the attitude of enlisted men who were obliged to endure the insults of the unrestrained and cocky youths.

> By and by, writing of young middies, I wish to inform my readers that we have on board two *infants*, in the dress of midshipmen, whose insulting conduct at times is almost unbearable. The morals of the ship, under the especial patronage of the officers, may be judged of when I state that these infants, for future patronage, are from *twelve* to *fourteen* years of age, and they will very often walk up to the seaman, and utter, with as much smartness as their gentlemanly elders, "You be d--d," and "You d--d son of a b---h."[80]

Rockwell saw the problem from a chaplain's point of view and advocated the policy of giving the early professional training to midshipmen on shore where they would be "under far more correct and efficient moral, social, and religious influence than on shipboard," and where "they would not, when young and reckless, be exposed to strong and overwhelming temptation to vice." He wrote:

> I have never heard a naval officer of age and experience speak in favor of placing boys as midshipmen on board our ships of war. They uniformly reprobate it, and say, that at least three fourths of them are ruined, or otherwise fail, and leave the service.[81]

Chaplain Rockwell supported the establishment of a naval academy for the "early scientific training of our naval officers."

IMMORALITY ON SOME SHIPS

Sharp criticism of immorality permitted aboard some naval vessels while in port was made by Chaplain Hervey H. Hayes as a result of his observations aboard the *Java* from May 1827 to September 1830 while this vessel was with the Mediterranean Squadron. About two years after his return to the States, Chaplain Hayes related what he had seen to a friend, Arthur Tappan of New York City, who evidently re-

[77] *Southern Literary Messenger*, Feb. 1843, pp. 73 ff.
[78] V:1:1 Apr. 1826.

832159—49——5

[79] *Sailor's Magazine*, June 1840, p. 307, and 5 *Stat.* 547.
[80] Murrell, *Cruise of the Frigate Columbia*, pp. 208–9.
[81] Rockwell, *op. cit.*, II:422.

quested the Chaplain to put his observations in writing. This the Chaplain did, but failed to sign his name to the paper. Tappan forwarded the account to Senator Theodore Frelinghuysen of New Jersey and told the Senator that the criticisms came from the pen of Chaplain Hayes.

Before submitting Hayes's unsigned report to the Navy Department, Frelinghuysen wrote to the Secretary, Levi Woodbury, requesting the address of Chaplain Hayes.[82] The Secretary replied that the Chaplain was stationed at the Navy Yard at Pensacola, Florida.[83] The Senator then wrote to Hayes requesting permission to submit his statement to the Secretary. The Chaplain was hesitant but finally acquiesced. The Senator sent the following letter to Secretary Woodbury together with Hayes's statement:

> The enclosed letter to a very reputable merchant from a Chaplain in the U. S. Navy was sent to me, some weeks ago, with a request that it might be laid before you. It disclosed such a surprising body of alleged facts that I was unwilling to send it to your department until the authority was furnished. The name of the writer has been given: and if you desire it, I will promptly inform you.
>
> I am fully persuaded that your frowns will be directed against licentious practises of the kind mentioned in the letter, & I believe also that the occasion will in your estimation fully justify this communication.[84]

In reading the following account by Chaplain Hayes, it should be remembered that he was writing to his friend Arthur Tappan:

> Dear Sir,
>
> When I saw you in N. York, you asked me whether it was still the practice to admit *abandoned females* into our public ships for two nights after arriving in port. I answered not for two nights only but for 2 after 2 indefinitely. You inquired to what extent this practice was tolerated, and what ships in the Mediterranean were free from this pollution at the time I was there. I had not time then to go into details and had some scruples about the propriety of doing so, but promised to inform you further at another time. In fulfilment of this promise I now write you.
>
> The picture that rises before my mind, whenever reflection turns back my attention to the Mediterranean & fixes it upon the scenes of which I now speak is too shocking and horrible for minute description. It would shock every decent person in the nation who has not been familiar with such scenes to read a true account of the loathsome brutality that disgraces some of our public ships.
>
> When *Com. Rodgers* commanded in the Mediterranean he prohibited the admission of abandoned females into his own ship & forbade their being tolerated in any ships under his command, but his successor no sooner drew sword there than all these bands were cut asunder—every obstacle removed—orders given to let them come on board, & night after night his ship was swarming with vileness & disease, equal to any thing that could be found in the lowest dregs of New York. Encouragement was given to other Commanders to follow this example, & soon a *Broad Pennant*, about twice the ordinary size, proudly fluttered over a squadron of *licensed American brothels*, decorated with the ensigns of the country. Their 13 stars had never shone more dimly, but to the Com. this seemed to be a real luxury of which he often spoke most humorously. Boat-loads of these creatures were hanging about the ship early in the p.m. & it was sometimes late in the morning before they got on shore. I have never seen the like in any men-of-war of other nations & all lying about us seemed to blush for us, as we seemed to have no shame for ourselves. I do not say that the ship's boats steered by Midshipmen were sometimes employed to bring them on board, but I believe this to be the fact.
>
> I think there was but one ship in the squadron which did not yield to the persuasions of the Com. to let the *sweet-hearts*, as he called them, on board. He had good principles & moral courage enough absolutely to resist all entreaty on the subject. He was branded as a superstitious Methodist—his religion cursed—his ship made a convict ship—a dozen criminals after punishment sent on board at a time to be *reformed* by Capt. S.—& to annoy him.[85]

Hayes claimed that more men "were disabled from duty from this cause [venereal disease] than all others put together on board the ship in which" he sailed. In picturesque and forceful figures of speech, he continued his exposé:

> I often pitied the policy that supported a most splendid ship of the line for nearly 2 years in the Mediterranean & at an expense of about $1,000 per day, apparently for little else than a magnificent charnel house. For 8 or 9 months together without our raising anchor the daughters of pollution like so many Turkey Buzzards were feeding on her carcass.[86]

The Secretary of the Navy was disturbed by the revelations and wrote the Senator stating that the "subject of certain immoralities on board our public ships has been referred to the Board now reviewing our Naval Regulations with a view to have some changes made in relation to this subject if deemed necessary." However, more than twenty years elapsed before a new edition of *Navy Regulations* was issued and this carried no new ruling on the subject. The Secretary, promising to investigate, wrote:

> The scenes described in the paper which you enclosed, are so gross & improper, that if the name of the writer of the letter is communicated, I will at once take measures to ascertain the truth of the Statements & punish if practicable those offending so highly against public decency, correct discipline, & sound morals.[87]

[82] Nav. Rec. Coll., IV:Dec. 1832:135.
[83] *Ibid.*, VI:Dec. 1832:135.
[84] *Ibid.*, IV:Feb. 1833:100.

4,a

[85] *Ibid.*
[86] *Idem.*
[87] Nav. Rec. Coll.,VI:19:154.

Senator Frelinghuysen then sent Chaplain Hayes's name to the Secretary who, under date of 7 March, wrote to the Chaplain saying:

> The object of this communication is to inform you of the willingness of this Department, at all times, to commence an investigation as to the conduct of our officers abroad, in relation to that subject—whenever you may think it proper to prefer to me your charges and complaints.[88]

"Nothing was farther from my anticipations," the Chaplain replied, "than that a copy of this communication would be sent to the Hon Secretary, but I am happy that such has been its destination." He added:

> I feel able to establish every statement it contains. They are notorious to every officer who was on that station at the time to which it refers. Indeed, it but half unvails the pollutions of the picture. Should I lift the vail from *cabins* and other places where officers openly carried on their shameless licentiousness, it would be enough to shock the public mind throughout all the land.[89].

After returning from the Mediterranean in the fall of 1830, Chaplain Hayes saw duty at the Norfolk and Philadelphia Navy Yards. In February 1832 he was sent to the Pensacola Navy Yard where he found Captain Dallas, who had been Commodore of the Mediterranean Squadron during the time the alleged immoralities took place. Hayes, commenting on his attitude aboard ship, said:

> I felt bound, as Chaplain of the Navy, to face the enormous evil in question & not to abandon it till the last means within my power has been used for its correction. I talked & preached against it on board, published it, & all to little effect.[90]

Hayes was very frank and at times tactless in his utterances. He often made direct references in his sermons to certain of his fellow officers. This naturally made him *persona non grata* with them. It appears evident that he was not given a warm welcome when he reported for duty at Pensacola for no quarters were provided. He claimed that he had to walk a total of twelve miles daily in order to perform his duties. Referring to his long daily journeys, he wrote, "As the way is loose sand, & the sun powerful, I found this put to trial my health & strength, but I punctually performed my duties notwithstanding . . ." Hayes also reported that the "Chapel at Pensacola is part of a joiners shop without any glass windows, & is utterly unfit for use in winter." He served at this station from 29 February 1832 to 7 January 1833, and looking back upon those months wrote: "My relation to the Yard was neither useful nor happy."[91]

Consequently, at the time he made his formal charge, Chaplain Hayes was more interested in conditions existing at Pensacola than those which he condemned in the Mediterranean Squadron nearly three years previous. He made charges of wasteful expenditure of Government funds at Pensacola, of excessive gambling, and of the desecration of the Sabbath by the promotion of hunting parties on that day.

Hayes, realizing the serious nature of his charges, sent in his resignation, effective at the expiration of his leave, 18 April 1833. However, he left the way open for continued service. "I shall at the same time," he wrote, "offer myself as a candidate for reappointment."[92]

The Secretary of the Navy wrote to Hayes asking whether he intended to make his charges against Captain Dallas public and whether he wished an official investigation.[93] Hayes replied that while he hesitated to cause any person to suffer, he was willing to face the issue of an investigation.[94]

The Navy Department then sent extracts of the Hayes letter to Captain Dallas at Pensacola. Evidently the Department made no inquiry into the serious charges of immorality on board the ships in the Mediterranean but rather asked Captain Dallas for an explanation of alleged irregularities which Chaplain Hayes claimed had occurred at Pensacola. Dallas circulated the charges among his officers who wrote rebuttals. These letters were sent to the Department by Captain Dallas together with his own explanation. John Boyle, Chief Clerk, who signed his name as Acting Secretary of Navy, acknowledged these letters on 11 May and stated:

> It was deemed a matter of justice to you as well as to the service, to make you acquainted with the complaint. The explanations furnished are entirely satisfactory. The Rev. Mr. Hayes ceased to be an officer of the Navy on the 18th ultimo: on that day his resignation was accepted.[95]

Although the Secretary of the Navy, Levi Woodbury, expressed a desire to correct any gross immoralities that may have been countenanced on board Navy vessels, but when his chief clerk, John

[88] *Ibid.*, I:20 3/4:148.
[89] *Ibid.*, III:Mar. 1833:128.
[90] *Ibid.*

[91] *Ibid.*, III:Apr. 1833:12.
[92] *Ibid.*, III:Mar. 1833:120.
[93] *Ibid.*, I:20 3/4:187.
[94] *Ibid.*, III:Mar. 1833:215.
[95] *Ibid.*, I:Apr. 1833:304.

Boyle, handled the case, attention was shifted from the major charges made by the Chaplain to minor ones. With the acceptance of the Chaplain's resignation, the matter appears to have been dropped. If Chaplain Hayes had confined his criticism to the charges first made, official action might have been taken. Nevertheless, it is evident that the Department was increasingly sensitive to such criticisms and that some measures were taken to prevent their recurrence.

While Chaplain Hayes may have exaggerated, it is apparent that his criticisms of moral conditions aboard certain ships were, in general, well founded. An editorial in the *American Spectator* stated: "Many of our national ships, while in foreign ports, instead of carrying with them that healthful moral influence which we might righteously expect, have been made the scenes of the most degrading and brutal vices."[96]

Chaplain Rockwell, who also served in the Mediterranean Squadron in 1835, five years after Chaplain Hayes' tour of duty, made similar charges. He wrote:

> In times past, it has been customary with our naval commanders, when in foreign ports, both of savage and of so called civilized and Christian nations, to permit hundreds of abandoned females to spend nights on board our national ships; thus converting them into floating brothels, and deeply disgracing the land from whence they came. The experiment was tried on a limited scale, by a base and profligate commander, on board two ships belonging to the station where we cruised; the one, just before our arrival, and the other, while we were lying the same port. So decided, however, was the opposition of many of the officers to this vile profanation of our country's flag, that the evil was soon checked and did not spread to the other vessels in the squadron.[97]

The schoolmaster on the *North Carolina* in 1825, Father Adam Marshall, gave a brighter picture. He wrote:

> The idea that a Man of War is a place of unbridaled [sic] debauchery is totally incorrect, at least as regards this ship. When ladies of respectability came on board it is always by special invitation, and they are accompanied by officers and treated with greatest respect.[98]

Father Marshall sailed under Commodore John Rodgers who is known to have forbidden the admission of "abandoned females" on his vessel. Chaplain Hayes happened to be under an officer of different moral standards.

[96] Reprinted in *Sailor's Magazine*, Nov. 1830, p. 90.
[97] Rockwell, *Sketches of Foreign Travel*, II:402-3.
[98] Terence King, S.J., "The First Priest in the U. S. Navy," *The Army and Navy Chaplain*, April-May 1942.

FLOGGING

One of the earliest protests raised by a naval officer against the degrading custom of flogging was made by a Navy chaplain in 1830. The December issue of *Sailor's Magazine* published a letter written from Chaplain Edward McLaughlin containing the following suggestions:

> Let there be no alcohol in any of its forms, but in the doctor's medical stores—let flogging be suppressed, and other modes of punishment substituted, and regulated by courts martial according to crime—give them the Bible . . . and back all by sending on board of every ship of the line, every frigate and every sloop of war, an enlightened, discreet, evangelical, and efficient chaplain,—let the schoolmaster in all cases, be a man of practical godliness; and then . . . change every ship's company into as moral, christian, and orderly a community as any of our country societies generally are. . . .[99]

The second known published protest against this practice was by Chaplain Charles S. Stewart in his book, *A Visit to the South Seas.* In one passage the Chaplain tells of the shocking sights that confronted him when he loitered at the gangway for meditation "as night gathers round."

> In one respect, however, I have proved the spot chosen to be most unfortunate; the gangway is the place of punishment, and twice, within the last two evenings, the keenest emotions I have known on board the Guerriere, have come suddenly upon me in the sound of the lash and the cry of some wretch suffering at my side. This mode of punishment is deemed by many indispensable on board a man-of-war; and it may be so—but as yet I am far from being reconciled, in feeling, to the necessity. To me there is an indignity and degradation in it, which seem inconsistent with the high-toned principles and spirit of Americanism; and independent of all other consideration, I never witness it without being tempted to ask Paul's question to the centurion, "Is it lawful for you to scourge a man that is a Roman?"
>
> But though there was sorrow in the case, and I felt every stroke almost as if across my own shoulders, it was not without a profit too. Only the moment before I had been gazing on the mild splendor of the southern cross, absorbed in associations inseparable from it, in the life and death of Him who hung upon the accursed tree; and the mind at once reverted powerfully to that scourging "endured for us," and to those "stripes by which we are healed."[100]

Others joined in the protest. Schoolmaster E. C. Wines, who began duty on the *Constellation* in July 1829, wrote of the arbitrary way in which punishment was inflicted. "I have seen," he declared, "a sailor taken to the gangway and receive a dozen for

[99] *Op. cit.*, p. 116.
[100] *Op. cit.*, I:30-1. This protest was reprinted in *Sailor's Magazine*, Aug. 1831, p. 367.

walking aft on the starboard side of the deck."[101] Wines here referred to that side of the sacred quarter-deck reserved for the captain and the higher ranking officers. He also told of how the innocent were often punished with the guilty.

> Men belonging to the same top, the same boat, the same mess, &c, will almost universally be flogged sooner than betray their companions; and as to the old maxim that it is better that six guilty persons should escape than that one who is innocent should be punished, it is here quite the reverse: it is, let scores of the innocent be flogged, rather than that one who is guilty should escape.[102]

Chaplain Charles Rockwell stated that, although the law forbade more than twelve lashes for any one offense, often commanders would order "two or three dozen at a time, for as many different offences." His revealing description of the punishment follows:

> The cats are commonly used in the presence of the whole ship's company, and, being applied to the naked back, each cord causes the blood to settle under the skin, or cuts to the flesh, according to the severity with which punishment is inflicted.
>
> Flogging with the cats is an exceedingly severe punishment, though some suffer much more from it than others. During our cruise, two men were sentenced, by a court-martial, to receive twenty-five lashes apiece, on board each of the four ships of the squadron, as an example and warning to all who might witness their sufferings. Having reached our ship, one of them received his twenty-five lashes, and passed on to the next to be punished there, while the other one was so much reduced, that after receiving ten lashes, our surgeon gave orders to desist, as, in his opinion the man's life would be endangered by inflicting more.[103]

Interesting revelations on flogging were made by Murrell in his *Cruise of the Frigate Columbia Around the World*. In frank, revealing words that surge at times with feelings of indignation and resentment, Murrell not only focused public attention on the evils of flogging but unhesitatingly mentioned the names of the officers on the *Columbia* who were most flagrant in the abuse of their authority.

Men were flogged for the most trivial offenses. Murrell tells of an old man who received twelve lashes for "talking on the fore yard, to another man, *upon a point of duty*."[104] Another sailor was given twelve for neglecting to close the door to the head.[105] Murrell himself received a dozen because he failed to mark properly a piece of his clothing, and again when he accidentally spilt ink on the deck. "At this time," he wrote, "I happened to have on a thin cotton frock, and every lash, to the number of twelve, fetched a piece of flesh from my back, leaving scars which I can show to the present day."[106] A man could be flogged for failure to polish his shoes; he could be flogged for the negligence of another; he could be flogged for drunkenness—it was a terrible punishment when used at the direction of a cruel, sadistic officer.

In the following extracts Murrell seems to speak for all the members of the crew:

> This morning there were two hundred and forty lashes served out in a short space of time. The boatswain's mate afterwards acknowledged to me, that his arm was never so fatigued before. These back rations were served upon twenty of the berth deck cooks, for the following trivial offence: It was "customary" on board of this ship, however "uncustomary" in other men of war, for the first lieutenant to go round the berth deck every morning, and inspect the men's "tin pots and pans," with a piece of white paper, with which he wiped the insides of the said pots and pans, and if the least soil came off on the paper, the owner was sure to receive a dozen lashes from this *molly coddle* of tin ware.
>
> Cleanliness on ship board is absolutely necessary, and should be enforced; but even clean tin will soil *white paper*, especially when dampened, as at sea; and cleanliness can be enforced on an American by a kind word, more effectually than by a dozen lashes.[107]

The author did not plead for the complete abandonment of flogging. He felt that there were offenses aboard ship which warranted this punishment and stealing was one of them. When two convicted thieves were flogged, he wrote that no one disapproved of this "deserved chastisement."

Murrell cited another flogging incident, however, which he felt was unjust.

> How different the case of a poor marine, whose back was absolutely cut to pieces, while our ship was lying at Norfolk, for an offence of the most trivial nature! The commodore on this occasion was absent from the ship, and the unfortunate marine was reported to his commanding officer, who, contrary to the rules of the service, ordered the poor fellow to strip, and not satisfied with using the cats, the usual instrument of punishment, whipped him over his naked back and shoulders with a *colt*. The colt is a piece of hard twisted rope, about the size of a man's fore finger, which not only scarified, but absolutely brought out a piece of flesh with every blow that was struck. Let me here appeal to the hearts of all, even to the most degenerate man: was not a scene like this sufficient to harrow up every feeling, and freeze the very blood in our veins;—to behold a wretch, destitute of any feeling for a

[101] Wines, *Two Years and a Half in the Navy*, I:21.
[102] *Ibid.*, I:92.
[103] Rockwell, *op. cit.*, II:407–8.
[104] Murrell, *Cruise of the Frigate Columbia*, p. 207.
[105] *Ibid.*, p. 198. "Head" in the Navy is a term for the toilet.

[106] *Ibid.*, p. 138.
[107] *Ibid.*, p. 55.

fellow-creature, to have power placed in his hands, so as to enable him to torment those under his command;—a wretch, deaf to all feelings of humanity, unworthily abusing that power, which perhaps would have been an honor in the hands of a humane and worthy officer?[108]

The books, written by Navy chaplains and members of the ship's company, drew aside the veil and revealed actual conditions on many vessels. They contributed to the rising tide of public sentiment which condemned the continuance of flogging. There is evidence, also, that the Navy Department was developing a tender conscience on this subject. As early as 26 September 1831 the following circular was issued:

The other subject, and our wishes upon it are, that till Congress deem it proper to alter the existing laws concerning punishments in the Navy, and whenever those laws allow a discretion in the choice of punishments, the first resort, in the case of offenses by seamen, is recommended to be always had to pecuniary fines, badges of disgrace, and other mild corrections, rather than to the humiliating practice of whipping; and that never on the same day, by punishing, under an Officer's own authority two offences at once, should the stripes, limited by law, be exceeded in number; or be inflicted otherwise than in the presence and under the sanction of the commanding officer of the vessel or station.[109]

A Navy Department directive in 1832, contained the following: "Flogging is recommended to be discontinued when practicable, by courts as well as officers; and some badge of disgrace, fine, etc., substituted when discretion exists."[110] Thus, sentiment was growing both among naval personnel and within the Department itself in favor of discouraging this form of punishment.

THE LAMBERT DIARY

The earliest known extant diary of a Navy chaplain is that of Chaplain Thomas R. Lambert, who was in the service from 1834 to 1856. He was aboard several vessels of the Mediterranean Squadron in 1836 and 1837, the period covered by the unpublished Diary. A generous portion of this work is devoted to a description of places visited in the Mediterranean lands. (See page 50, footnote 69.)

The Diary contains numerous items which refer to the chaplain's activities. Much incidental information of interest is revealed in such entries as the following:

Oct 2, 1836. We had no service in consequence of watering the ship.

Oct 9, 1836. Not feeling very well. I did not have any service.

Oct 23, 1836. I read the regular Service for to-day, but in consequence of a stormy breeze, I had no sermon.

Nov 20, 1836. . . . the weather would not admit of service.

July 11, 1837. Had no service today, owing to one of Comd Elliott's freaks. Nothing to prevent it, being a very fine day—Comment is unnecessary.

It is evident that the holding of Divine Services was dependent upon such factors as the state of the weather, the state of the Chaplain's health, the work of the ship, and pleasure of the commanding officer. Chaplain Lambert referred several times to the valued assistance rendered by the ship's band in the Sunday services.

On 12 April 1837, while the Squadron was at Port Mahon, Chaplain James Everett of the *Constitution* died. Chaplain Lambert was requested to officiate at the funeral held ashore and gives the following detailed description of the service:

April 13 1837. Today, the flags of all the Men of War in the harbor were hoisted half mast, for the Rev. James Everett of the Constitution, at the usual hour about 8 o'clock in the morning, & continued so until his body was interred. I was appointed to read the funeral service, & repaired on board about half past ten o'clock where I found all the officers assembled. At a little before 11 o'clock the procession moved from the ship to Fort St. Philip, consisting of 20 boats with each an ensign half mast. The Band in the first boat, the corpse in the 2nd, myself & the Rev. John Falls, the chaplain of H.B.M.S. Rodney, & the Bearers, in the 3d Boat—his messmates & marines in the 4th Boat—then followed the other Boats. As soon as they had all shoved off from the Ship, the Band struck up "Lord dismiss us with thy blessing" in a solemn & impressive strain, & the boats pulled minute strokes to the shore. Here the procession formed, & the corpse was received by the Marines which had perviously landed, with arms reverse & 3 rolls of the drum. We then proceeded to the grave, preceded by the guard of Marines. Then followed all the officers of the English, French, & our ships, & the civil and military authorities of the place. When we had arrived at the burial place, the Marines formed a line & the procession passed by them to the grave. They then formed near the grave, & after I had finished the service, they fired three volleys over the body & we returned to the Boats. . . . Mr. Everett was upwards of 50 years of age, & was the oldest chaplain in the service. He had been in the service nearly nineteen years.

Commodore J. D. Elliott on 19 April ordered Chaplain Lambert to take over the duties of chaplain aboard the *Constitution*. At first Lambert was happy with his new assignment, but soon difficulties, such as those previously quoted from his Diary of 11 July 1837, arose. The Chaplain promptly submitted a request to the Navy Department through his Com-

[108] *Ibid.*, pp. 35–6.
[109] Nav. Rec. Coll., V:1:289.
[110] *Rules of the Navy Department, Regulating the Civil Administration of the Navy of the United States*, p. 4.

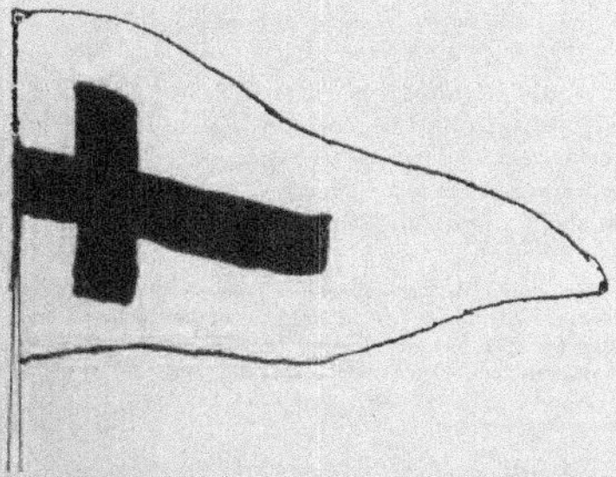

First known illustration of the church pennant from Journal kept by Acting Midshipman P. C. Johnson, USS *Ohio*, 1840–1848. Courtesy, Naval Records Collections.

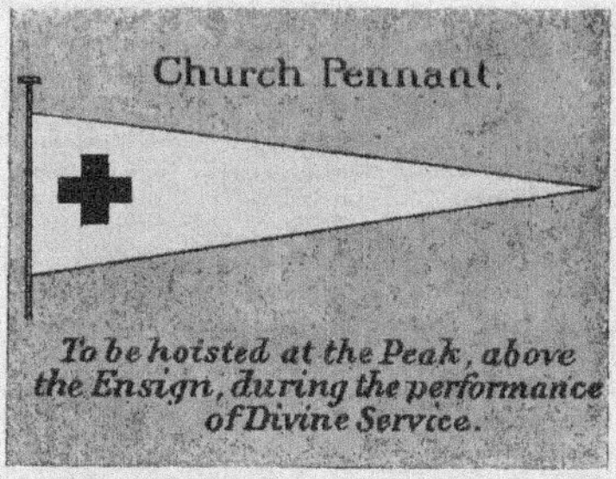

First known official illustration of the church pennant. From *Navy Signal Book,* 1858.

manding Officer asking to be detached from the Mediterranean Squadron,[111] but Commodore Elliott did not give it a favorable endorsement. Several days later, without warning, Chaplain Lambert was ordered to the sloop *Shark*. Under contemporary naval regulations, sloops did not rate chaplains. Lambert described his transfer in the Diary:

> This order I received while the Ship was getting under weigh, & when I left her she was 6 miles from the Shark, which distance I had to pull in an open boat in a heavy sea, & I did not reach the Shark until after dark. In leaving the Constitution which was going 6 or 8 knots, & which was not hove to, I lost nearly one trunk filled with clothes and books.

Chaplain Lambert registered a formal protest with the Department over this assignment. Much to his satisfaction he was soon ordered back to his former ship, the *United States.*

After Commodore Elliott returned to the United States, he was tried by a court-martial on thirteen charges filed by disgruntled subordinates. During the trial, Chaplain Lambert was called as a witness and referred to his transfer from the *Constitution* to the *Shark,* to the discredit of the Commodore. The Court sentenced Elliot to four years' suspension from active duty.[112]

[111] Nav. Rec. Coll., III: July 1837: 99.

[112] It is interesting to note that Commodore Elliott was in command of the *Niagara* in the Battle of Lake Erie, 10 September 1813. His failure to come to the assistance of Perry when the Flagship *Lawrence* was being heavily attacked precipitated a controversy which raged in and out of the Navy for decades.

THE CHURCH PENNANT

The use of the church pennant[113] in the United States Navy, like many other naval customs, was undoubtedly appropriated from the British Navy. The Library of Congress contains a manuscript on "The Church Pennant" by Commander H. H. Lippincott, ChC, USN, Retired, which gives the following account of the history and use of the pennant in the British Navy:

> The present Church Pennant (British Naval) is a survival of the old "common pennant" which went out of general use in 1864. The use of a pennant to signify that the ship's company was at prayers appears to have been instituted by Rodney during the American war, circa 1780, Art. X of his additional sailing instructions provided:
>
> "In order that the performance of divine service may meet with as little interruption as possible, the ships are to hoist a common pennant at the mizzen peak before they begin the same, and to keep it flying until they have finished."
>
> This was adopted by Admiral Arbuthnot when Commander-in-Chief of the North American Station in 1781, by Kempenfelt in 1782, and by Howe in his signal book of 1790 and became the established practice.[114]

No reference to a church pennant was made in the first Navy Department *Signal Book* to come into general use, which was adopted in August 1813. The earliest known reference to the existence of this

[113] The church pennant should not be called the church flag. In naval terminology a flag is square or rectangular, while the pennant is triangular.

[114] From a letter to Lippincott from Engineer Commander H. S. Brown, Royal Navy, acting Naval Attache to the British Embassy in Washington, D. C., for 1922.

pennant in the United States Navy is found in a manuscript Signal Book by Commodore Charles Morris which bears the label: "circa 1827."[115] The following appears in this book:

> *Divine Service,* there will be time for—When the ships company are engaged in the performance of Divine Service, the ship will wear a pendant at the mizen top Gallant Mast head.

Even though the signals prepared by Morris in 1827 were never officially adopted, the mention of a pennant to indicate "the performance of Divine Service" is evidence that it was being used at this time.

At the opening of the nineteenth century the merchant seamen of both England and America were a spiritually neglected class. "Floating Chapels" or "Ship Chapels," which were called "Bethels," were started in London and vicinity about 1814. In 1818 the British and Foreign Seamen's Friend Society and Bethel Union was formed to expand this type of evangelism. A flag was adopted consisting of the word BETHEL in large white letters on a blue background. Some of these flags were embellished with designs such as a red star in a corner or a dove bearing an olive branch.[116]

The Bethel flag first appeared in the United States on a vessel entering New York harbor in March 1821. Under the leadership of John Allan, who had been commissioned by the Bethel movement in England to carry the idea to America, a group was soon at work in New York.[117] The zealous members of this organization supplemented the activities of the Navy chaplains in ministering to the spiritual needs of naval personenl.

The *Sailor's Magazine,* the official publication of the Seamen's Friend Society, is sprinkled with references to the Bethel movement in the years following 1821. It is clear that the Bethel flag was often raised over British and American naval vessels. The following are typical quotations:

> 1835 June 21st This evenlng we had a prayer meeting on board the United States ship Franklin, lying in the North river.

> July 13th This day we hoisted the bethel flag on board the English brig Cambria.[118]

It may be assumed that after 1821 the Bethel flag was usually hoisted whenever the zealous members of the Seamen's Friends Society went aboard to conduct religious services. The January 1837 issue of *Sailor's Magazine* carried the following letter:

> Brooklyn, Dec. 21st, 1836.
>
> Mr. Editor, In accordance with my promise I would now give you some faint description of the blessed work of grace that prevails amongst the Sailors of the U. S. Navy, on board the Hudson, Receiving Ship, at Brooklyn. For some weeks past the good work has been in steady progress onwards.
>
> W.G.[119]

Missionaries of the Seamen's Friend Society carried on the work in foreign ports. One of these workers, writing from Rio de Janeiro on 1 March 1839, stated: "During the year past we have omitted no occasion of holding services under the Bethel flag on board English and American vessels in the harbour."[120]

A number of Bethel missions, often located on an abandoned ship tied to a dock, were established in the principal port cities of the eastern seaboard. The Bethel flag was invariably associated with such missions. The June 1831 issue of the *Sailor's Magazine* reported the publication of a hymnal containing 620 hymns, which sold for 37½¢ and which had been compiled especially for these Bethels and mariner churches. Several churches still in existence in some of the larger eastern sea port cities are direct descendents of these Bethel missions.

The movement spread to foreign ports frequented by American ships. The famous pioneer missionary to China, Robert Morrison, raised the Bethel flag on an American merchant vessel at Whampoa near Canton in 1822. By the end of the third decade the Seamen's Friend Society had sent representatives to such distant places as Havre, Marseilles, Honolulu, and Canton. During the following decade work was commenced in other places.

Chaplain Charles Rockwell, who was sent to Marseilles as a missionary of the Seamen's Friend Society in 1834, and who transferred to the naval chaplaincy, referred to the Bethel flag as follows:

> Our Saviour, also, from a vessel's deck preached the gospel to the multitudes who thronged the shore; thus, as it were, raising the Bethel Flag on the sea of Galilee. . . .[121]

---

[115] Nav. Rec. Coll. The presence of a letter from Morris under the front cover, dated 19 October 1848, suggests the possibility of the latter date. However, the January 1880 issue of *Naval Institute Proceedings* contains some autobiographical material of Morris which shows that he was the author of the *Signal Book* of 1813. In 1827, when Morris was a commissioner of the United States Navy, he was at work on an improved and revised edition of the 1813 signals. Such evidence, combined with the general appearance and style of the manuscript volume, confirms the 1827 label.

[116] An article entitled "History of the Bethel Flag" is contained in the 1 June 1846 issue of *The Friend of Temperance and Seamen,* Honolulu. See also *Sailor's Magazine,* Sept. 1835, p. 10; Nov. 1839, p. 90; and Jan. 1836, p. 135.

[117] *Sailor's Magazine,* Sept. 1835, p. 10.

[118] *Ibid.,* p. 33.

[119] *Op. cit.,* p. 159.

[120] *Sailor's Magazine,* July 1839, p. 356.

[121] *Sketches of Foreign Travel,* I:28.

The United States Navy had no official regulation governing the use of a church flag or pennant during these years under review. There is evidence that the Bethel flag was used at times aboard naval vessels. For example, a writer in the *Army and Navy Chronicle* for 21 September 1837 expressed his disapproval of the use of the Bethel flag on United States naval vessels in the following words:

> Even the very sight of the ships, when the unassuming peak-pendant is hoisted for manly prayer, excites a deeper devotional feeling in the spectator than where he sees the arrogant and pharisaic Bethel flag flying.

The first known reference to the actual use of the church pennant in the United States Navy is found in an entry in the Diary of Chaplain Thomas R. Lambert for Sunday, 21 August 1836. On that morning the *United States,* on which Lambert was chaplain, sailed into the harbor of Athens, Greece, at Piraeus, and found the other vessels of the Squadron —the *Constitution, Potomac,* and *John Adams.* "The salutes & the working of the ship prevented us from having service," wrote Lambert, "but was pleased to see the church flag hoisted on board the Potomac."

Such incidental references to the position of the church pennant as "at the missen peak," "the mizen top," and the "peak-pendant" imply that if the national emblem were flown at the same time, the church pennant took precedence in position. Naval traditions have a strange tenacity. Customs never regulated by official directives have persisted through the years with startling uniformity. Although no known reference to the church pennant flying above the national emblem appears before the forties, it seems reasonable to assume that this was a custom of long standing.

ITEMS OF GENERAL INTEREST

A number of items of incidental or general interest, some of the smaller mosaics in the historical picture of the chaplaincy, belong to the third and fourth decades of the nineteenth century.

The first extensive distribution of Bibles to United States naval personnel was in 1820 and 1821 when the American Bible Society responded to an invitation from the Secretary of the Navy to furnish a copy of the Scriptures to every petty officer and seaman. According to the Society's *Annual Report* of 1821, approximately 3,500 Bibles had been made available and of this number 1,650 had already been distributed. The Secretary of the Society wrote:

> Suitable directions have been issued from the Navy Department in relation to the distribution and preservation of the books; and the Managers have been highly gratified from learning that the officers are active to circulate among their crews, and that the men are very thankful to receive, the precious volume of Divine Truth.[122]

The "suitable directions" are found in a Navy Department Circular dated 3 January 1821 which reads:

> The Agent of the American Bible Society, John Nichie, Esq of New York is requested to forward to you copies of the Holy Bible.
>
> On the receipt of the same, you will distribute them, as they may be required, among the Petty Officers & Seamen under your Command, as well as those on board vessels in ordinary, as in the Yard, and direct the Purser to charge each man with the value, and on their discharge from the service, should the Bible be returned in good order, necessary wear excepted, you will cause them to be credited to the same.
>
> You will be particular in having it explained to the men, that it is not intended to charge them with Bibles delivered to them, but merely to secure the careful use of them.[123]

While Chaplain Charles S. Stewart was on duty at the New York Navy Yard, 1835–39, he took an active part in the Naval Lyceum which had been established there in 1833. During 1836 and 1837, the Chaplain was editor of a bimonthly, the *Naval Magazine,* which contained articles on professional subjects, accounts of cruises, and general naval news. Its friends claimed that it was the first periodical established in America which was devoted exclusively to naval matters.[124]

No uniform was prescribed for Navy chaplains prior to the Naval General Order of 1 May 1830 which read:

> Plain black coat, vest and pantaloons; the pantaloons to be worn over boots or shoes, or black breeches, silk stockings with shoes; coat to have three black covered buttons under pocket flaps and on the cuffs.[125]

Prior to this date chaplains probably wore their civilian clerical garb. The order of 1830 also forbade chaplains, schoolmasters, and clerks from wearing swords. On 12 November 1838 a General Order permitted chaplains to wear the official naval buttons which other officers wore. These buttons had the emblem of the eagle.[126] The wearing of vestments at Divine Services was optional.

[122] *Op. cit.*, p. 16.
[123] Nav. Rec. Coll., V:1:246-7.
[124] *Ibid.*, MS. History New York Navy Yard, p. 75.
[125] *Ibid.*, Uniform File.
[126] Ibid., V:1:339. Old Navy buttons on display in a museum case on the *Constitution* at Boston have the eagle's head facing the right.

Chaplains had no supervision other than that of their commanding officers at this period, and it was not until 1860 that they reported their activities to the Navy Department. The Department had no way of knowing whether or not commanding officers were granting chaplains the opportunity of carrying out their customary duties. Rockwell described how some chaplains were handicapped by their commanding officers:

> Those in command have it also in their power to prevent or arrest all efforts for the moral and religious improvement and those under them, as to efforts in favor of temperance and other plans of reform, and may throw constant obstacles in the way of compliance with such laws as require public religious worship in the Army or Navy. Thus in our own Navy, while no ship smaller than a frigate is allowed a chaplain, there has been more than one instance in which a chaplain has been on board a ship during a whole cruise of two or three years, and yet, in all that time, has been called upon by the captain to preach but a single sermon.[127]

The first Roman Catholic priest known to have served in the Navy was Adam Marshall, S. J., who is listed as schoolmaster aboard the *North Carolina* for the period beginning 22 December 1824. Marshall's Diary is now in the library of Georgetown University, Washington, D. C. The last entry, written by another hand, was:

> Sept. 20, 1825. At 2:30 A.M., the Revd. Adam Marshall, Schoolmaster, departed this life. At 10 A.M. called all hands to bury the dead and committed the body of Revd. Adam Marshall to the deep.

It is altogether probable that the ship's Chaplain, John W. Grier, an Episcopalian, read the burial service on this occasion.

While Marshall undoubtedly gave first consideration to his teaching duties, it is possible that he also served as the unofficial chaplain to all Catholic personnel who were on the *North Carolina*. It has been claimed that Father Marshall was the first Catholic priest to serve in either branch of the armed forces of the United States.[128]

No history of the Chaplain Corps of the United States Navy would be complete without paying tribute to that numerous company of spiritually minded officers who, in the absence of a chaplain, have conducted Divine Services or read prayers for the men under their command. The following extracts, taken from the log of the *Vincennes* at a time when no chaplain was aboard, are typical:

> July 5 Read prayers and mustered crew.
> July 12 at 11 performed divine Service. Read the Articles of War and mustered the crew.

Other officers, who may not actually have conducted Divine Services for naval personnel, were known for their religious convictions. Among these should be mentioned Matthew Fontaine Maury, the outstanding naval scientist of the nineteenth century.[129] This versatile naval officer after spending more than a dozen years at sea became interested in and wrote widely on subjects such as oceanography, mineralogy, geology, astronomy, and exploration. His writings are studded with quotations from the Bible which this versatile student knew so well. He had definite ideas about the relation between his beloved science and his beloved Bible and felt that the two did not conflict if each were rightly interpreted. Answering his critics who objected to his confirming scientific doctrines with Biblical quotations, he said: "The Bible is true and science is true. . . . They are both true; and when your men of science, with vain and hasty conceit, announce the discovery of disagreement between them, rely upon it the fault is not with the Witness or His records, but with the 'worm' who essays to interpret evidence which he does not understand."

Maury was eager to place in the hands of every ship's captain charts that would indicate the condition of winds, calms, and drift of currents in all seas of the world at all seasons of the year. Such charts would provide a dependable substitute for the trial-and-error method of navigation upon which ship's masters then relied. He figured that since God had ordered the universe, He had also decreed order in the universe. He was convinced that constant laws of God prevailed amid the howling of winds and the rush of ocean currents. This idea of divine order and design in the universe occurs again and again in his writings and thinking, as a motif recurs in music. After the most painstaking scientific research, his famous *Wind and Current Charts* were developed which have become a great boon to mankind in saving lives, time, and money.

Shortly before the Civil War appeals were made to Maury to devise a method of preventing collisions at sea between east-bound and west-bound ships due to fogs off the Newfoundland banks. Maury naturally turned to the Bible for guidance and found in the Eighth Psalm the words "paths of the sea." He pondered these words carefully and then cried out, "If

[127] Rockwell, *Sketches of Foreign Travel*, II:122.

[128] Terence King, *The Army and Navy Chaplain*, April–May 1942.

[129] The material on Maury is drawn from C. L. Lewis, *Matthew Fontaine Maury, Pathfinder of the Seas*.

God says 'paths of the sea,' they must be there, and I will find them." And find them he did. After patient and prolonged research in ocean currents, prevailing winds, fogs, and drifting icebergs, he charted two tracks across the Atlantic, each about twenty-five miles wide, one for east-bound and one for west-bound vessels. The Navy used these lanes at once and finally in 1898 all transatlantic steamship companies adopted them by written agreement.

Although not formally joining a church until late in life, this "Pathfinder of the Seas" received much of his inspiration from Holy Writ. His philosophy may be summed up in his own words: "When principle is involved be deaf to expediency," and "He only is truly brave who has the courage to do right. This is the highest quality of bravery that a military, or any other man can possess."

Frequently the commanding officers of naval vessels without chaplains request local clergy to conduct Divine Services on board their ships in this country. When in foreign ports American missionaries sometimes acted in this capacity on United States naval ships. The *Sailors Magazine* in July 1840 printed a letter recounting the services rendered by the Reverend Justin Spaulding, missionary of the Seamen's Bethel at Rio de Janeiro.

> Dear Brother,—The Independence is about leaving us, which makes us all quite sad; she takes with her the finest set of officers perhaps you ever saw. . . .
>
> Constantly on the Sabbath, when in port, we have endeavoured to minister to them and the crew in holy things. There are on board between six and seven hundred souls. What an invaluable cargo! At half past 10 o'clock A.M., all hands are mustered on deck for service. Even when they have no clergyman they attend prayers in the form of the Protestant Episcopal Church. In no house of worship have I ever observed attention to the services more uniformly correct, and sometimes to the word spoken, marked. Every Sabbath, we have carried with us on board a quantity of moral and religious reading, which, after the regular services, we have distributed among the sailors and which they appeared eager and thankful to obtain. Sometimes the commodore [Nicholson] himself would come and assist in circulating them. . . .[130]

IN SUMMARY

During the years 1821 to 1830 inclusive, a total of thirty chaplains saw service in the United States Navy for varying periods. Twelve of this number had been appointed prior to 1821. Since the Navy Department during these years adhered rather consistently to the general policy of limiting the chaplains on duty to nine most of them must have served only a few years. Such a turn-over reflected the dissatisfaction of many chaplains over matters such as compensation. It also revealed certain weaknesses of the Navy Department's method of selecting chaplains. The Department was extremely lax in regard to physical fitness and age of applicants. For example, Burgess Allison was seventy years old when he was commissioned.

While reforms were taking place during the years 1820–1840, it was in the latter half of this period that the effect of these changes became evident. Gradually greater care was being taken in the selection of chaplains, their duties were more carefully defined, their status was improved, and their pay liberalized. Evidence of a happier situation for chaplains during this latter decade may be noted from the fact that only nine new men were needed to maintain the quota compared with the eighteen new men of the previous decade. More and more, consecrated men of ability were looking upon the naval chaplaincy as a life career. Such an attitude naturally increased the effectiveness of their ministry in the Navy. At the close of 1840 there were thirteen chaplains on duty, among whom were such able men as Charles Samuel Stewart, Walter Colton, and George Jones. Not only did they rise above their contemporaries in ability and accomplishments, but they rank among the greatest of Navy chaplains in the entire history of the Corps.

The year 1840 in a very real sense marks for the chaplains the end of one era and the beginning of another. The days of rugged pioneering were over. A number of well-defined principles and official naval regulations which related to chaplains had been established, largely by the trial and error method. The opening years of the next decade brought enlarged opportunities with a greatly increased Corps. The chaplaincy was beginning to come into its own as a vital and integral part of the United States Navy.

[130] *Op. cit.*, July 1840, pp. 353–4.

CHAPTER SEVEN

"LIMITED TO TWENTY-FOUR"

1841 – 1860

For about twenty years the number of chaplains in the United States Navy had been limited arbitrarily, with more or less consistency, to nine. This total included the sick and those on leave. In 1842 the limit was raised to twenty-four and fixed by law. For seventy-two years, or until 1914, this quota of twenty-four remained, even during the Civil War and the Spanish-American War.

The fixing of a limit of twenty-four was a somewhat fortuitous result of other circumstances. A war scare with Great Britain in 1841 had inspired a large increase in the appointment of midshipmen. In that year alone a record number of 219 were received into the service. It so happened that three Secretaries of the Navy—James K. Paulding, George E. Badger, and Abel P. Upshur—held office successively during 1841 and each was under the usual pressure to make midshipmen appointments. By August 1842 the number of midshipmen had increased to the unprecedented total of 490.

To a lesser degree there had also been an increase in other ranks and ratings. Eleven new chaplains were appointed in 1841, which added to the thirteen already on duty, brought the total to twenty-four. All but one of the eleven were appointed by Secretary George E. Badger who served from 3 March to 13 September 1841. It is interesting to note that Badger appointed seven chaplains on 8 September, less than a week before he vacated his office. These were: Bartow, Gillett, McKenny, Newell, Stockbridge, Talbot, and Moses B. Chase. This was the second appointment for Chase, who had previously served from 1819 to 1822.

In August 1842 Congress was laboring over an appropriation bill for the Navy. To halt any further increase of officer personnel, the appropriation bill limited the number of midshipmen to the total on duty as of 1 January 1841, and of other officers to the number on the rolls as of 1 January 1842. Thus, the quota of chaplains was fixed at twenty-four. The provision regarding chaplains read: "There shall be in the Navy, for the public armed vessels of the United States in actual service not exceeding twenty-four chaplains who shall be appointed by the President with the advice and consent of the Senate."[1]

Fitch Waterman Taylor, 1841-1864.
Who also served in the Civil War.
(From photograph)

During the decade, 1841–1850, twenty-six new chaplains were commissioned and ten additional in the following decade. The Secretary of the Navy, in his 1843 report, stated that from 1 January to 30 June 1843 there were 10,843 enlisted men in the Navy. This meant an average of 460 men for each of the twenty-four chaplains. On 1 January 1842 the United States had in commission two ships of the line, the *Ohio* and the *Delaware*. These vessels, together with six frigates, called for the services of

[1] During World War II, there was still on active duty one chaplain who entered the service in 1913 under this rule. He was Chaplain Thomas Berthier Thompson, who until his retirement on 6 March 1946 was the senior chaplain of the United States Navy.

eight chaplains. It was the custom to assign at least one chaplain to each of the six squadrons, even though the squadron might not have one of the larger vessels. Also, chaplains were usually assigned to the various exploring expeditions during these years. There was a chaplain on duty at the Boston, Portsmouth, Washington, Philadelphia, Norfolk, and Pensacola Navy Yards, and after 1857 one was usually on duty at the Mare Island Navy Yard. Since the establishment of the Naval Academy at Annapolis in 1845, a chaplain has always been assigned there.

The first chaplain to complete thirty years of continuous service in the Navy, which, however, included several long leaves of absence on partial pay, was Addison Searle, who entered the Navy in April 1820 and died at sea 2 August 1850. While he was serving in the *Constitution* at Leghorn Roads, Italy, on 21 May 1822, Lord Byron visited the ship. It so happened that on the same day a young man by the name of George Bancroft, who was destined to became a great historian, a Secretary of the Navy, and the founder of the Naval Academy, also visited the ship.[2]

A number of interesting reminiscences of Chaplain Searle remain. He was a large man, weighing between 250 and 300 pounds. Perhaps this excessive weight contributed to his continued ill health. Once a naval captain wrote to the Secretary of the Navy:

> Do pray indulge Mr. Searle with a leave of absence. He is sick and miserable, and will make all so around him. A ship full of officers is no place for a sick person of any clique, but above all things for a *sick Parson*.[3]

Chaplain Searle died on board the *Cumberland* in his fifty-ninth year while the ship was en route from Jaffa to Alexandria. It was noted in the ship's log that "at 7:45 p.m. the Rev. Addison Searle (Chaplain) departed this life." Carpenters were set to work to make a large coffin to accommodate the body. Another entry reads:

> At 11:50 a.m. Called "all hands bury the dead", hoisted the Colors at half mast: read the Service and Committed to the deep the body of the Rev. Addison Searle, Chaplain. The Marine Guard Firing Three Vollies at the Ceremony—At 11:50 finding the Coffin not to sink, wore Ship, hove to and sent a boat to attach more weights to it.

None of the thirty-six chaplains who entered the service during the years 1841–1860, rose to the stature of any of that great trinity—Charles Samuel Stewart, Walter Colton, and George Jones—who continued in service from the previous decade. Perhaps the best known chaplain to enter the service in these years was Fitch Waterman Taylor, who wrote two books decribing his experiences at sea.

Chaplain Taylor saw service in the Mexican War aboard the *Cumberland*, Flagship of the Gulf Squadron. An account of his experiences is found in his book, *The Broad Pennant*, which appeared in 1848. The book does not, however, treat the war alone. Its account of the Chaplain's activities during the conflict reveals little of particular interest, for it was a war in which the Navy played a relatively minor part.

A note of incidental interest is found in *The Broad Pennant*, however, regarding the term "padre" which is often used today as a term of affection for chaplains of all denominations. Chaplain Taylor was in contact with Spanish speaking people in Valparaiso where the natives referred to their Catholic clergymen as "padres." Chaplain Taylor wrote that one of the women of the city could not believe that he was a "padre" because he had no tonsure.[4] This is the earliest known instance of a naval chaplain being referred to as "padre."

THE APPOINTMENT OF CHAPLAINS

The policy of requiring chaplains to be ordained became official with the adoption of the *General Regulations* of 1841 which contained the following article:

> No person shall be appointed a chaplain in the Navy, who shall not be a regularly ordained or licensed clergyman, of unimpeached moral character, not exceeding thirty years of age.[5]

Even though this regulation stipulated that chaplains should not be more than thirty years of age when appointed, the fact remains that most of those selected during the years 1841 to 1860 were in their forties. The age stipulation was repeated in the regulations prepared by the Secretary of the Navy, Isaac Toucey, in 1858. Congress on 1 June 1860 raised the age limit to thirty-five. This law reads: "A chaplain shall not be less than twenty-one nor more than thirty-five years of age at the time of his appointment."[6]

---

[2] Log of *Constitution* and notes of Chaplain W. W. Edel.

[3] Nav. Rec. Coll., II:June 1833:92. Letter from Capt. H. E. Ballard dated 24 June 1833.

[4] *Op. cit.*, p. 71. The use of the term "padre" for chaplain by Captain Marryat in his *Mr. Midshipman Easy* shows that this use of the term was well known in 1897.

[5] *Gen. Reg.* 1841, p. 15. These regulations were issued "By Authority of the Navy Department." In 1853 the Attorney General ruled, however, that the Naval Department had not authority to issue such general regulations.

[6] R. S. 1396.

Naval chaplains were not always appointed during these years on the basis of their spiritual and professional qualifications. Sometimes they were selected because they enjoyed the favor of important government officials. A good example of this is the commissioning of the eloquent and able Orville Dewey shortly after he was called to the Unitarian Church in Washington which President Millard Fillmore attended. Dewey rather apologetically accounted for this appointment in his *Autobiography:*

> At the time that I was invited to Washington, I received, in February 1851, a document from the Government, which took me so much by surprise that I supposed it must be a mistake. It was no other than a commission as chaplain in the Navy. I wrote to a gentleman in Washington, asking him to make inquiry for me, and ascertain what it meant. He replied that there was no mistake about it, and that it was intended for me, I then concluded, as there was a Navy Yard in Washington, and as the President, Mr. Fillmore, attended the church to which I was invited, that he intended by the appointment to help both the church and me, and I accepted it. On going to Washington I found that there was a chaplain already connected with the Navy Yard, and on his retirement some months later, and my offering to perform any duties required there, being answered that there was really nothing to be done, I resigned the commission.[7]

The official accounts do not harmonize, however, with Dewey's belated explanation. The records show that he accepted his commission on 5 February 1851 and did not resign until 9 February 1853. The chaplain he was supposed to succeed at the Navy Yard was Joel Newton, who was transferred on 21 May 1851. Dewey was officially listed as Chaplain of the Washington Navy Yard from this date to February 1853, but by his own admission did nothing. There is little wonder that critics arose who claimed certain chaplains were making a sinecure of their office. It should be pointed out however, that Dewey's *Autobiography* did not appear until 1883, thirty years after he resigned. It is possible that he had forgotten the performance of routine duties at the Yard which were sufficient to justify the acceptance of a chaplain's salary for two full years. When Dewey resigned, Chaplain Mason Noble succeeded him at the Yard.

The arbitrary method of selecting chaplains often brought into the service men ill-prepared to perform the exacting duties of this office. Christian laymen as well as ministers urged reforms. There is evidence that Army and Navy officers were likewise concerned. As early as 4 November 1857 a number of clergymen and interested laymen met in Baltimore to discuss what could be done about this matter. Commodore Matthew Perry served as chairman. A participant stated that

> it was determined to memorialize the Hon. Secretaries of War and Navy, John C. Spencer and Abel Upshur, requesting them to appoint no one as chaplain who did not submit the recommendation of the highest ecclestiastical authority of the Church to which the applicant belonged.[8]

It was almost sixty years before this practical suggestion came into being.

SHOULD THE CHAPLAINCY BE ABOLISHED

During the latter part of the forties, a growing criticism of the chaplaincy was climaxed in the submission to Congress of a number of memorials signed by citizens of several states. These petitioners prayed "that the office of chaplain in the army, navy, at West Point, at Indian stations, and in both houses of Congress be abolished." They claimed that the employment of chaplains by the Federal Government was unconstitutional; that it portended a union of church and state; and that the office had frequently been dishonored by the appointment of unworthy men.

These memorials were referred to the House Judiciary Committee which reported on 13 March 1859. After answering certain of the objections, the report defended the sailor's right to Divine Service as follows:

> Portions of our naval forces are at all times in some distant coast—often on stations where not only the language of the people is strange, but their rites are abhorrent. He often finds himself where no holy days or Christian service would ever remind him of his distant home in a Christian land, were it not for the provision the Government has kindly made for him. Would it not be a privation to deprive him of it? Could it benefit him to do so? If he heard no sound of the gospel in language familiar to his ears, and in forms well suited to his understanding, would he be a better soldier or more obedient sailor?
>
> . . . .
>
> The spirit of Christianity has ever had a tendency to mitigate the rigors of war, if as yet it has not been entirely able to prevent it; to lead to acts of charity and kindness; and to humanize the heart. It was true philanthropy, therefore, to introduce this mitigating influence where, of all other places, its fruits were to be more beneficially realized, namely into the Army and Navy, and to abolish it, in this Christian age of the world, would seem like retrograding rather than advancing civilization. While much good and no perceptible evil has resulted from the practice; while no constitutional prohibition exists in relation to it, and no tendency to a "religious establishment" is discernible under it; while diversity of truth is tolerated as freely as the constitutional requirement, in the minister, as well as in those for whom he officiates; and while the

[7] Dewey, *Autobiography*, p. 108.

[8] Johnson, *An Address*, p. 23.

expense is so small as not to be felt by any one, your committee do not think it necessary to interfere with the office of chaplain, as it exists at present, in the Army and Navy.[9]

This did not silence the critics, for about four years later the House Judiciary Committee made a similar report in reply to another set of memorials from citizens who wanted the chaplaincy abolished. The Committee reported that the members "have had the subject under consideration, and, after careful examination, are not prepared to come to the conclusion desired by the memorialists." It was pointed out that chaplains were in the Army before the adoption of the Constitution; that the First Congress had appointed chaplains; that the expense of the chaplaincy was slight; that the need for religious guidance even in Congress was great; and that the exercise of religion was necessary for the "safety of civil society."[10]

The opponents of the chaplaincy were so persistent and active that they finally aroused Lorenzo Dow Johnson, evidently a member of an Episcopal Church in Washington, D. C., to speak and write in favor of the continuance of government chaplains. Johnson issued two pamphlets on the subject. The first appeared in 1856 and was entitled: *Chaplains of the General Government.*[11] The second was published the following year and bore the caption: *An Address to the Pastors and People of these United States on the Chaplaincy of the General Government.* Johnson's avowed purpose was to educate the public and to urge members of Congress to take "right action on the subject." His pamphlets constitute the first endeavor to give a short but accurate history of the chaplaincy. The 1856 pamphlet gave the names of thirty-four chaplains who had served in the Navy. The 1857 pamphlet listed seventy-five chaplains, some of whom were the unordained men from the ship's company who served in this capacity.[12]

Johnson stated that the leading spirits in the movement to abolish the chaplaincy were "those who avowed their disbelief in all revealed religion." Point by point he met their objections and answered them with indisputable facts and realistic logic. Then Johnson with equal frankness pointed out certain glaring weaknesses which then existed in the chaplaincy.

First there was the matter of the appointment of naval chaplains.

The manner in which they obtain their appointments is widely different from that in the army, namely, by seeking it, just as any other office is sought, in the gift of the government. The applicant produces his recommendations and his friends as vouchers, and like those who seek to be elected chaplains to Congress, he brings into requisition all the influence he can command, and if his appliances are more effective than those of others, he gets the appointment. Once obtained it is an appointment for life.[13]

The chaplain like other office-seekers got his job "through political considerations." After receiving his appointment, he usually found "himself released from nearly all those influences brought to bear upon the pastor of a church . . ." Johnson added:

A naval chaplain can do all that the ship's rules require of him, and yet have very little love for doing good. He can make himself an acceptable chaplain without being acceptable to God. After defining his rank and privileges among the officers of the ship, all the rest is left to himself, that is, government acts on the presumption, that being a minister, he will always be right.[14]

Johnson then asked several searching questions. Why were so few of the twenty-four chaplains "doing service in sea-going ships?" Why were so many on "leave of absence" or "waiting orders?" He repeated the criticism regarding one naval chaplain who was said to have had but two years sea duty in twenty-eight years' service.[15] Johnson added: "By the complaining public who are memorializing Congress to abolish the chaplaincy, many of these men are regarded as holding mere sinecures, instead of doing all they can, as ministers of Christ, to save souls committed to their care."[16]

Johnson stated that there were in 1857 "about eighty-six vessels of war, with crews numbering from fifty, to five, six, and seven hundred men." He then commented:

Cruising on every ocean they pass through the extremes of heat and cold, and the unhealthy climates of every latitude; in which some sicken and die, and are buried in the sea;—and but for a chaplain they would hear no prayer when sick, nor hardly have a Christian burial when dead. Long months, yea years even, would pass, without hearing a sermon in a language they could understand. Who will deny that the navy opens many an important field for the labors of a faithful Christian teacher. One

[9] 31st Cong., 1st sess., *House Reports,* Vol. I, No. 583. Rep. 171.

[10] 22nd Cong., 1st sess., *House Reports,* Vol. II, No. 743. Rep. 124.

[11] This rare pamphlet is found in the Toner Collection, Library of Congress.

[12] Johnson, *op. cit.,* p. 29.

[13] *Ibid.,* p. 17.

[14] *Ibid.,* p. 19.

[15] The reference is possibly to Chaplain Timothy J. Harrison, commissioned in 1829, who had but two and a half years sea duty up to 1857.

[16] Johnson, *An Address,* p. 18.

who has an aptness to teach, and a love for doing good, will find in the American Navy, a great work to do. In this far-reaching navy there are only twenty-four chaplains.

After pointing out certain weaknesses in the chaplaincy, Johnson offered constructive suggestions. He proposed that

> There be appointed a Board of Commissioners for government chaplains, composed of some eight, ten or twelve clergymen, chosen by as many denominations; then, like those men who apply to enter the medical department of the army or navy, who are required to obtain the approval of a Board of Surgeons, so let all candidates for the chaplaincy pass an examination of this Board, from which a certificate of approval should be made necessary, to constitute the applicant eligible to an appointment.[17]

Johnson advocated reforms nearly sixty years ahead of his time. No Chief of Chaplains with the prerogative of examining the credentials of an applicant was appointed before 1917. The various denominations were not called upon until 1913 to endorse their clergy seeking chaplains' appointments.

Johnson also demanded that a chaplain be appointed on the basis of his ability as a spiritual leader rather than by "skillful wire-pulling." "Then, and not till then," pointed out Johnson, "will he enjoy the respect due to his position." He had one other important suggestion to make for the improvement of the chaplain's work. He wrote:

> Then let it be the duty of all government chaplains to make an annual report to this Board, containing statistical facts and the general results of their work, whether it be at a fort, on a campaign, at hospitals, navy yards, on board receiving ships, or on a cruise at sea.[18]

It may be that this suggestion lay back of the regulation of 1860 which required all chaplains to make annual reports of their activities.

In a circular letter directed "To the Pastors of Every Christian Church in these United States", dated 16 September 1857, Johnson solicited financial support for publishing his material in pamphlet form, copies of which were to be sent to members of Congress and to accredited clergymen in the United States. The existence today of the pamphlets is proof that his goal was achieved at least in part.

The fact that efforts to persuade Congress to abolish all government chaplains were abandoned was due, no doubt, in part, to the activity of Johnson.

REGULATIONS AND DIRECTIVES GOVERNING CHAPLAINS

The new regulations issued in 1841 contained for the first time instructions to chaplains relative to visiting the sick. This reference recalls the difficulties Chaplain Edward McLaughlin had with the doctor in charge of the Naval Hospital of Portsmouth in 1827. The new regulations read:

> He is to perform divine and funeral services, at all times, when required so to do by his commander.
>
> He shall be very attentive to the requests of all sick persons who may desire his attendance, and shall, although not requested, visit all such as may be dangerously ill, and offer them such consolation or admonitions as they may require.
>
> When a person shall be appointed to instruct the boys of the vessel, he shall frequently attend, to see that he performs his duty properly, and that the boys attend regularly. He shall report to the commander those who may be particularly deserving, and all who may be idle or negligent.[19]

The question of chaplains' pay was considered by Congress in 1852. It will be recalled that in 1835 their compensation had been fixed at $800 per annum while on leave or awaiting orders and $1,200 while on active duty. They were relatively the lowest paid naval officers of comparative status. The Act of 1852 rectified this situation somewhat and fixed chaplains' pay at $1,000 a year while on leave or awaiting orders and $1,500 while on active duty.[20] Finally the Act of 1860 placed chaplains on the same basis as lieutenants. This gave them $1,200 a year while on leave or awaiting orders, and $1,500 a year during the first five years of sea duty and an increase of $200 each two years thereafter.[21] For example, a chaplain who had had seven years of sea service drew $1,700 and, after he had been at sea eleven years, $2,100.[22]

Another ruling which referred to the chaplains' status is found in the proposed naval regulations submitted by the Secretary of the Navy on 6 December 1858. Article 12, chapter II, reads:

> Chaplains and chief engineers who have now been commissioned twelve years as such shall have precedence with commanders and with surgeons and paymasters who take precedence with commanders under Article 27.

Chaplains did not receive "relative rank" by law until 1871 and actual rank was not given until 1899. However, in the proposed regulations of 1858 the trend is clear. No corps of the Navy took so long to crystallize into a well defined unit as did the Chaplain Corps. For years chaplains were the poorest paid naval officers and for decades they did not receive

[17] *Ibid.*
[18] *Idem.*

[19] *Gen. Reg.*, 1841, p. 132.
[20] Callan and Russell, *Laws of the United States relating to the Navy and Marine Corps*, p. 380.
[21] This marks the beginning of "fogies" for chaplains. A fogey is a military term signifying an increase in pay due to longevity.
[22] R. S. 1394.

the same rights and privileges for longevity as other officers.

THE CHAPLAIN'S UNIFORM

As has been stated, the first attempt to regulate chaplains' uniforms came in 1830 when the following uniform regulation for chaplains appeared:

> Plain black coat, vest and pantaloons, the pantaloons to be worn over boots or shoes, or black breeches, silk stockings with shoes, coat to have three black covered buttons under pocket flaps and on the cuffs.

Thus the chaplain had a choice between "pantaloons" and "black breeches," the latter with silk stockings. At first the buttons were plain but in 1841 the official eagle button was authorized.

The *Regulations* issued in 1841 introduced certain radical changes which authorized chaplains to wear practically the same uniform as that of other naval officers. Their coat was described as of "dark blue cloth, with rolling collar of black velvet, in other respects like the undress coat of the lieutenants."[23] The coats were double-breasted with a row of nine buttons, three inches apart on each side. The *Regulations* also permitted "three buttons on the cuffs and pocket flaps, and one in the middle of the skirts." Since the sword was an official part of the uniform and since nothing was said to the contrary, it would appear that chaplains were permitted to wear swords. Accounts of Perry's expedition to Japan in 1853–1854 mention that Chaplain Edmund C. Bittinger, a member of the expedition, carried a sword.[24]

During the winter of 1843–1844, certain chaplains in the vicinity of Washington and Philadelphia met to discuss the possibility of changes in the uniform. They felt that the uniform required by the *Regulations* of 1841 was "an unsuitable one for ministers of the gospel," and hoped to restore the former uniform. Writing on 8 December 1843 to the Secretary of the Navy, Chaplain John Robb pointed out various inconsistencies in the regulations and commented as follows:

> By the present regulations, Chaplains are required to wear a uniform coat, cocked hat, white vest, blue pantaloons in winter, and white in summer. This dress, however, is not worn by the Chaplains stationed at Washington, Philadelphia, New York, and Boston, and there is but one, as far as can be ascertained, on shore service who wears any part of the uniform. They still adhere to the former dress—a plain black suit, which is certainly the more appropriate one of the two, and which is the dress worn by Chaplains in the British Navy. Besides the unsuitableness of the present dress to the sacred calling of the minister of God, the Chaplain is subjected to additional expense in providing himself with this unifrom when upon sea service; for although the Captain of the ship who disapproves of such a dress, (and some of them do), may not require him to wear it, the Commodore of the Squadron may. The Chaplain at Boston informed me that the officers upon that station thought the dress prescribed by the regulations of 1841 altogether unfit for a Chaplain, particularly Commodore Downes, and that even Com. Nicholson, one of the Commissioners of the Navy, who recommended it did not now justify it.[25]

Walter Colton, in a letter of 6 January 1844 to the Secretary of the Navy, wrote:

> I have been requested by several Chaplains of the Navy to address you in reference to our Uniform. We are anxious to have the color changed from blue to *black*, the coat to have a black velvet collar, with Navy buttons, the same as now,—this uniform to be worn in foreign ports and on state occasions. We then want the favour of your permission to officiate on the Sabbath either in the black silk gown such as is generally worn by the clergy, or the plain black coat. Should uniformity in your opinion render it expedient, the use of the silk gown can be made obligatory when attached to cruising vessels. My own humble opinion is in favor of making the use of the gown in this situation peremptory. It is in harmony with the etiquette of a man of war, is an appropriate badge of a Chaplain's office and is a graceful garment for a public speaker. The Navy button will look well on black and will secure a Chaplain in foreign ports the respect due to his commission. A very few of the Chaplains may perhaps be averse to the gown, but nine tenths of them, I think, are in favor of it,—all indeed, so far as my knowledge extends, whose opinions are based on a practical experience of the usages of a man of war.[26]

The Secretary of the Navy, evidently impressed by these appeals, issued the following directive under date of 20 January 1844:

> Chaplains shall wear a Black Coat, with a black velvet collar, and the navy button now in use. (They need not, however, provide themselves with new coats until those they now have are worn out.) While performing religious services on the Sabbath, or on other occasions, on board vessels of war or at yards and Shore Stations, they shall wear the Black Silk Gown usually worn by clergymen.
>
> David Henshaw[27]

The wording of this regulation was too exacting to win universal approval. While most chaplains did not object to the order that "Chaplains shall wear a black coat," a few objected vigorously to the mandatory rule that while "performing religious services . . . they shall wear the Black Silk Gown." At once there were objections. Chaplain Colton, a Congregationalist, wrote the following explanation of the new ruling for *The Independent North American:*

[23] *Gen. Reg.*, 1841, p. 198.
[24] Hawks, *Narrative of the Expedition*, pp. 418, 475.

832159—49——6

[25] Nav. Rec. Coll., III:Dec. 1843:59.
[26] *Ibid.*, III:Jan. 1844:40.
[27] *Ibid.*, V:2:61.

I see that several religious papers ascribe the introduction of the gown into the Navy as the costume of the Chaplains to *sectarian* purposes in the Department. The facts are these. At a meeting of several Chaplains of the Navy, I was appointed to correspond with the Department in reference to a change in our uniform, and to prefer a request that when performing religious services, the silk gown usually worn by the Clergy might be substituted for the *blue coat and navy button.* The Department acceded to the request, and issued a circular to that effect. Now, if any one supposes this originated in a spirit of sectarianism, or that there is anything in it which squints at Episcopacy, I can only say it had its sources mainly with those who do not belong to that persuasion, and who can find more important features of distinction than those which lie between a blue and a black coat. Or, if any one thinks the blue coat and navy buttons a more appropriate costume for a Clergyman while performing religious services, than a plain black gown—so be it. There is no accounting for tastes. Or, if any one thinks a Chaplain in the Navy should have no uniform at all, no evidences of his office, no outward tokens of his profession, I leave him to the *simplicity* of his idea.

This is written as an act of justice to the late head of the Navy Department, and would be anonymous, could it have, in that shape, the authority which may, perhaps, be given it by the name of its author.[28]

The Navy Department was quick to change the rule regarding the mandatory use of the gown, and on 23 April 1844 issued the following clarification:

The Regulation of the 20th January, 1844, prescribing a uniform for Chaplains in the Navy, is so modified, that, in performing Divine Service, the Chaplain may, in his discretion, wear the black gown, a plain black coat, or the uniform coat prescribed by that Regulation.

J. G. Mason

New regulations appeared in 1852 which changed the double-breasted coat to single-breasted, and eliminated one row of the "nine large Navy buttons in front." Collars and cuffs were to be "of black velvet, without embroidery." The wearing of the gown at Divine Service was still optional.[29]

A circular, issued by the Navy Department on 3 March 1853, modified this regulation by substituting "black covered buttons" for the "navy button." The new order also authorized an "Undress Black Frock Coat, single breasted, with one row of nine black covered buttons." Another change read: "In performing Divine Service, the Chaplain may wear the black gown & *white cravat,* or the uniform prescribed in the Regulations."[30]

[28] Copied in *Army and Navy Chronicle,* 21 March 1844; p. 363.

[29] *Regulations for the Uniform and Dress of the Navy and Marine Corps of the United States,* p. 10.

[30] Nav. Rec. Coll., V:2:281. Italics (the author's) indicate addition of a new provision.

From this series of regulations and modifications it is evident that the whole subject of the chaplain's uniform was under considerable discussion. Strangely enough no objection seems to have been raised to the chaplain's wearing a sword. The question at issue was: Should the naval chaplain dress as a civilian clergyman or should he be completely identified with the service by wearing an officer's uniform? The latter view prevailed. Chaplain Charles W. Thomas, writing to the Secretary of the Navy on 31 January 1855, summarized his own convictions as well as those of many of his brother chaplains when he stated:

My humble opinion is that the uinform prescribed in the book of Regulations is in every way suitable and appropriate. The more we are identified with the service the greater will be our influence over both officers & men, and I think we should not be above expressing in the outward dress that identity which I trust we all feel inwardly. This is the opinion also of others of the grade with whom I have conversed.[31]

TO BE OR NOT TO BE LITURGICAL

Many of the non-liturgical church people of the United States were concerned about the trend among naval chaplains toward the use of the gown and the Episcopal prayer-book in Divine Services. Some said that the Navy department had appointed a disproportionately large number of Episcopal chaplains. Others claimed that Navy regulations required the non-Episcopal chaplains to use the Episcopal liturgy. What did the *Regulations* of 1818 mean when it stated that a duty of a chaplain was "to read prayers at stated periods"? Were the Baptist chaplains, for instance, required to *read* their prayers or could they, according to the custom of that denomination, offer an extemporaneous prayer?

These matters were brought to the attention of the members of the House of Representatives, who requested an explanation from the Secretary of the Navy. Isaac Toucey, then Secretary, wrote to the Speaker of the House saying in part:

Sir: I have the honor to acknowledge the receipt of the resolutions of the House of Representatives dated January 13, 1859, requesting the Secretary of the Navy, during the present session, "to communicate to this House the number of Chaplains appointed in any branch of the Navy service since eighteen and thirteen; the religious denomination to which each person so appointed was attached, so far as it can be ascertained; whether chaplains by any navy regulations, or any act of commanders of vessels or stations are required to use a particular

[31] *Ibid.,* III:Jan. 1855:200.

uniform or clerical dress, including a gown, or to read prayers, or to comply with any particular forms or ceremonials of divine service; and whether there is any evidence on file in the department tending to show that non-Episcopal ministers are required by officers of the navy to use the Episcopal liturgy."

Secretary Toucey then submitted a list of sixty-one chaplains who had been appointed between 12 May 1815 and 11 September 1856. An analysis of his classification appears in column A of the following chart. Column B is a classification of the same chaplains according to data more complete than was evidently in the possession of Secretary Toucey.

	A	B
Baptist	1	6
Congregational	3	5
Methodist	8	9
Episcopal	10	24
Presbyterian	6	11
Unitarian		2
Universalist		1
Unknown	33	3
	61	61

The fact that the Secretary did not know the regilious affiliation of thirty-three chaplains, including two who were on duty at the time, speaks eloquently of the inadequate records in the Navy Department. The system of including in a "jacket" all papers relevant to each officer was not instituted in the Department until on the eve of the First World War. Toucey's list did not include chaplains who were appointed from the ship's company, and it omitted at least five, who had been regularly appointed, including such men as Charles Rockwell and Orville Dewey.

The data which the Secretary furnished indicated that ten of the sixty-one, or about fifteen percent, were Episcopalians. The correct record would have shown twenty-four, or approximately forty percent. On the basis of actual facts, the critics had good reason to charge the Department with discrimination.

Regarding the use of a particular dress, the Secretary wrote: "The commanding officer of a vessel or station has no authority to establish the uniform or dress of any officer of the Navy." Toucey referred to the *Uniform Regulations* of 1853 which permitted but did not compel the chaplain to wear the gown at Divine Services.

As to the ruling that the chaplain was "to read prayers at stated intervals," the Secretary wrote: "the department is not aware that this has ever been construed other than to offer prayers at stated periods. However this may be, to put at rest any doubt, an order has been recently issued which establishes this to be the true construction."[32]

The new order regarding prayers is dated 17 January 1859, the same date as the Secretary's reply to the Speaker of the House, and is worded as follows:

> It is understood that the Navy Commissioners' Regulations of 1818, requiring Chaplains "to read prayers at stated periods", have heretofore been construed to require them to *offer* prayers, and such will hereafter be the construction.[33]

This discussion undoubtedly led to the adoption of the following naval regulation, on 1 June 1860, which is still (1946) in effect: "Every chaplain shall be permitted to conduct public worship according to the manner and forms of the church of which he may be a member."[34]

BUGLE CALLS FOR CHURCH

From early days the bugle has supplemented the spoken word aborad ship. Available records do not indicate with exactness when the bugle calls for church and hammocks were introducted into the Navy. It is possible they were used during the period under review. It appears that they were used first in the Army, then in the Marine Corps, and finally were adopted by the Navy. There are two church calls known to have been used in the Navy. Chaplain E. L. Ackiss heard the earlier call sounded as late as 1920 by a bugler who had just reported aboard his ship, the *New Hampshire.*

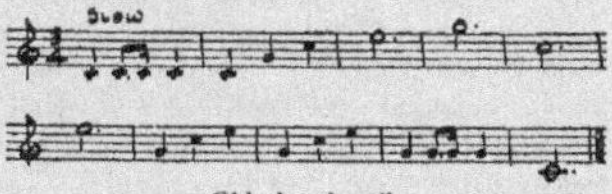

Old church call.
Upton, *Infantry Tactics,* 1874, p. 404.

Present church call.
Ship and Gunnery Drills, 1927, p. 330.

[32] 35th Cong., 2nd sess., *House Reports,* Exec. Doc., No. 59.
[33] Nav. Rec. Coll., V:2:343.
[34] R. S. 1397; *Nav. Reg.,* 1920, art. 1245 (10).

Today the ship's bell is often tolled aboard ship during or immediately after the bugle call for church. This custom is said to have been taken over from the British Navy. According to tradition, Lord Nelson tolled the bell of his ship when going into battle for the souls of those who were about to die. In all probability the tolling of the bell to announce the hour of Divine Service antedates the bugle call in the United States Navy. The earliest known mention of this custom appears in Herman Melville's *White Jacket* and refers to a Sunday in 1843 when the author was aboard the *United States.* Chaplain Colton who served on the *Congress* states in his Journal for Sunday, 7 December 1845: "At eleven o'clock, the tolling of the ship's bell announced the hour of worship."[35] Chaplain Taylor also refers to this custom aboard the *Cumberland* for 22 March 1846.

THE CHURCH PENNANT

The first reference to a church pennant flying above the national emblem is contained in the *Proceedings of the Court of Inquiry appointed to Inquire into the Intended Mutiny on board the United States Brig Somers.* The incident referred to occurred in 1842 when the *Somers* was returning to the United States from a trip to the African coast. Commander Alexander Slidell Mackenzie, believing that the safety of his ship was in danger, acted promptly and confined to irons three alleged ringleaders, including Acting Midshipman Philip Spencer, son of the Secretary of War. A court of the ship's officers found them guilty of "attempted mutiny" and they were hanged at the yard-arm. Commander Mackenzie's report aroused a nation-wide controversy and a court of inquiry was held to determine all of the facts involved. The court exonerated Commander Mackenzie.

In the course of the inquiry Mackenzie described the events on the Sunday following the execution. According to custom, the crew was assembled and the laws governing the Navy were read. Then Mackenzie spoke briefly on "the lessons to be drawn from the fate of those who had suffered." He stated:

> In conclusion, I called on them, as they had given three cheers for their country, now to give three cheers for God—as they would do by singing his praise. The colors were then hoisted, and above the American Ensign was raised the Banner of the Cross—the only flag that ever floats above it from any vessel under my command. The 100th Psalm was sung, after which the crew dispersed.[36]

This testimony suggests not only that the church pennant was raised above the national emblem during Divine Worship but also that it was customary to fly it even when no chaplain was aboard to conduct such a service.

The first known illustration of the church pennant is found in a Journal kept by Acting Midshipman P. C. Johnson on the *Ohio* during 1846–1848. His drawing shows a Latin cross with the short arm parallel to, and the top of the long arm beginning with, the base of the pennant.[37] No printed Navy signal books are known to be in existence for the years 1813 to 1858.[38] The *Signal Book* of the latter date contains the first known official illustration of the pennant, picturing a blue Greek cross on a white background. There is no text to describe its use. A slightly different illustration of the Navy's church pennant appeared in a French signal book also published in 1858. On the page showing United States Navy flags is the picture of the church pennant with the Latin cross, the long arm of which is horizontal, with the short arm extending completely across the pennant.[39] (See illustrations page 57)

MELVILLE'S DESCRIPTION OF DIVINE SERVICES

A rather comprehensive idea of the position of the chaplain and of religion on board a man-of-war during this period may be obtained from Melville's somewhat autobographical narrative, *White Jacket.* Melville served aboard the *United States* during the 1841 to 1844 tour of duty of Chaplain Theodore B. Bartow, an Episcopalian. While the author did not mention the Chaplain by name, the following description, taken from the work, without a doubt applied to him:

> No Sundays on shipboard! You may as well say there should be no Sundays in churches; for is not a ship modeled after a church? has it not three spires—three steeples? yea, and on the gundeck, a bell and a bellfry? And does not that bell merrily peal every Sunday morning, to summon the crew to devotions?
>
> At any rate, there were Sundays on board th's particular frigate of ours, and a clergyman also. He was a slender, middle-aged man, of an amiable deportment and

[35] Colton, *Deck and Port*, p. 67.

[36] *Proceedings of the Court of Inquiry,* 1843, p. 14, (pamphlet).

[37] Nav. Rec. Coll. McCauley and Blake, *Tables of Allowances of Equipment,* published in 1854, includes an allowance of one "Church and Meal Pennant" for each ship. Three sizes of the pennant are specified (p. 30): 24′ x 4½′ for ships of the line; 22′ x 3¾′ for frigates, sloops, and side-wheel steamers; and 20′ x 3½′ for sloops and side-wheel and screw steamers, 2d class.

[38] Several manuscript Signal Books by Commodore Charles Morris for the years 1827–48 are in the Nav. Rec. Coll.

[39] See McCracken, "The Church Pennant," *Naval Institute Proceedings,* Aug. 1930, p. 718.

irreproachable conversation; but I must say, that his sermons were but ill calculated to benefit the crew. He had drank at the mystic fountain of Plato; his head had been turned by the Germans; and this I will say, that White-Jacket himself saw him with Coleridge's Biographia Literaria in his hand.

Fancy, now, this transcendental divine standing behind a gun-carriage on the main-deck, and addressing five hundred salt-sea sinners upon the psychological phenomena of the soul, and the ontological necessity of every sailor's saving it at all hazards. He enlarged upon the follies of the ancient philosophers; learnedly alluded to the Phaedon of Plato; exposed the follies of Simplicius's Commentary on Aristotle's "De Caelo", by arraying against that clever Pagan author the admired tract of Tertullian—*De Praescriptionibus Haereticorum*—and concluded by a Sanscrit invocation. He was particularly hard upon the Gnostics and Marcionites of the second century of the Christian era; but he never, in the remotest manner, attacked the everyday vices of the nineteenth century, as eminently illustrated in our man-of-war world. Concerning drunkenness, fighting, flogging, and oppression—things expressly or impliedly prohibited by Christianity—he never said aught.[40]

The writer complained that this particular chaplain talked only of "harmless, non-committal abstrusities."

Church attendance was compulsory on the *United States*, and the attitude of the men is reflected in the following description by Melville:

The accommodations of our chapel were very poor. We had nothing to sit on but the great gun-rammers and capstan-bars, placed horizontally upon shot-boxes. These seats were exceedingly uncomfortable, wearing out our trouwsers and our tempers, and, no doubt, impeded the conversion of many valuable souls.

To say the truth, men-of-war's men, in general, make but poor auditors upon these occasions, and adopt every possible means to elude them. Often the boatswain's-mates were obliged to drive the men to service, violently swearing upon these occasions, as upon every other.

"Go to prayers, d--n you! To prayers, you rascals—to prayers!" In this clerical invitation Captain Claret would frequently unite.

At this Jack Chase would sometimes make merry. "Come, boys, don't hang back," he would say; "come, let us go hear the parson talk about his Lord High Admiral Plato, and Commodore Socrates."

Melville tells of how "a remarkably serious, but bigoted seaman" objected to compulsory church attendance. Respectfully approaching his captain, he said: "Sir, I am a Baptist; the chaplain is an Episcopalian; his form of worship is not mine; I do not believe with him, and it is against my conscience to be under his ministry. May I be allowed, sir, *not* to attend service on the half-deck?" The captain haughtily replied: "You will be allowed, sir, to obey the laws of the ship. If you absent yourself from prayers on Sunday mornings, you know the penalty."[41]

Melville felt that a strict enforcement of the compulsory attendance at Divine Services did not nurture a receptive attitude for spiritual truths. Not only was there reluctance on the part of some, which at times turned into resentment, there was also ridicule and derision. The author mentions the Bibles which were distributed to the members of the crew and tells of the inspection held on Sunday morning.

Although, by the regulations of the Navy, each seaman's mess on board the Neversink was furnished with a Bible, these Bibles were seldom or never to be seen, except on Sunday mornings, when usage demands that they shall be exhibited by the cooks of the messes, when the master-at-arms goes his rounds on the berth-deck. At such times, they usually surmounted a highly-polished tin-pot placed on the lid of the chest.[42]

However, there is another side to the story and Melville does not overlook it. He speaks of the liberality with which seamen contributed to such causes as the building of a seaman's chapel in China or paying the salary of a colporteur in Greece. Melville felt that a religious captain often made "a far better chaplain for his crew than any clergyman could." He wrote:

This is sometimes illustrated in the case of sloops of war and armed brigs, which are not allowed a regular chaplain. I have known one crew, who were warmly attached to a naval commander worthy of their love, who have mustered even with alacrity to the call to prayer; and when their Captain would read the Church of England service to them, would present a congregation not to be surpassed for earnestness and devotion by any Scottish Kirk. It seemed like family devotions, where the head of the house is foremost in confessing himself before his Maker. But our own hearts are our best prayer-rooms, and the chaplains who can most help us are ourselves.[43]

This same work presents a sympathetic picture of Divine Worship. The following is a description of the procedure at morning and evening prayers:

Shortly after breakfast the drum beats to quarters; and among five hundred men, scattered over all three decks, and engaged in all manner of ways, that sudden rolling march is magical. . . .

---

[40] *Op. cit.*, pp. 147–8.

While Melville's works are usually classified as fiction, his picture of man-of-war life is drawn from personal experience on the *United States*. In *White Jacket*, he tried to cloak the frigate with anonymity by calling her the *Neversink*. It is felt that Melville has authentically portrayed shipboard life in his day.

[41] *Ibid.*, pp. 148–9.

[42] *Ibid.*, p. 150.

[43] *Ibid.*

First Naval Academy Chapel, 1853–1859.

The sailors run to and fro . . . to gain their respective stations in the shortest possible time. In three mintues all is composed. . . .

At a sign from his finger, the brass band strikes up the Portuguese hymn. This over, from Commodore to hammock-boy, all hands uncover, and the Chaplain reads a prayer. Upon its conclusion, the drum beats the retreat, and the ship's company disappear from the guns. At sea or in harbour, this ceremony is repeated every morning and evening.

By those stationed on the quarter-deck, the Chaplain is distinctly heard; but the quarter-deck gun division embraces but a tenth part of the ship's company, many of whom are below, on the main-deck where not one syllable of the prayer can be heard. This seemed a great misfortune; for I well knew myself how blessed and soothing it was to mingle twice every day in these peaceful devotions, and, with the Commodore, and Captain, and smallest boy, unite in acknowledging Almighty God.[14]

There was no public address system then to carry the voice to every part of the ship. The Portuguese Hymn is better known today as "Adeste Fideles."

[14] *Ibid.*, pp. 269–72.

GEORGE JONES AND THE NAVAL ACADEMY

As has been pointed out, naval chaplains were more or less concerned with the education of midshipmen ever since the days of the "Academy" conducted by Chaplain Robert Thompson at the Washington Navy Yard. They were also required by regulation to teach, or supervise the instruction of, the boys in the Navy.

Among the Navy officers who were active in advocating the establishment of a naval school for the training of midshipmen was Chaplain George Jones whose letter of 30 January 1839 to the Secretary of the Navy stressed this imperative need. Jones also wrote an appeal for such an institution which was published in the *Naval Magazine,* and he discussed this subject with Secretary Upshur. It was natural, therefore, when the Naval School was formally opened in the fall of 1845, that Chaplain Jones should be one of the eight members of the first academic board. Jones was made the first head of the Department of English Studies, which then included both history and geography. These duties seem to have taken precedence over his usual chaplain's responsibilities. He was first of all the professor of English and then chaplain. In similar manner Navy Surgeon John A. Lockwood served both as professor of chemistry and medical officer.[45]

Chaplain Jones was not detached from the *Brandywine* until September 1845. When he arrived at Annapolis he found that no living accommodations had been reserved for him. Consequently he had to find quarters in the town until the following year when the old quartermaster's office was rebuilt to provide a home. Divine Services were held in the quarters formerly used by the Army.

Jones, writing from the Naval Academy on 17 March 1847, told the Secretary about the establishment of a museum in the newly erected lyceum building. "I have presented all that remains to me of collections which I have made in the Mediterranean & in other seas," he stated, and added, "I also spent the last vacation in travelling, at my own expense, for the purpose of procuring information & geological specimens for the school."[46] Thus was laid the foundation of the splendid museum now at the Academy at Annapolis.

Following the establishment of the lyceum, the Navy Department sent a collection of historical naval relics, including the famous DONT GIVE UP THE SHIP flag, to the School for safe keeping. These were displayed and cared for by Chaplain Jones. Jones was also active in promoting the interests of the library, being sent to New York in 1851 to make purchases of books.[47]

In 1850 the Naval School was reorganized and was thereafter known as the Naval Academy. At this time Jones was appointed the first chaplain of this institution, the office then being separated from any professorship. It should be noted, however, that subsequent to that date chaplains often served as professors.

When the Navy took over Fort Severn, the quarters which the Army had used as a chapel soon proved inadequate for Divine Services. A new chapel was built in 1853 and was dedicated by Chaplain Theodore B. Bartow, Jones's successor, on Sunday, 5 February 1854. This building was rectangular in shape. In 1860 a portico in classical Grecian style was added.

The rules for the government of the Naval Academy issued in 1850 (reissued in 1855) provide that a chaplain and certain other specified officers be attached to the Academy. Attention is also given to Divine Service in the rules:

> 1. Divine service shall be performed on Sunday, and shall be attended by every person attached to the Academy.
>
> 2. Any Midshipman or Acting Midshipman who shall behave indecently or irreverently while attending divine service, or shall use any profane oath or execration, or profane the Sabbath, shall be dismissed the Naval Service or otherwise less severely punished.
>
> 3. Fifteen minutes before the breakfast hour, the Midshipmen and Acting Midshipmen shall assemble in the chapel for prayers.

When strong criticism arose regarding compulsory attendance at Divine Services, the Secretary of the Navy in 1859 modified the earlier Naval Academy regulations as follows:

> The 2nd Section of Chapter 19, of the existing Regulations of the U. S. Naval Academy, approved Jan'y 25, 1855 which requires the attendance at Divine Service on Sunday of all persons attached to the Academy, is thus modified.
>
> Officers will be excused by the Superintendent from such attendance upon their declaration, in writing, that they cannot conscientiously attend.
>
> Acting Midshipmen will be excused for the same reason, upon the written request of their parents or guardians.
>
> In all such cases the Superintendent will prescribe such Regulations as will insure from those excused a decent observance of the Sabbath during the performance of Divine Service.
>
> Acting Midshipmen will at all other times during the day conform to the prescribed hours of Study.[48]

The custom of having morning prayers before breakfast has continued at Annapolis throughout the years. At first this brief service was held in the Chapel, but with the growth of the Academy, especially during wartime, it was transferred to the mess hall. An interesting entry in the log of the Officer of the Day made during the examinations held at the end of the second academic year of the Naval School reads: "At 11:10 the Midshipmen assembled in the Lyceum for examination in Gunnery. The session was opened with prayer by Chaplain Jones." It is recorded that Chaplain Jones held

[45] Sturdy, "The Founding of the Naval Academy," *Naval Institute Proceedings*, Oct. 1935, pp. 1367 ff.

[46] Nav. Rec. Coll., III:Mar. 1847:189.

[47] Robert, "The Naval Academy Chapel," *Naval Institute Proceedings*, Oct. 1945, p. 1136.

[48] Nav. Rec. Coll., V:2:344.

the same service each day during the examination period.[49]

CHAPLAINS AT WORK

The primary responsibilities of naval chaplains have always been religious although a few chaplains have concentrated their efforts almost exclusively upon education or other collateral activities. The effectiveness of their religious activities, then as now, was conditioned not only by the attitude of their commanding officers, as already pointed out, but also by the physical surroundings under which they worked, the material equipment provided for them, the rules of the naval service, and many other factors which seldom concern clergymen in civilian life.

While few descriptions of accommodations for holding religious services on shore stations for this early period are accessible, it appears that makeshift arrangements in such buildings as were available, were the rule. Chaplain Fitch Taylor described the chapel at Pensacola: "The Chapel is a prettily arranged upper room, in one of two octagonal buildings, which are ornamental to the yard."[50] Mention is made in 1851 of the facilities at the Boston Navy Yard as follows: "The Carpenters completed the enlargement of the Chapel, in the steam house near the Dry Dock, by removing the Office of the Timber Inspector."[51] It was not until after the turn of the next century that a chapel was built for naval personnel, aside from the chapel at the Naval Academy.

Contemporary evidence indicates that chaplains frequently appealed directly to the Secretary of the Navy for supplies and equipment. Chaplain Theodore B. Bartow, in December 1841, inquired of the Navy Department whether the Secretary was "empowered to make any allowance to purchase Sunday School books for the boys on shipboard." Bartow said he felt certain that if the Secretary could not do this Congress would grant him the right. He added: "For the nation responds to the truth 'that the care of the children is its first duty'." He was informed that the Navy Agent at Norfolk would provide them.[52]

Other chaplains were, on occasion, less successful. In 1843 Chaplain Mortimer Talbot requested five or six dozen Books of Common Prayer which he felt could be procured for about three dollars a dozen. He was turned down with the comment that the "table does not include it."[53] Four years later Chaplain Stockbridge requested a stove for the vestry of the chapel at the New York Navy Yard, together with "permission to have a fire in it when necessary.[54] In forwarding this letter to Washington the commanding officer gave the following reasons for his disapproval.

> . . . it would require a man to attend to it, besides the danger from fire should it be left unattended. . . . The attendance of Mr. Stockbridge is only required on Sundays and at funerals; hence, there is no occasion to incur an extra expense for coal and an attendant, for his Study.[55]

Evidently the Commanding Officer of the New York Navy Yard felt that the Chaplain's religious duties were confined exclusively to the chapel on Sunday, when heat was provided. No provision was made for conferences in a suitable office during the week. Stockbridge was informed that "the Dept. cannot give the authority" for a stove.

Chaplains often appealed to the Navy Department to secure personal favors. Writing from the *Potomac* while it lay off the Pensacola Navy Yard in 1847, Chaplain Rodman Lewis requested a transfer to the Yard. The position had been vacant since the death of Chaplain Charles Henry Alden a few months earlier. Lewis pointed out that his home and family were at this station since he had previously served there more than four years. His Commanding Officer, Captain Aulick, forwarded his request with the following endorsement:

> Forwarded. This ship not being allowed a Chaplain by the regulations of the service—Mr. Lewis is being therefore a supernumerary on board.—I respectfully recommend that he be detached.

The Secretary of the Navy, in granting this request, was careful to point out that the Navy Department had the authority to determine the complements of ships. His endorsement read:

> Let the Rev'd. Mr. Lewis be detached—inform Capt. Aulick that this order is given because the Chaplain does not appear to be useful in his profession on board the Potomac—No officer can be regarded as a supernumerary on board a ship, who is ordered to her on duty by the Navy Dept. Done 18 Jany. 1847.[56]

The experience of Chaplain Edmund Bittinger when he was sent to the *Macedonian* about a decade later was parallel. He found himself in a situation comparable to that of a clergyman in civil life who

[49] Robert, "Extracurricula," *Naval Institute Proceedings*, Apr. 1945, p. 386.
[50] Taylor, *Broad Pennant*, p. 197. Date is before 1848.
[51] Nav. Rec. Coll., History Boston Navy Yard, II:312.
[52] *Ibid.*, III:Dec. 1841:131.
[53] *Ibid.*, III:Apr. 1843:141.
[54] *Ibid.*, III:Oct. 1847:312.
[55] *Ibid.*, II:Sept. and Oct. 1847:192.
[56] *Ibid.*, III:Jan. 1847:19.

reported to a parish where he was neither expected nor wanted. In the service, however, his only recourse was an appeal to the Secretary of the Navy. "Commodore Stringham has just informed me," wrote the Chaplain, "that as he considered me a *'Supernumerary'* there is no room for me on board of the 'Macedonian' . . . I hope the Department will at its earliest convenience settle this question so that I may know what relation I am to sustain to the ship."[57] In forwarding it to the Secretary, the Commodore commented: "I have the honor to state, that, on referring to the Allowance book, I find, that Chaplains are only allowed to Flag Ships. I have therefore given to the Pass'd Ass't Surgeon the state room, to whom I think it properly belongs."[58] In his reply, the Secretary said in part: ". . . you are informed that the Act of the Department, in ordering Chaplain Bittinger to the Macedonian has, in this case, modified the Complement Table. You will assign him the room that is usually occupied by Chaplains on board of Flag Ships."[59]

Chaplains continued to be disturbed over the fact that they were sometimes handicapped in their duties by their commanding officers to whom they were exclusively accountable. Chaplain Stockbridge expressed the feeling of many when he wrote the following to the Secretary in 1853:

> . . . . My experience of nearly twelve years in the service is, that the Chaplain gets on the most smoothly with his superiors who does the least for Christ & his cause. Hence it is that Chaplains who have entered the service in hope of doing good soon find their zeal quenched. And, moreover, while doing duty in sea going ships, their personal piety is subject to the most unfavorable social influences. Besides these considerations, there is no ecclesiastical power to which Chaplains are accountable. Are they not on the whole about as faithful as we can expect them to be under the existing system?[60]

As the result of the agitation of Chaplain Stockbridge, Lorenzo Johnson, and others during this period, Congress finally acted on 1 June 1860 and attached a rider to an appropriation bill which read: "Chaplains shall report annually to the Secretary of the Navy the official services performed by them."[61]

That chaplains aboard ships were gradually experimenting in new fields of usefulness during this period may be seen in the work of Chaplain Vernon Eskridge. Sprague's *Annals of the American Pulpit* describes one of his activities:

> He applied for and obtained in 1851 an appointment as chaplain in the Navy. In March, 1852, he was ordered to the United States frigate Cumberland, then fitting out for the Mediterranean Station. He engaged in this work with great zeal, and in a short time some fifteen or twenty of the men, including several of the officers of the squadron, professed their faith in Christ, and a little society was established on shipboard, called "The Church of Christ."[62]

This is perhaps the first time a Navy chaplain tried the experiment of establishing a church organization aboard ship. Chaplain W. W. Edel in his "Navy Chaplains from 1775 to 1917" gives the following account of this zealous man:

> While the ship was in the Mediterranean, an epidemic of Cholera broke out on board, and Chaplain Eskridge was ever constant in his attentions to the sick and dying. Before long he too was stricken and after a long siege of illness he fought his way back to health, although he was but the shell of his former self. At the end of the three year cruise he returned to Norfolk, where his family were waiting for him, and was granted a leave of absence in which to recuperate. But even while he rested in the first days of his leave, word was brought to him that the smallpox had broken out in Portsmouth, just across the river. Bidding farewell to his family, and paying no heed to the advice and protestations of his friends, the devoted Chaplain went out among the sufferers and ministered unto them. Early in the morning and late at night he tended the men of the Navy and their families, going from house to house to nurse the sick, pray with the dying and say the last rites over the dead. He was attacked by the disease, but although scarcely able to stand he refused to remain in the house and struggled to go on with his work. On the 21st of September 1855 this gallant and devoted man completed his duties upon this earth and went on to report to One who once said, "Greater love hath no man than this—."[63]

A report of the launching of the first *Princeton* at the Philadelphia Navy Yard on 10 December 1843 refers to the offering of a prayer by a civilian clergyman on that occasion. While religious ceremonies at the laying of the keel, the launching, and the commissioning of a vessel go so far back in human history that their origin is lost, this is the first known record of a prayer at the launching of an American war vessel. A newspaper account of the occasion follows:

> Just before the vessel was released Captain Stockton, U. S. Navy, who was in charge, assembled those on board and a prayer was offered by the Rev. Doctor Suddards:
>
> 'Eternal God, Creator of the Universe, Governor of Nations. Humbly we prostrate ourselves before Thee and ask Thy blessing. Most humbly we beseech Thee with Thy favor to behold and bless Thy servant the President of the United States and all the officers of the Government. May the vessel about to be launched be guarded by Thy gracious Providence and care. May it not bear the sword

57 *Ibid.*, III:May 1858:194.
58 *Ibid.*, VII:1 of 1858:133.
59 *Ibid.*, I:57:315.
60 *Ibid.*, III:May 1853:208.
61 R. S., 1398.

62 *Op. cit.*, VII:736.
63 *Op. cit.*, pp. 10–1.

in vain, but as the minister of God be a terror to those who do evil and a defense to those who do well. Graciously bless its officers and men. May love of country be engraven upon their hearts. Remember in mercy both arms of our National defense, and may virtue, honor and religion pervade all their ranks. Bless all nations and individuals on the earth and hasten the time when the benefits of holy religion shall have so prevailed that none shall wage war again for the purpose of aggression and none shall need it as a means of defense. All of which blessings we ask in the name of Him who taught us to say: "Our Father who art in Heaven". . . ."[64]

A good portion of a chaplain's time aboard ship was spent with the crew in an advisory capacity. As Chaplain Taylor said, "They feel that they can speak to him as they cannot to a watch-officer."[65] Frequently the men requested the chaplain to talk over their problems with them during the night watches.[66] Often the chaplain's room was available for conferences and interviews. "Three seamen," wrote Chaplain Colton, "came into my state-room to converse with me on the subject of religion. I encourage them in their good resolution."[67] Chaplain Jones felt that his most pleasant moments aboard his ship were spent "with men or officers, in night watches, or in state-rooms conversing on the subject [religion[68]].

"Seamen are prefectly accessable," wrote Chaplain Rockwell, and they "will rarely refuse to answer a question of the most personal nature, if your manner is such as to gain their confidence."[69] Indeed they seemed to take "peculiar pleasure in dwelling even in the darker portions of their past history."[70] He admitted, however, that sailors were given to "yarning," but he added that a chaplain "soon learns how to detect them in this, and when they are alone with one whom they respect, they do not attempt it."[71] To win a man's confidence chaplains knew that it was necessary to show an interest in him. "Listen kindly to the sailors tale of woe, and from that time forward you became his friend and may hope to do him good" was the way Rockwell summarized it.[72]

A part of each chaplain's day was devoted to calls on the sick and injured in sick bay, a compartment forward which housed these unfortunates. Perhaps at no time did the chaplain find sailors more amenable to spiritual guidance. "If you would get at the true character of the sailor," wrote Chaplain Colton, "you must visit him in his sickness. His better feelings then gush out over the asperities of his lot. . . ."[73] Chaplain Rockwell noted that the "heart is open and tender" at such times.[74]

On these visits the chaplain offered what comfort he could, talked to the patients about their loved ones at home and often prayed with them. These calls were sometimes heart-rending. Chaplain Colton, referring to a conversation with a very sick man said, "He spoke to me of his mother and his sisters, and tears filled his eyes." Chaplain Taylor, referring to a particularly trying visit wrote, "I returned to my room and wept over the scene of the sadness. . . ."[75]

It was to the chaplain that many a tar confided his last words and confessed, as a relief to his own spirit, his sins and shortcomings. Chaplain Taylor tells of an urgent request to come to sick bay. "I have been sent for by one who had declared himself an athiest, and endeavored to spread his opinions among his messmates, but on his dying cot desired to make a public declaration of his folly."[76]

What type of men made up the chaplain's parish on the old pre-Civil War frigates? Certainly no civilian parish ashore contained the diverse elements that were found aboard a frigate of the sailing-ship days. "A man-of-war is a world in miniature," said one chaplain, "in which every different kind of temper and disposition is to be found. . . ."[77] Chaplain Jones noted, "we have here, men from all climates, speaking all languages and of fierce bold habits."[78] Twenty-three foreign nationalities were represented on board the *Constitution* in 1844.[79] The sailors of the *Vincennes,* according to Chaplain Stewart, were "gathered principally from the South American coast—wanderers upon the earth, without a country or a home" and only a "small portion of them were shipped in the United States."[80] It was the feeling of Chaplain Rockwell that most sailors were of no nation, "but change from the employ of one to

---

[64] Benham, *Ships of the United States Navy and their Sponsors,* 1797–1913, pp. 151–2. It is now the general rule to have a prayer either by a chaplain or a civilian clergyman at the launching of a naval vessel the size of a destroyer escort or larger, and usually a prayer is offered at the commissioning.

[65] Taylor, *A Voyage Round the World,* II:277.

[66] Stewart, *A Visit to the South Seas,* I:118–9; Rockwell, *Sketches of Foreign Travel,* II:391.

[67] Colton, *Deck and Port,* p. 375; Stewart, *op. cit.,* I:121.

[68] Jones, *Sketches of Naval Life,* II:236.

[69] Rockwell, *op. cit.,* II:391.

[70] *Ibid.*

[71] *Idem.*

[72] Rockwell, *op. cit.,* II:428 ff.

[73] Colton, *Deck and Port,* p. 305.

[74] Rockwell, *op. cit.,* II:428 ff.

[75] Colton, *Deck and Port,* p. 42; Taylor, *A Voyage Round the World,* II:276.

[76] Taylor, *op. cit.,* II:276.

[77] Stewart, *op. cit.,* I:32.

[78] Jones, *op. cit.,* I:22.

[79] Snow and Gosnell, *On the Decks of "Old Ironsides,"* p. 301.

[80] Stewart, *op. cit.,* II:286.

that of another, just as convenience, or caprice, or higher wages may induce them to do so."[81] Of the officers aboard his ship Chaplain Colton said, "We present, perhaps, in our assembled capacity, as great a variety of intellectual, moral, and social habit, as any group of the same size, ever yet convened on flood or field."[82]

Many reasons prompted men in the United States to enter the Navy. "Disgrace, or misfortune, or difficulties at home," said Chaplain Jones, "drive them to thoughts of sea," and he added that for the unruly the "severe discipline of a man-of-war, may restore a healthy character to the individual."[83] Groups similar to those described made up the motley crew the chaplain called his parish.

Chaplains themselves had difficulty in evaluating their own services aboard the men-of-war of this early period. Chaplain Colton felt that the capstan of a man-of-war was an ideal pulpit from which to influence an audience. He admitted, however, that the impressions on sailors were less permanent than on other men.[84] Chaplain Stewart at one time was convinced that no class of men were more open to convictions of truth than seamen or more susceptible to religious impressions, yet writing from another ship some time later, he reported, "I have long felt great discouragement, as to any decisive benefit to them [sailors] from the discharge of the service of my station. . . ."[85] Samuel Leech, who served aboard ship as an enlisted man, later wondered that sailors who saw so much peril should have treated religion with so much neglect and added "when danger is imminent, they send up a cry for help; when it is past, they rarely return a grateful thank-offering."[86]

THE EXPLORING EXPEDITIONS

During the forties, a number of events focused public interest in the United States upon the Pacific and had far-reaching effects on the Navy and incidentally on naval chaplains. An agreement with Great Britain was reached in 1846 which settled the Oregon question. In that year the American flag was raised over California; the discovery of gold there two years later gave a mighty impetus to an emigration which had already begun. American trade with China was second only to that of Great Britain. The whaling industry in the Pacific was bringing in its shiploads of oily wealth. All these factors and more combined to make the citizens of the United States vividly aware of the Pacific.

Beginning as early as 1838 and continuing through the forties and the fifties, several exploring expeditions were sent to the Pacific. A number of scientists and artists accompained these cruises. Usually there was at least one chaplain attached to each expedition.

The first of these expeditions sailed from Hampton Roads under the command of Lieutenant Charles Wilkes on 18 August 1838. The Squadron consisted of the sloops *Vincennes* and *Peacock,* the brig *Porpoise,* and two small schooners. On board the *Vincennes* was the newly appointed chaplain, Jared L. Elliott. After visiting some of the islands of the South Pacific, Wilkes, with the *Vincennes* and the *Peacock,* sailed into the Antarctic Ocean in December 1839 and a month later sighted land which he called the Antarctic Continent.

The Expedition surveyed some 280 islands of the Pacific, and in 1841 some of its personnel made an overland exploring journey through the Oregon Country and down into California.[87] The *Vincennes* arrived at San Francisco on 24 August 1841 and departed on 31 October. It appears that sometime between those dates Chaplain Elliott was detached from the ship. He returned to New York and reported to the Secretary of the Navy on 7 March 1842. His resignation was accepted on 18 October.

Available records regarding the part of Chaplain Elliott had in this Expedition are scanty. An examination of the log of the *Vincennes* shows that Divine Service was held with fair regularity on Sunday mornings, even after it was clear that the chaplain had been detached. The log of the *Peacock,* which had no chaplain, and the journals of some of her officers show that Divine Service was also held aboard this vessel almost every Sunday morning. The *Peacock* was wrecked off the mouth of the Columbia River in Oregon on Sunday morning, 18 July 1841. The last entry in her log included the statement, "Performed Divine Service as usual."[88]

A service was held on board the *Vincennes* 10 August 1840 at which Chaplain Elliott preached a

[81] Rockwell, *op. cit.*, II:386.
[82] Colton, *Ship and Shore*, p. 23.
[83] Jones, *op. cit.*, I:64–5.
[84] Colton, *Sea and Sailor*, p. 53.
[85] Stewart, *op. cit.*, I:35; II:Dec. 11, 1829.
[86] Leech, *Thirty Years from Home*, p. 82.

[87] Alden and Westcott, *The United States Navy*, p. 125.
[88] The Journals of George Foster Emmons of the *Peacock* and other original material belonging to this expedition are in the Coe Collection, Yale University. The original logs of the *Vincennes* and the *Peacock* are in the National Archives. Joseph Clark, seaman and corporal of Marines, a member of the Wilkes Expedition, wrote a narrative of the voyage. His book *Lights and Shadows of Sailor Life*, printed in Boston in 1848, contains a few references to Chaplain Elliott.

memorial sermon for two members of the Expedition who had been killed a short time previously by the natives. This sermon was printed in full in the 7 November issue of the *Polynesian* published at Honolulu and is the earliest known printed sermon of a Navy Chaplain.

Chaplains Joel Newton and N. C. Fletcher were assigned to the *Columbus* and the *Vincennes* in 1845. These vessels made up the squadron under the command of Commodore James Biddle which was sent to Japan. Fletcher, who was ordered to the sloop *Vincennes,* was unhappy about the living accommodations aboard his ship. On 2 June 1845 he wrote to the Secretary of the Navy:

> It was never intended for "Sloops" to be allowed Chaplains, consequently there is no accommodations for them; they must be crowded in with the luggage, with the black stewards and waiters and no "State Room", to flee to for repose. If there is an officer in the Navy who needs some little privacy it is the Chaplain, and yet all the room I could be spared was about 4 feet square and that in the open "Ward Room". I submitted to this for a time and should have done so perhaps for the cruise had I not received *"orders"* through Com. Biddle to perform the duties of "Professor of Mathematics" in addition to my duties as Chaplain. This being a distinct office from the Chaplaincy (and one for which the law provides) I deemed altogether wrong, and an order to which I could not, *nor would not,* submit.[89]

Fletcher carried his complaint to Commodore Biddle who said that nothing could be done unless the doctor issued a certificate saying that the Chaplain's health was such that he could not make the cruise. The Chaplain secured the certificate, and when the crusty Commodore still delayed he resigned. The *Vincennes* sailed without a chaplain.

Commodore Biddle sailed into Yedo (Tokyo) Bay in July 1846, seven years before Perry's expedition. The *Columbus* and *Vincennes* were the first United States naval vessels to enter those waters.[90] While the Japanese authorities treated the Commodore with kindness and supplied the needs of the Squadron, no one was permitted to land. According to the report of the Secretary of the Navy for 6 December 1847: ". . . to the offer of friendy intercourse, the unchangeable answer was, 'Go away, and do not come back any more'."[91]

Chaplain Henry Wood was on the *Powhatan* in 1859, when that ship carrying the United States Commissioner, J. E. Ward, lay off Taku, near Tientsin, China. This was during the Anglo-French attack on the Taku forts, precipitated by the Powers' unsuccessful efforts to negotiate a new treaty with China. Since the Americans had not engaged in actual fighting, Ward decided to go to Peking in advance of the British and French. He took with him the Reverend W. A. P. Martin, D.D., a missionary well versed in the Chinese language, and Chaplain Henry Wood of the *Powhatan.* There was no character in the Chinese written language for "chaplain." The nearest symbol was that which is translated: "Jossman." A flag with this character was carried over Chaplain Wood's sedan chair on the trip inland, described by the Chaplain as being "thru heavenly weather over hellish roads." This was the first official United States delegation to visit Peking, and Dr. Martin and Chaplain Wood were the first Protestant ministers to enter the forbidden city.

WITH PERRY IN JAPAN

Two naval chaplains accompanied Commodore Matthew Galbraith Perry, the younger brother of Oliver Perry, on his expedition to Japan in 1853 and again in 1854. They were Chaplains Edmund C. Bittinger who was aboard the *Susquehanna* and George Jones in the *Mississippi.*

On 14 March 1854, while the American Squadron was at anchor off Yokohama, Chaplain Bittinger launched out on an exploring expedition of his own, causing considerable excitement in local Japanese official circles as well as giving concern to Commodore Perry. American officers were not permitted to penetrate into the interior, however, Chaplain Bittinger was either ignorant of or indifferent to this fact. Wearing his sword, with which he sometimes threatened those who tried to stop him, Chaplain Bittinger started out to walk to Tokyo. An excited Japanese official notified Commodore Perry, who at once ordered guns to be fired as a recall signal for the wandering Chaplain. Messengers with written orders were also dispatched to find him and escort him back. The following account of the incident is given in Perry's official report:

> The American officer, whose intrusion had created so great an excitement, was Mr. Bittinger, the chaplain of

[89] Nav. Rec. Coll., Book of Resignations, 1841-1850.

[90] Charles Nordhoff, crew member of the *Columbus* and fourteen years old at the time of the expedition, wrote an account of the voyage. It is found in his book *Man of War Life* which appeared in 1857 as a part of *Nine Years a Sailor,* and as a separate edition in 1858. The same story edited by the grandson of the same name appeared in 1940 under the title, *In Yankee Windjammers.* Nordhoff makes only incidental mention of the presence of the Chaplain aboard the *Columbus.*

[91] See also Luce, "Commodore Biddle's Visit to Japan in 1846," *Naval Institute Proceedings,* Sep. 1905, pp. 555 ff.

the steamer *Susquehanna.* While taking a walk on shore, this gentleman's curiosity prompted him to extend his observations somewhat beyond the usual circuit of some four or five miles, within which the Japanese authorities had contracted the movements of their visitors. Starting from Yoku-hama, opposite to where the squadron was anchored, the enterprising investigator pushed on to the town of Kanagawa, some three miles further up the bay, where he was accosted by some of the Japanese officials and the interpreter, Gohatsiro, who urgently solicited him to return. He was not, however, to be so easily balked of his purpose, and continued his journey, followed by the Japanese officers, who dogged his steps at every turn until he reached Kamasaki. Here there was a river to cross, and he tried to prevail upon the Japanese boatmen to ferry him to the opposite side, but they refused in spite of bribes and threats, in the course of which the chaplain, if the Japanese accounts are to be believed, drew his sword. He now pursued his way higher up the river with the hope of finding a place that might be forded, and had just reached a very promising looking crossing, the depths of which he was about trying, when the messenger, who had hurried in rapid dispatch, from the steamer *Powatan,* accosted him with the written order of the Commodore. "He," thus reported the Japanese authorities, with their usual minuteness of description "read it, walked four steps further, read it again, then suddenly returned, and intimated his intention of going back to the ship."[92]

Chaplain Bittinger's conduct in a land where the officials viewed with suspicion all foreigners was tactless to say the least. It is to be remembered that he was without an interpreter and had no means of conveying his desires except by the sign language and his sword. The Japanese interpreter, Yenoske, who figures in the incident, had received his knowledge of English from Ranald MacDonald, a half-breed American Indian from the Oregon country who had deliberately cast himself adrift off the coast of Hokkaido in 1848. Held as a prisoner by the Japanese until he was taken aboard an American naval vessel at Nagasaki in 1849, MacDonald was used as a teacher of English. Thus when Perry forced his presence upon the unwilling officials of Japan, they had several individuals, including Yenoske, who were able to serve as interpreters. The account of Chaplain Bittinger continues:

The chaplain, in the course of his wanderings, had an opportunity of seeing one of the largest towns of Japan, that of Kanagawa, which, with its numerous wide streets, and its crowded population, had quite an imposing appearance. He penetrated into several of the dwellings and temples, and by his pertinacious perseverance, succeeded in obtaining, in one of the shops, some Japanese money in exchange for American coin. The native authorities seemed particularly worried in regard to this last matter, as it was so great an offence against their laws. The Japanese, in their report of the occurrences, stated that the American officer had gone into a shop by the roadside and asked the keeper to allow him to see some coins. The Japanese shopman complied with the request, but as he seemed somewhat chary in the display of his treasure the chaplain insisted upon seeing more, which demand was also granted. Scales were now asked for, which being brought, the chaplain took out some silver pieces, and weighing them in one balance against the Japanese gold and silver coins, mixed indiscriminately in a heap, in the other, transferred the latter to his pockets and left his American coin to console the shopman for the loss of his Japanese change. . . . On the next day Yenoske brought back the sum of three dollars and a half in American silver coin, which had been left in compulsory exchange with the Japanese shopman, and stated that six pieces of gold, six of silver, and the same number of copper, were in possession of the chaplain. Yenoske requested that the Japanese money should be returned, and was told it should be restored.

When a marine aboard the *Mississippi* died, Commodore Perry requested permission of the Japanese authorities to purchase land for an American cemetery. After considerable discussion, and with great reluctance, the Japanese finally consented "to allow the burial to take place at Yoku-hama, at a place adjoining one of their temples, and in view of the squadron."

George Jones, as chaplain of the *Mississippi* and the senior chaplain of the Expedition, was responsible for the religious part of the last rites. He was aware of the old antipathy of the Japanese toward Christianity. In 1636 the Japanese government had had difficulties with Catholic missionaries and native Christians. The Japanese authorities felt that the Christian missionaries were the advance agents of foreign imperialism against which the best defense was absolute seclusion. An Imperial edict was issued in June 1636, which stated in part: "So long as the sun shall warm the earth, let no Christian be so bold as to come to Japan; and let all know that the king of Spain himself, or the Christian's God or the great God of all, if he violate this command, shall pay for it with his head."[93]

Being somewhat apprehensive as to what might happen when conducting Christian burial services ashore, Chaplain Jones appealed to Commodore Perry for instructions. "Do exactly as you always do on such occasions," advised the Commodore, "no more, or less."[94] On the day of the funeral, a Japanese official went aboard the *Mississippi* to accompany the party ashore. Among those in the party was Perry's interpreter, the veteran printer-missionary to

[92] Hawks, *Narrative of the Expedition,* I:360.

[93] Hawks, *op. cit.,* I:475.

[94] *Ibid.*

Used by permission—The Leatherneck, July 1945
First Marine Burial in Japan.

China, Wells Williams, who had acquired a knowledge of the Japanese tongue from some Japanese shipwrecked on the Oregon coast in 1834.[95]

An entry in the log of the *Mississippi* for 9 March 1854 stated:

> At 2:45 called all hands to bury the dead—read the funeral Services over the body of Robert Williams (Private Marine) and sent the body ashore for interment with an escort under the command of Captain Slack.

The account of the funeral reads:

> The flags of every vessel in the squadron were hoisted at half mast as the boats pushed off. The body was borne to a very picturesque spot at the foot of a hill, at a short distance from the village of Yoku-hama. The chaplain, Mr. Jones, was robed in his clerical gown, and on landing was received in the most courteous manner by some of the Japanese authorities, who showed none of their supposed repugnance to the Christian religion and its ministers. Crowds of people had also gathered, and looked on with great curiosity, but with decorous respect, as the funeral procession moved slowly along to the sound of the muffled drum. The road lay through the village, and its inhabitants came out from their houses and open shops to behold the novel scene. The place chosen for the burial was near a Japanese place of interment, with stone idols and sculptured headstones, and as the procession came up a Buddhist priest, in robes of richly embroidered silk, was observed already on the ground.[96]

For the first time in more than two hundred years, the Japanese Government permitted a Christian service to be held on Japanese soil. Moreover, this was the first known Protestant service ever conducted in Japan by an ordained clergyman. The ceremony for the deceased marine is unique also because of the presence of a Buddhist priest, who, uninvited by the Americans officiated according to the rites of his faith.

The account of the double ceremony is as follows:

> Mr. Jones read the service of the Protesant Episcopal church, and while he was officiating the Buddhist priest sat near by on a mat, with an altar before him, on which was a collection of scraps of paper, some rice, a gong, a vessel containing saki, and some burning incense. The service having been read, the body lowered, and the earth thrown in, the party retired from the grave. The Buddhist priest then commenced the peculiar ceremonies of his religion, beating his gong, telling his rosary of glass and wooden beads, muttering his prayers, and keeping alive the burning incense. He was still going through his strange formulary when the Americans moved away. . . .[97]

Chaplain Jones, mindful of the significance of the observance of Christian rites ashore in Japan, sought to answer the question: "What are the prospects of now presenting Christianity to the Japanese mind?" Jones wrote:

> Apart from governmental influence, I think there would be no great difficulty in introducing Christianity; *but the government would interfere most decidedly.* I performed funeral services on shore four times: once at Yoku-hama, twice at Hakodaki, and once at Simoda; in every instance in the presence of the Japanese, and, in most, when large numbers were collected. They always behaved well. Japanese officers were present, with their insignia, on all occasions. I thus became known among the people everywhere as a *Christian clergyman,* or, to follow their signs for designating me, as 'a praying man'. Instead of this producing a shrinking from me, as I had supposed it would, I found that I had decidedly gained by it in their respect, and this among officials as well as commoners.

The following epitaph was inscribed on the tombstone[98] at one of the American graves:

> Sleeping on a foreign shore,
> Rest, sailor, rest! thy trials o'er;
> Thy shipmates leave this token here,
> That some, perchance, may drop a tear
> For one that braved so long the blast,
> And served the country to the last.[99]

Chaplain Jones wrote the third volume of Perry's official report, entitled *Observations on the Zodiacal Light from April 2, 1853, to April 23, 1855.*

Jones also contributed to the second volume of the official record. This comprised a series of reports by

[95] See Drury, "Early American Contacts with the Japanese," *Pacific Northwest Quarterly,* Oct. 1945.
[96] Hawks, *op. cit.,* I:354. See Drury, "First Marine Burial in Japan," *The Leatherneck,* July 1945, pp. 64–5.

[97] Hawks, *op. cit.,* I:355.
[98] *Ibid.,* I:446.
[99] *Ibid.,* I:476.

various officers of the Expedition. Among the several by Jones were: "A Geological Exploration, etc., of the Island of the Great Lew Chew," "A Visit to the Coal Regions of the Island of Formosa," and "A Mineral Spring near Hakodaki." The Island of the Great Lew Chew is better known today as Okinawa.

WALTER COLTON

Walter Colton, one of the best known Navy chaplains of the nineteenth century, launched "a gallant fleet of books" which spread his reputation throughout the reading public. Colton's unique experience as governor of a large part of California in the early days of its history under the American flag, added to his other achievements, won for him a prominent niche in California's hall of fame.

Colton was born in Rutland, Vermont, 9 May 1798. He received his B.A. degree from Yale in 1822 and was graduated from Andover Theological Seminary in 1825. He was ordained in the Congregational Church. For three or four years Colton served as Professor of Moral Philosophy and Belles-Lettres in the Scientific and Military Academy at Middletown, Connecticut. He became editor of the *American Spectator* and *Washington Chronicle* in Washington, D. C., in 1828 and as such attracted the attention of President Jackson who offered him an appointment as a naval chaplain. Colton accepted it and remained with the Navy until his death in Philadelphia on 22 January 1851.[100]

After Colton's first duty on the *Vincennes* in the West Indies, he went to the Mediterranean aboard the *Constellation*. Out of this epxerience came the two books which proved him to be an extremely observant traveler: *A Visit to Athens and Constantinople*[101] and *Ship and Shore*.

Upon his return from the Mediterranean in 1834, Colton appears to have been without regular naval duties for several months. During the winter of 1834 and 1835, he lobbied on behalf of the proposed law to increase the pay of naval officers, including chaplains. Congress approved this law in March 1835. Colton was attached to the Navy Yard at Boston that same month, where he remained until June 1837. In the spring of that year he was appointed Historiographer and Chaplain to the South Sea Surveying and Exploring Squadron and carried on special studies for a year preparing himself for this work. However, the delicate state of his health at that time caused him to request detachment from the Expedition. The billet was then given to Chaplain Jared Elliott. Colton edited *The Colonization Herald* for a few months in 1837.[102]

After a short tour of duty in the *Macedonian*, Colton was assigned to the Navy Yard at Philadelphia and in 1844 to the Naval Asylum (now called the Naval Home) in that city. While in Philadelphia, he joined his old associates and, with the written consent of the Navy Department, helped to edit the *Independent North American*. Some of Colton's editorials were reprinted in *The Sea and the Sailor* in 1851. He was a forceful writer who sought to put his Christian principles into practical every-day use. However, a new Secretary of the Navy brought a change of policy. On 4 August 1842 Secretary A. P. Upshur wrote to Colton saying: "Your connection with a political newspaper is considered inconsistent with your position as a Chaplain in the Navy."[103] Colton, faced with the alternative of resigning one or the other, elected to stay in the Navy.

While serving as chaplain at the Naval Asylum, Colton faced a number of practical problems which invariably, at some time or another, challenge the initiative of all Navy chaplains. Foremost was excessive drinking on the part of certain pensioners. During long years of sea service, these sailors had received their daily ration of grog. Now in their old age and with little to do, it was easy to indulge in drink. Colton was eager to provide adequate and satisfying substitutes. There was a reading room, but it had few books and periodicals. Even Divine Services were dull since the institution had no musical instrument and consequently no singing.

On 28 May 1845 Colton wrote to the Secretary of the Navy requesting $435 to purchase an organ, some periodicals, and a die. The Secretary granted the request but wrote to the Governor of the Asylum requesting a more detailed statement. The Governor wrote in reply:

> In compliance with which I have the honor to state that the Chaplain wishes to obtain for the Chapel a parlor organ, the cost of which will be about two hundred and fifty dollars—to which will be added the salary of an Organist, amounting to about one hundred dollars per annum.
>
> For the Temperance Society of the Asylum a Die, cut with appropriate inscriptions from which medals will be struck for presentation to those pensioners who become

[100] Colton, *Sea and Sailor*, pp. 365 ff.

[101] This volume was edited and reissued in 1851 by his friend, the Rev. Henry T. Cheever, under the title *Land and Lee in the Bosphorus and Aegean*, after the death of Chaplain Colton.

[102] Colton, *Sea and Sailor*, pp. 378–9.

[103] Nav. Rec. Coll., I:33:112.

members of the association. The Die will cost about Fifty Dollars, and each medal from ten to twelve cents—and for the Reading Room some magazines and papers from different parts of our country, a large number of which could be secured very economically by purchasing the "*Exchange List*" of one of the large Philadelphia Papers, which could be bought for about thirty five dollars, per annum, and would embrace a much greater variety of reading matter than could possibly be obtained for that sum in an other way.[104]

This appears to be the first time the Navy Department provided an organ for a chaplain. This instrument, known as a Packard Organ, manufactured by the Fort Wayne Organ Company, Fort Wayne, Indiana, is still in use at the Naval Home.[105]

The years 1845 and 1846 were years of national crisis in which the United States Navy played an important part in Californian waters. In June 1845 Secretary of the Navy Bancroft sent secret instructions to Commodore John D. Sloat of the Pacific Squadron to seize San Francisco and to blockade other California ports in case war broke out between the United States and Mexico. Commodore R. F. Stockton was sent out in the fall of 1845 in the *Congress* to relieve Commodore Sloat. Walter Colton was the chaplain on this historic voyage.

Colton, at the age of forty-seven, had married Cornelia Baldwin of Philadelphia. In 1845, a year later, he received orders for duty aboard the *Congress*. He bade farewell to his wife and started on a cruise that was destined to keep him away for nearly four years.

On Sunday, 16 November, less than three weeks after sailing, Colton betrayed his homesickness for his wife in the following entry in his Journal:

> You who cannot leave your wives and children for a week, without intelligence from them, go to sea with the prospect that we have, of not hearing from them for a year. The truth is, none but old bachelors and hen-pecked husbands should go to sea. The latter flies from persecution, the former from that wretchedness which a sight of real domestic happiness inflicts. The bliss of Eden made even Satan more wretched than he was before.[106]

Writing to one of his brothers from Norfolk shortly after reporting for duty on the *Congress*, Colton declared:

> We have the noblest frigate in the service—admirable officers—and as fine a crew as ever trod a deck. I have been here almost two weeks, and have not seen one sailor intoxicated, nor one punished for any offence; and—what is still more remarkable—I have not heard any profaneness, either among the crew or officers. I came on board a thorough teetotaller, and such shall remain. No one here shall drink even wine under the countenance of my example. I am anxious to have evening prayers—have proposed it to Commodore Stockton, and he has it now under consideration. I intend to devote myself thoroughly to my appropriate duties. We have four hundred souls on board, all told. I now intend to keep a journal, which I can use on my return, if God permit.[107]

The Journal which he started grew into two books —*Deck and Port* and *Three Years in California*. These volumes are rich in source material on the early days of the American occupation of California.

Chaplain Colton was fortunate in having Commodore Stockton as his Commanding Officer for he was given every encouragement in his work. Drinking and profanity were discouraged; grace was said at meals in the wardroom. Colton's Journal records many instances of the Commodore's interest in and presence at Divine Services.

Commodore Stockton, on one occasion, spoke to his officers and men on the importance of Bible study. "He rebuked the idea that religion was out of its element among sailors," wrote Colton, "and told them that of all classes of men they were the one that most needed its restraining influences and glorious promises and denounced as insane a disposition to trifle with its precepts." Colton summed up his convictions regarding the value of such a testimony in these words: "Such remarks as these, coming from the commander of a ship or squadron, will do more to sustain a chaplain in the discharge of his difficult duties than any privileges which can be conferred upon him through the provisions of law."[108]

Writing to the Secretary of the Navy on 30 October 1845 when the *Congress* was about to sail for the Pacific, Colton states:

> We weighed anchor from Hampton Roads this morning at day light. When we had cleared the Capes all hands were called and Commodore Stockton delivered a brief and very pertinent address to the officers and crew. Prayers were then offered for the Divine protection, the broad pennant saluted and the ship cheered. A spirit of cheerfulness and alacrity in duty pervades all hearts. The shore is fading from our view and how many of us will tread it again is known only to Him with whom are the issues of life and death.[109]

Colton's Journal is rich in references to chaplains' duties and to his own interesting experiences. The

[104] *Ibid.*, II:May–June 1845; 157.

[105] The 1916 *Regulations, Origin, History & Laws of the United States Naval Home* (p. 47) states that the organ was purchased by order of Secretary George Bancroft. The date of the printing of this pamphlet overlaps the service of the present organist at the Naval Home, Mrs. Alice S. Harrison, who has been playing the organ at Divine Services held there since 1909.

[106] Colton, *Deck and Port*, p. 41.

[107] Colton, *Sea and Sailor*, pp. 392–3.

[108] Colton, *Deck and Port*, p. 324.

[109] Nav. Rec. Coll., III:Oct. 1845:291.

following is his account of the Divine Service held on Sunday, 26 October, on the eve of their departure:

The wind still ahead. This being the sabbath, we had divine service. The crew were attentive: not the rustle of a hand or foot disturbed the stillness; the speaker's voice only broke the silence of the deck. The text was the injunction of the prophet, "Go up now, look towards the sea." The object of the speaker was to sketch the stern magnificence of the ocean as illustrating the majesty of God; to exhibit the effects of an ocean life on the social and moral character of man; and to inculcate the great lesson, that into whatever climes we may penetrate, through whatever seas we may pass, we cannot escape from the presence of the Deity.[110]

Colton took a deep interest in the ship's library which was heavily stocked with "between three and four hundred volumes." He had received many of the miscellaneous and religious books from the Presbyterian Board of Publication, the Sunday School Union, the American Tract Society, and Commodore Stockton himself; and the supply of Bibles from the American Bible Society. The Chaplain realized the importance of a splendid collection such as the one with which he was provided, and remarked:

No national ship ever left a port of the United States more amply provided with books suited to the habits and capacities of those on board. This desideratum has been supplied, so far as the crew is concerned, with comparatively little aid from the department. The government furnishes the sailor with grog to burn up his body, a christian liberality with books to save his soul.[111]

On 9 November, the third Sunday at sea, Colton wrote:

I distributed tracts to-day to the crew—to all who came to me for them: and few remained behind. It would have encouraged the hearts of those who supply these sources of salutary instruction, to have witnessed the eagerness with which our sailors took them. In a few minutes there were three or four hundred men on the decks of our ship reading tracts; each catching some thought which lures from sin, and throws its clear and tender light on the narrow path which leads to heaven.[112]

Divine Services were held on Sundays throughout the voyage with consistent regularity often in spite of strong winds and rolling seas. Where other commanding officers might have found an occasion to omit the Sunday services, Commodore Stockton agreed with his Chaplain that such should be held if at all possible. Colton gives us the following picture of a Divine Service:

Sunday, Dec. 7th. At eleven o'clock, the tolling of the ship's bell announced the hour of worship. The officers took their accustomed station on the starboard-quarter; the marines on the poop-deck; the crew on the larboard-quarter, stretching back to the waist and circling the main-mast to the opposite side; the band and singers between the after-hatches; Mr. Ten Eyche and Mr. Turrel, with their families, forming a group between the officers and marines. The commodore, being informed by the captain that the crew were assembled for worship, appeared and took his station on the left of the officers. The chaplain then took his station at the capstan, which was covered with a large flag, when the band played the impressive air to the words, "O come and let us worship."[113]

On 10 May 1846 Colton commented on certain advantages a chaplain had over his brother minister ashore. He wrote:

Sunday. . . . Divine Service: officers and crew all present. Subject of the sermon, the temptations of the sailor. A chaplain in the navy has one advantage over his brethren on land. He has his parishioners in the most compact of all possible forms, and every one present when he officiates. In making his official visits he has not to ride around among five hundred families located at all points of the compass. He cannot stir without coming in contact with them. But he has this disadvantage; in the vicissitudes of a sea life they are extremely apt to break away from his constraining influence. They may be brought back again, but it is too often through the deepest self-inflicted humiliation.[114]

In his Journal for 12 July 1846 Colton refers to a prayer meeting which he was then conducting for an hour every other evening. Although the attendance was voluntary, the meeting filled a large store-room. "A good number of our sailors are earnestly seeking religion," he wrote, "and several hope they have found it." This was a service in which the men themselves took part by testimony and prayer.[115]

On 27 July 1846 he wrote from Monterey:

We have had for two or three months past an increased attention in our ship to the subject of religion. It began in my Bible-class, but spread beyond that number among the crew. As the interest deepened, I established a prayer-meeting, which has been held three times a week in the store-room, an ample and convenient apartment for that purpose. Here you will find at these meetings some sixty sailors on their knees at prayer; some thirty of them, it is believed, have recently experienced religion; the rest are inquirers, and come to be prayed for. Among the subjects of the work are some of the most efficient seamen in our ship, but who have hitherto led a thoughtless life. Those who give evidence of having experienced a change of heart are called upon to pray. Their prayers have no finished sentences, but they are full of heart and soul. When they speak in their exhortations it is with great directness and

[110] Colton, *Deck and Port*, p. 16.
[111] *Ibid.*, p. 19.
[112] *Ibid.*, p. 33.

[113] *Ibid.*, p. 67.
[114] *Ibid.*, p. 300.
[115] *Ibid.*, p. 382.

832159—49——7

force. It would affect you to tears to hear these rough, hardy sailors speak in these meetings of their sins, of the compassion of Christ, and their new-born hopes. Almost every evening some new one, the last perhaps expected, comes in, and, kneeling down, asks to be prayed for. These meetings have no opposition among the officers, and very little, if any, among the men. There has been a great change in the Navy within a few years on this subject. We can now have Bible-classes and prayer-meetings on board our men-of-war, and find among our officers many who will encourage them, and not a few who will give them their efficient aid.[116]

Colton summed up his convictions regarding the effect of a strong religious program aboard a naval vessel in the matter of discipline:

The effect of this on the discipline of the ship is too marked to escape observation. There is no disobedience and no punishment. Each performs with alacrity the duties of his station. It would seem as if we might throw every instrument of correction and coercion overboard; their requirement, for the present at least, has ceased. Give me the religious sentiment in a crew, and you may sink your handcuffs, cats, and colts in the depths of ocean.[117]

The United States flag was first raised in California at Monterey on 7 July 1846 by Commodore Sloat. On 15 July the *Congress* arrived at Monterey, the capital of California, and Stockton took over command of the Pacific Squadron. Since the United States was then responsible for the civil administration of California, some provision had to be made for the establishment and maintenance of civil authority. Commodore Stockton selected Chaplain Colton to be Alcalde at Monterey, a district which extended 300 miles along California's coastline. The office of alcalde was a Mexican institution which combined the duties of sheriff, judge, prosecutor, coroner, and governor. Colton took up temporary quarters in the home of Thomas O. Larkin, the American Consul at Monterey.[118] He was the first Protestant clergyman to settle in California.

At Monterey Colton played the role of a benevolent dictator for nearly three years. After serving as alcalde by appointment for about two months, he was elected to that office on 15 September 1846 by the citizens of Monterey. Through his fair and impartial decisions, Colton soon won universal respect among all classes. It is reported that the poor almost worshipped him and the rich knew that he had no "itching palm."[119]

[116] Colton, *Sea and Sailor*, p. 395. See also *Deck and Port*, p. 371.
[117] Colton, *Deck and Port*, p. 383.
[118] The Larkins established the first white American home on the Pacific Coast and their house, which was built in 1835, is still standing.
[119] Colton, *Sea and Sailor*, p. 396–7.

Chaplain Colton established the first American press in California. He received financial aid from Commodore Stockton and purchased an old abandoned press "which had been used by a Roman Catholic monk in printing a few sectarian tracts."[120] Colton examined the press with enthusiasm. "Mice had burrowed in the balls," he wrote, "there were no rules, no leads, and the types were rusty and all in pi." Why not start a newspaper—an American newspaper—in California? Colton found Robert Semple, an emigrant from Kentucky, who wore a "buckskin dress, a foxskin cap," and was "true with his rifle, ready with his pen, and quick at the type-case," who agreed to assist. The two became partners and on 15 August 1846 published the first issue of the *Californian*. It was printed on sheets of paper about the size of foolscap which had been used to wrap tobacco.[121]

The first issue of the first American newspaper in California told about the declaration of war between the United States and Mexico. One half of it was in Spanish, the other half in English. "It produced quite a little sensation," wrote Colton.

On 4 September this Navy chaplain impanelled the first jury to be summoned in California. There was no appeal from his court. In his Journal Colton noted: "There is not a judge on any bench in England or the United States, whose power is so absolute as that of the alcalde of Monterey."[122]

The Spanish natives loved to gamble, a pastime which Colton, with typical New England puritanism, resolved to stop. In October he issued an ordinance against this vice but observed later that this only drove the gamblers out of the settlement into the bushes. Some were caught. Once Colton took a "file of soldiers" and surrounded a saloon. He caught fifty in his net, including the Alcalde of San Francisco, and fined each twenty dollars.[123]

With the money received from fines and with the aid of the prisoners incarcerated in the town's jail, Colton built a town hall of white stone quarried from a nearby hill. The building measured about seventy by thirty feet and had two stories—the lower

[120] The press used was taken to California in 1834 by Thomas Shaw of Boston for A. V. Zamorano. It may have been used by priests attached to the San Carlos Church in Monterey or at the Carmel Mission about four miles from the village. The Church had been built in 1795 and was one of the four presidio churches in California erected by the Franciscan fathers; the Mission had been founded by the Franciscan friar, Junipero Serra in 1771.
[121] Colton, *Three Years in California*, p. 32.
[122] *Ibid.*, p. 55.
[123] *Ibid.*, p. 195.

for a school, and the upper for public assemblies. It took more than a year to construct. The building was completed in the summer of 1847, and was known throughout the community as Colton Hall. It was there in September 1849 that the first Constitutional Convention met. Colton Hall, a fitting memorial to a great Navy chaplain, is still standing at Monterey and is still used as a municipal office building.

Colton's duties ashore were almost entirely secular. There were not enough Protestant American adults in Monterey to provide even a small audience. The native population was Catholic. Occasionally, naval vessels without a chaplain called at Monterey and Colton went aboard to conduct Divine Services.

On Sunday, 2 August 1846, Colton wrote in his Journal:

> I officiated to-day on board the Savannah. It is much to the credit of the officers of this ship that though without a chaplain, they have had, during a three years' cruise, their religious services regularly on the Sabbath. Four of their number, two lieutenants, the surgeon, and master, are professors of religion, and exert a deep influence through their consistent piety. Their Sabbath exercise has consisted in reading prayers selections from the Scriptures, and a brief, pertinent sermon. They have had, also, their Sabbath-school.[124]

Here is a striking testimony to the faithfulness of certain naval officers in conducting Divine Service when no chaplain was aboard.

The following entry from Colton's Journal for Sunday, 21 February 1847, tells of his interest in the distribution of Bibles and religious literature:

> The American Tract Society has sent me out, by the Lexington, a large box of their publications. Nothing could be more timely. I have not seen a tract circulating in California. Emigrants are arriving, settling here and there, without bringing even their Bibles with them. The same is true of the United States troops. All these are to be supplied from home, and by those two great institutions which are now throwing the light of life over continents and isles. It remains for the Missionary Society to do its duty, and dispatch to this shore the self-denying heralds of the Cross.[125]

Colton started a school in the hall he had erected. Writing to his wife in April 1847, he announced that he had been appointed Judge of the United States Court of Admiralty. He wrote:

> You don't know, I suppose, what this court is, so I will explain it: when a vessel of any kind is captured by our men-of-war, she is considered a prize. But before she becomes really so, it is necessary that she should be tried and condemned; if it is found that she belongs to individuals of a neutral nation that have not been trading with the enemy, she is liberated; but if she belongs to the enemy or those who reside among them, she is condemned. Now, to decide this question is my office; there is an appeal from my decision to that of the Supreme Court of the United States, if the owners choose. I have just condemned the schooner William and her cargo: they are both worth about twenty thousand dollars.[126]

Colton said that seven prize cases were brought before his court at Monterey during his three year service there.[127]

Chaplain Colton was among the first to send to Eastern newspapers the news of the discovery of gold in the Sacramento Valley in January 1848. His letters appeared in the *Independent North American* and the *New York Journal of Commerce.*

Colton sent a messenger to the mines to confirm the discovery and, according to his Journal, the messenger returned with a bag of nuggets. An eager crowd of Monterey's townsfolk gathered around Colton and his messenger to examine the specimens. "All admitted they were gold," wrote Colton, "except one old man, who still persisted they were some Yankee invention, got up to reconcile the people to the change of flag."[128] The gold excitement swept the village of its men, except those who were in prison or in uniform. "I have only a community of women left," bemoaned the Chaplain, "and a gang of prisoners, with here and there a soldier, who will give his captain the slip at the first chance." Indeed, one whole platoon of soldiers deserted for the gold fields, leaving "only their colors behind."[129]

On 18 October 1849 Colton was relieved of his duties as alcalde and returned to Philadelphia by way of Panama. He was again attached to the Philadelphia Navy Yard where he passed away on 22 January 1851.

On 10 December 1942 a liberty vessel, the *Walter Colton,* was launched by the Kaiser Shipbuilding Company at Richmond, California, in honor of this Navy chaplain who played such an important role in the early days of California's history.

It was Walter Colton, the patriot and at the same time the spiritually-minded Navy chaplain, who penned the quotation:

THE AMERICAN PEOPLE LOVE VALOR, BUT THEY LOVE RELIGION ALSO. THEY WILL CONFER THEIR HIGHEST HONORS

[124] *Ibid.*, p. 21.
[125] *Ibid.*, p. 181-2.

[126] Colton, *Sea and Sailor*, p. 399.
[127] Colton, *Three Years in California*, p. 407.
[128] *Ibid.*, p. 247.
[129] *Ibid.*, pp. 248-9.

ONLY ON HIM WHO COMBINES THEM BOTH.[130]

A CHAPLAIN'S LIFE ABOARD A SAILING VESSEL

The life of a Navy chaplain on a frigate presented a strange contrast, not only to that of a contemporary clergyman in a parish on shore, but also to that of later generations of chaplains who have gone down to the sea in mighty 45,000 ton ships. It is the purpose of this section to picture the chaplain back in the days of sail at work in his new home, amidst his nautical surroundings, sharing with his shipmates all of the difficulties and experiences of life within "wooden walls" under billowy clouds of sail on voyages that often kept them away from their homeland for two years and more.

When the new chaplain went aboard his man-of-war for his first voyage, he was known as a "greenhorn" and immediately had to set to work to "learn the ropes," as the expression went. To save himself embarrassment, a vocabulary, unlike anything on land, had to be acquired quickly; for example—deck, bulkhead, galley, sick bay, ladder, had to be substituted for floor, wall, kitchen, hospital, and stairs. Even the method of determining time aboard ship differed from that back home. The timepiece was a half-hour glass in charge of the sentry stationed at the cabin door. Time was divided into watches and reckoned by bells each half hour. Hence the new chaplain should never ask the question "what o'clock?" but "How many bells is it?"[131]

Many customs aboard ship seemed strange to him as he went about his duties in his new surroundings. He now had to request permission from the deck officer before he could go ashore and had to be particular to report his return aboard to the same officer. He soon learned how to deport himself at the various ship evolutions and drills and also what parts of the vessel were reserved for the various ranks and grades of the personnel aboard. The sacredness of the quarter-deck was impressed upon him. Sometimes he learned to his own embarrassment that loud talking and "boisterous conversations" were forbidden here—that sitting on the quarter-deck underway or being out of uniform there were not small offenses. This area which was appropriated exclusively to officers—no seaman was allowed to be seen on it except on duty—had to be approached by lifting or touching the hat.[132]

One of the first things that impressed the new officer was the "ceaseless noise and systematized confusion" which prevailed in his new home—the man-of-war. He had to become accustomed to the "noise of the men climbing and hauling the ropes, and answering to the orders from deck; the creeking of timbers; the rustling of the canvas; the heavy plunging of the vessel; and above all, the loud roar of winds and waves" which combined to produce a congregation of sounds he had never dreamed of. If he tried to write during the day "the lively conversation of the wardroom officers in one ear, the prattle of the pantry boys in the other; the echoing tread of sailors overhead" all conspired to prevent concentration.[133] He had to learn to sleep under the most trying conditions and in the midst of all sorts of bustle and din. There was little privacy even for the officer. The sailor had none and for him it was impossible to retire from the crowd; there was no such thing as seclusion, "no going beyond the sound of cursing and blasphemy."[134]

On the other hand, the chaplain was impressed with the great cleanliness of his new home for everything on a man-of-war required unremitting attention. Without it the ship would have become intolerable and sickness and disease would have followed.[135] At inspections, captains had the habit of wearing white gloves, which if soiled, resulted in dire punishment.[136] As one chaplain remarked, "there is not a house to be found on shore, cleaner in all its parts."[137]

Order was the first great rule on board, and that to which all others bent.[138] Everything had its place and had to be kept in it for one had to be able to lay his hands on a given object in any part of the ship in the dark. As one contemporary noted, "From day to day, from week to week, from month to month, and from year to year, the same stroke of the bell is followed by the same whistle, the same call, and the recurrence of the same duties."[139]

---

[130] *Ibid.*, p. 21. The words "They also love religion" appear on the corner stone of the David Adams Memorial Chapel, N.O.B., Norfolk with the date MCMXLI. A memorial tablet to Colton has been erected in the narthex of the Chapel.

[131] Jones, *Sketches of Naval Life*, I:97; Wines, *Two Years and a Half in the Navy*, I:27.

[132] Jones, *op. cit.*, I:3; Wines, *op. cit.*, I:29–30.

[133] Colton, *Land and Lee*, p. 8; Wines, *op. cit.*, I:87.

[134] Murrell, *Cruise of the Frigate Columbia*, p. 190-1.

[135] Wines, *op. cit.*, II:132; Colton, *Deck and Port*, p. 40.

[136] Jones, *op. cit.*, I:107.

[137] *Ibid.*, I:106.

[138] Wines, *op. cit.*, I:27.

[139] *Ibid.*

Captains were literally monarchs of all they surveyed and their authority, for the time being, was all but absolute.[140] They seldom interfered with the active work of their ships and many felt they succeeded best when they confined themselves to their cabins, only rarely showing themselves "to vulgar eyes." All in all the government on board a man-of-war was despotic—a far cry from the democratic atmosphere of the agricultural villages from which many chaplains came.[141] One of the first lessons the new chaplain learned was that he had superiors and that he must bow to the etiquette of rank.[142]

Many frigates of sailing ship days were overcrowded—often five hundred men were packed into two decks where they lived almost like kenneled dogs.[143] Men from almost every country under Heaven were crowded in these restricted quarters.[144] One observer pointed out, "They bring with them no principles of amalgamation, but, on the contrary almost as many apples of discord as there are individuals in the crew. Nothing but necessity can bind into one harmonious whole so many discordant elements, and that necessity must result from a system of discipline, stern in its nature and prompt in the execution of its penalties."[145]

The chaplain's part in the discipline of the ship's company, while different from that of the line officer, was very real and important. He did not speak the language of implicit authority for he did not command and govern. His duty was to be helpful, affable, and kind. "The stern character of Naval discipline," wrote one captain, "renders this a rare quality in the intercourse between officers and seamen, and indeed it is seldom admissible: but the chaplain comes among them in the character of a friend, not a stern superior."[146] The chaplain worked with the men as their personal friend and adviser; it was "his province to reason men into what is right, and to dissuade them from what is wrong."[147] He appealed to men's motives, to their conscience and their sober judgment, and if his influence was rightly used he greatly aided in the discipline of the ship.

The chaplain's parish was a most uncomfortable place in which to live and work when the ship pitched and rolled violently in a hostile sea. As one officer described such a scene, "The ocean for many days in succession appeared like a vast expanse of moving mountains. Nothing could surpass its dark and angry sublimity."[148] Under these circumstances the gun-deck was constantly shipping seas; it was hazardous to walk about, even with the aid of life lines; and below decks the air was foul because all ports were shut in. The description continues,

> Let the reader figure to himself barrells, trunks, books, and china—rolling, sliding, falling, and breaking around him—and he will have some idea of my situation. . . . The waves had become mountainous in size and giants in strength; and the ship, as if wearied and vexed by their angry power, seemed alternately to seek a dwelling place in the heavens above and in the sanctuaries of the deep.[149]

Chaplain Colton, who spent much of his time aboard ship writing, told of the problems involved even in this mild occupation. He said that his words were often written "on a table lashed down to prevent its being capsized, in a chair secured with laniards against the force of the ship's lurch, and the manuscript tacked to its place to escape the fate which befell the Sibylline leaves."[150] Chaplain Stewart, on one occasion, referred to an "almost sleepless night, from the excessive rolling of the ship", an experience which must have been somewhat common aboard the small sailing men-of-war.[151]

Normally, little sympathy was wasted on the seasick individual. Old tars had a habit of recommending, with feigned solicitude, "salt water, with fat bacon and molasses as a grand specific."[152] Wines described the feeling that came over one when afflicted with this disease. "You loath everything you see, or hear, or taste, or touch, or smell, or your own life in the bargain."[153] The only encouraging thing that one could hold to was the certainty that seasickness was not fatal in its effects and that sooner or later health would be restored.[154] Jones pointed out that a chaplain who was subject to seasickness was at a distinct disadvantage in carrying out his duties aboard ship.[155]

Food aboard ship, especially after months at sea, sometimes presented a hardship to chaplains, many of whom had lived in agricultural areas where it was abundant. Ships often stood to sea with six months' provisions and four months' water on board.[156]

[140] Wines, *op. cit.*, I:39–40; Jones, *op. cit.*, I:40.
[141] Wines, *op. cit.*, I:37.
[142] *Ibid.*, I:23.
[143] Hollis, *The Frigate Constitution*, p. 18.
[144] Wines, *op. cit.*, I:37.
[145] *Ibid.*
[146] Jones, *op. cit.*, II:238–9.
[147] Rockwell, *Sketches of Foreign Travel*, II:428; Colton, *Land and Lee*, p. 238.

[148] Wines, *op. cit.*, I:93.
[149] *Ibid.*, p. 88.
[150] Colton, *Land and Lee*, p. 8.
[151] Stewart, *A Visit to the South Seas*, II, 10 Feb. 1829.
[152] Jones, *op. cit.*, I:17.
[153] Wines, *op. cit.*, I:82.
[154] Reynolds, *Voyage of the Potomac*, p. 24.
[155] Jones, *op. cit.*, II:238.
[156] Colton, *Deck and Port*, pp. 18, 384.

Officers usually augmented their slender supply of fresh provisions by carrying chickens, ducks, sheep, pigs, etc., in apartments on deck conveniently near the galley.[157] Normally the fresh provisions were exhausted after several weeks. Salt meats, dried vegetables, cheese, and ship's biscuit were the principal items of diet of seafaring men until about the middle of the nineteenth century. Weavils usually reached the biscuits, which had been baked on shore, long before they appeared on the table.[158] "Our water was so bad both in taste and smell," complained Wines, "that I generally held my breath till I had drunk off all I wanted, to avoid, as far as possible, the unpleasant sensations occasioned to the olfactory and gustatory nerves."[159] Chaplain Colton, no doubt, expressed the wish of many officers when he wrote "How I long for a cup of milk! even a glass of good pure water would be a luxury."[160]

The chaplain lived with the other wardroom officers—the lieutenants, purser, surgeon, master and marine officers—in small compartments dignified with the name state rooms. These rooms were approximately five by seven feet and contained "neat little bureaus which were furnished by the government." Some officers provided carpets for the deck and fitted up their state rooms in "elegant style."[161]

The chaplain was treated with respect aboard ship and, like other officers, had his rights and privileges. The starboard side of the quarter-deck in port, for example, was available to him and he could freely use the starboard gangway when leaving or returning to the ship. He rated a hand salute from the men, was greeted as "Sir," and had the honor of being attended by two side boys when he was "piped over" the side by the boatswain's mate.[162] He, as other officers, had boys assigned to him who acted as personal servants and were at his beck and call. On many frigates the barber or one of his mates daily visited the chaplain's and other officers' rooms and twice a week "gave a general overhaul to the men's chins."[163] Chaplain Jones warned other chaplains that they must know what is due them and must require it from others, or they would not be respected aboard ship.[164]

Steam was introduced into the United States Navy during the latter thirties but won its way only gradually, and in certain quarters most reluctantly, over sail. With the final ascendency of steam, conditions aboard naval vessels greatly changed. The protracted voyages were over, since a steam vessel had always to be within reach of a coaling station; ships became larger and roomier; improved methods of caring for food were introduced. Life aboard ship became radically different for the chaplains of the post-sailing-ship era.

DISTRIBUTION OF BIBLES

The American Bible Society continued its activities in distributing Bibles and Bible selections to naval personnel after 1820, the year the Secretary of the Navy had invited the Society to engage in this project. Upon request they were sent direct to commanding officers of ships or shore stations or to chaplains.

In 1841 the American Bible Society again sought official endorsement of this plan and directed the attention of the Secretary to its willingness to supply copies of the Bible to naval personnel. The Secretary issued instructions to the commandants at the several Navy Yards that they should "supply the crews of public vessels going on a cruise, with the Bible, one copy to each mess."[165] The *Annual Reports* of the Society frequently mention its contacts with the Navy. In 1844 the *Portsmouth* was given fifty-two Bibles and 510 Testaments. In 1845 Chaplain Colton received 100 Bibles for the *Congress*. Chaplain Wood of the *Powhatan* received 300 Testaments on the eve of the departure of that vessel for China. From 1820 through 1860 the reports show that "at least 4,618 Bibles and 8,361 Testaments in English, French, Spanish, Portuguese, Arabic, German and Irish were distributed to or through chaplains and other officers of the U. S. Navy."[166]

THE END OF FLOGGING

While it is not possible to assess fully all of the factors which brought about the abolition of flogging, it is reasonable to believe that the writings of such eye-witnesses as E. C. Wines and W. M. Murrell, added to that of Chaplains McLaughlin, Rockwell,

[157] Bell, *Room to Swing a Cat*, pp. 132, 137.
[158] *Ibid.*
[159] Wines, *op. cit.*, I:23 ff.
[160] Colton, *Sea and Sailor*, p. 394. Perhaps in no respect does the contrast between sea life in sailing ship days and modern sea life stand out more clearly than in the comparison of food and drink. It was in the years immediately preceding the Civil War that primitive refrigerating systems were introduced on board ship. See Bell, *op. cit.*, p. 132.
[161] Jones, *op. cit.*, I:3; Wines, *op. cit.*, I:35.
[162] Stewart, *op. cit.*, I:119, 208; Taylor, *The Flag Ship*, I:21.
[163] Wines, *op. cit.*, I:15, 99.
[164] Jones, *op. cit.*, II:238.
[165] *Annual Report*, 1842, p. 54.
[166] CoC., Author's File, letter from Dr. Gilbert Darlington, Treasurer, American Bible Society, 4 June 1945.

and Stewart, undoubtedly contributed to the final result.

Two other important books which appeared during this period under review further pricked the public conscience. The first of these books was Richard Dana's *Two Years before the Mast,* 1840, which was to the Merchant Marine what the other works were to the Navy.

Dana gives (Chapter XV) a vivid description of a flogging which took place on the *Pilgrim.* He tells of one unfortunate man, stripped and tied to the gratings, crying out "Oh, Jesus Christ! Oh, Jesus Christ!" as the cat-o'-nine tails slashed across his back. Dana wrote:

> "Don't call on Jesus Christ," shouted the captain; "he can't help you. Call on Captain T.........., he's the man! He can help you! Jesus Christ can't help you now!"
>
> At these words, which I never shall forget, my blood ran cold. I could look on no longer. Disgusted, sick, and horror-struck, I turned away and leaned over the rail, and looked down into the water.

That night Dana found it difficult to go to sleep as he lay in his hammock in the forecastle. Nearby were the two men who had been flogged, groaning with pain. Dana vowed that if God should ever give him the means, he "would do something to redress the grievances and relieve the sufferings of that poor class of beings. . . ."[167] Dana's book, which contained this protest, was more widely read than he anticipated. His resolve was fulfilled. A "voice from the forecastle" had at last found an audience.

The other book which contains an exposé of the inhumanity of flogging was written by a man who had served as a member of the crew aboard a United States vessel. It was *White Jacket* by Herman Melville, which has been previously mentioned. Melville's vivid descriptions fairly pull the reader into the group of seamen assembled on the deck where the floggings took place.

Once Melville was threatened with the cat for an offense of which he was innocent. Of his reactions he wrote:

> My blood seemed clotting in my veins; I felt icy cold at the tips of my fingers, and a dimness was before my eyes. But through that dimness the boatswain's mate, scourge in hand, loomed like a giant, and Captain Claret, and the blue sea seen through the opening at the gangway, showed with an awful vividness. I cannot analyse my heart, though it then stood still within me.[168]

Like Dana, Melville resolved to do all that he could to arouse public opinion. "You see a human being, stripped like a slave; scourged worse than a hound," he wrote, "and for what? For things not essentially criminal, but only made so by arbitrary laws." Melville appealed to his readers: "Are we not justified in immeasurably denouncing this thing? Join hands with me, then; and, in the name of that Being in whose image the flogged sailor is made, let us demand of legislators, by what right they dare profane what God himself accounts sacred."[169]

In *Memories of a Rear-Admiral,* S. R. Franklin comments as follows on the effect of Melville's book:

> Melville wrote a book, well known in its day, called *White Jacket,* which had more influence in abolishing corporal punishment in the Navy than anything else. This book was placed on the desk of every member of Congress, and was a most eloquent appeal to the humane sentiment of the country. As an evidence of the good it did, a law was passed soon after the book appeared abolishing flogging in the Navy absolutely, without substituting any other mode of punishment in its stead; and this was exactly in accordance with Melville's appeal. He said: "Abolish it at once, even if you substitute nothing for it; but abolish it."[170]

Instances of flogging were more numerous during 1840–1849 than in the thirties.[171] During the forties, four sentences are on record of 200 lashes each and forty-one of 100. Usually a part of the sentence was remitted in these instances of severe penalty.

An enlightened public opinion was beginning to express itself through members of Congress as early as 3 June 1844, when Senator Hale introduced an amendment to the Naval Appropriation Bill calling for the abolition of flogging in the Navy. The amendment was not passed, but for more than six years Senator Hale kept the issue before his fellow legislators.

On 29 January 1850 the Secretary of Navy, William Ballard Preston, sent a circular to the higher ranking officers of the Navy soliciting views on the following question:

> Can corporal punishment be dispensed with consistently with the interests and efficiency of the Naval Service? If so, what punishment should be substituted?[172]

Eighty-six replies were received. Only three voted to abolish flogging. These were Commodores Conner

[167] Dana, *Two Years Before the Mast,* pp. 107-8. Yet in the closing chapter of his book, Dana refers again to flogging and reluctantly gives the following conviction (p. 380): "Yet, when the proposition is made to abolish it entirely and at once; to prohibit the captain from ever, under any circumstances, inflicting corporal punishment; I am obliged to pause, and, I must say, to doubt exceedingly the expediency of making any positive enactment which shall have that effect."

[168] *Op. cit.,* p. 263.
[169] Melville, *op. cit.,* pp. 132-5.
[170] Franklin, *op. cit.,* p. 64.
[171] "General Courts Martial and Courts of Inquiry 1799-1861."
[172] Nav. Rec. Coll., V:2:167-8.

and Stockton, and Lieutenant W. D. Porter. One other officer, who did not answer the questionnaire but who later was known to be in favor of the reform, was Commodore Uriah P. Levy. Among the statements approving flogging is the following:

> The seamen of our Navy have never applied for the abolishment of corporal punishment. On the contrary, as wherever they have expressed themselves upon the subject, they have advocated its continuance.[173]

A common argument was that many of the men who composed the average ship's company were reckless or slothful and that they could not be made to do their duty without the fear of corporal punishment.

One of the most influential advocates of the proposed change was Commodore Robert Stockton who was insistent that corporal punishment by the lash "can be and ought to be dispensed with." He felt that the American sailor ought to be regarded as an American citizen and that notwithstanding the peculiar habits incident to his mode of life and training, he has about him many noble characteristics of his countrymen."[174] Chaplain Colton only once or twice referred to flogging aboard the *Congress* which Stockton commanded.[175] In his reply to the questionnaire Stockton stated that he felt the power of moral suasion and the appeal to higher motives were more effective "than the dread of a degrading punishment." Commodore Stockton's influence in favor of the reform had far-reaching effect.

The vigorous protests voiced by naval surgeons who were required to stand by and observe flogging were additional factors in its eventual abolition.[176]

In the final analysis the abolition of flogging in the United States Navy can be claimed as another victory for the Christian church. The naval chaplains and such Christian-minded officers as Commodores Stockton and Levy were only a part of a larger movement which was sweeping across America. Finally, members of Congress moved by essentially Christian convictions said in effect: "This thing shall no longer be permitted." On 28 September 1850 Congress outlawed the practice in both the Merchant Marine and Navy.

GROG OR NO GROG

Mention has been made of the activities of chaplains in promoting temperance aboard naval vessels. When Chaplain George Jones served in the *Columbia,* he induced approximately two-thirds of the men and nearly half of the officers to sign the temperance pledge. Seeking a substitute for spirituous liquors, Jones turned to coffee. He wrote to the Secretary of the Navy and offered to procure at his own expense coffee and sugar "for the night drink for the deck watches," if the Secretary "would furnish conveniences for having it prepared."[177] Secretary Upshur accepted this arrangement in 16 February 1842:

> Sir, Your letter of the 10th inst has been received. In reply you are informed that the necessary apparatus for the preparation of Coffee or Coco, for the purpose mentioned by you, may be procured, if the measure be approved by Captain Parker.[178]

It seems reasonable to assume that this marks the beginning of the widespread habit of coffee drinking aboard naval ships. In this same year, coffee, cocoa, and tea, together with dried fruit, pickles, and cranberries were included in the food ration. It is significant that the grog ration was reduced at the same time.[179]

This reduction, however, did not always bring about completely satisfactory results. An observer in the *Constitution* remarked that while temperance had been steadily advancing, he felt that it had received a set-back in the law of 1842 which cut the grog ration in half. A much larger number of men were now drinking their grog, he said, since the money allowance for those who gave it up had dropped from six and a quarter cents a day to two cents a day. The immediate effect of this law "was to send back to the grog-tub a great many who had previously withdrawn." He approved the new law; nevertheless, he felt that the old money allowance should be continued, for the government would be compensated for the additional outlay "both in quiet and good order on board ship, and in the popularity of the service."[180]

[173] See Nav. Rec. Coll., "Corporal Punishment and Spirit Rations, Reports of Officers, 1850," with pamphlet "Naval Discipline and Corporal Punishment" by John J. Almy, bound with letters, p. 12. Sailors aboard the *Germantown,* on the other hand, expressed their gratitude to Senator Hale for his consistent effort in favor of the reform by presenting him with a medal. See Nav. Rec. Coll., Letter of H. K. White, 18 Jan. 1922.

[174] Nav. Rec. Coll., MS., "Corporal Punishment and Spirit Rations."

[175] Colton, *Deck and Port,* p. 23. Colton wrote: "Corporal punishments are opposed to the spirit of the age; but he would be worthy a monument who could invent an adequate substitute on board a man-of-war."

[176] Roddis, *A Short History of Nautical Medicine,* p. 95.

[177] Robert, "The Naval Academy Chapel," *Naval Institute Proceedings,* Oct. 1945, p. 1136.

[178] Nav. Rec. Coll., 1:32:66.

[179] Bradley, "Sketch of History of Medical Corps, Navy," *Journal of the Association of Military Surgeons,* May 1902, p. 506.

[180] *Southern Literary Messenger,* Feb. 1843, p. 73.

The wardroom presented another side of the liquor problem. Chaplain Joseph Stockbridge, who entered the service in September 1841 and had his first duty aboard the *Independence*, was distressed to find that he together with other officers who did not indulge in intoxicating liquors were obliged to pay a proportionate part of the cost. On 11 May 1843 Stockbridge addressed a letter to the Secretary of the Navy saying in part:

Without doubt you are acquainted with the fact that in most Wardroom messes wine and other intoxicating drinks are used at dinner and on other occasions; but it may not be known to you that as a general rule, this is done at the expense of each individual in the mess,—that total abstinence men are taxed to support a practice against which they are conscientiously opposed of which they feel themselves pledged to suppress by every honorable means. As a total abstinence man I have protested against being taxed for wine and other intoxicating liquors drunk in the mess to which I belong, and hitherto the only response I have received has been, that I must submit to the will of the majority, who forget that the minority has any rights at all. I feel that in this manner my rights are invaded.[181]

The Secretary replied that the dispensing of liquors "in the messes of officers on board vessels of war is one which belongs properly to the mess itself." He declined to interfere.

The same circular sent out by the Secretary of the Navy on 29 January 1850 to sound out opinion among high ranking naval officers on the matter of flogging included this question: "Can the issue of the spirit ration be dispensed with and what substitute will answer in its stead?" Twelve out of the eighty-four who replied said that the spirit ration could be eliminated. The three officers who voted for the abolition of flogging were also in favor of giving up grog.

Captain J. Downes wrote:

When I commanded the frigate Macedonian in 1819, being on the West Coast of Mexico where no spiritous liquor could be obtained, the crew were about three months without their spirit ration. The effect of the absence of spirit from the ship was most happy, as, from the day that the last of the spirits was served out, until the ration was restored, the men appeared perfectly contented: there was no quarreling & very little, if any, occasion for punishment. I am perfectly satisfied that nine tenths of the offences committed by seamen on board our ships of war are caused by the spirit part of their ration.[182]

Commodore Stockton, writing on 6 February 1850, was of like opinion, claiming that:

The good of the service would be promoted by dispensing with the spirit ration. The habitual use of ardent spirits feeds the appetite for Stimulus, and the appetite naturally grows stronger by indulgence—And the best substitute for it, in my opinion, an increase of pay.[183]

Chaplains were very much concerned about the liquor problem. Temperance education and the promotion of temperance societies had proved but a partial answer. Several chaplains wanted the daily issuance of grog aboard ships abolished. Colton wrote: "The whiskey ration is a curse to the service, and a damning blot on our national legislation."[184] Chaplain Taylor summed up the opinion of his colleagues when he wrote:

Strange that the government of the United States will not dispense with the spirit ration in the Navy, as well as to have done it in the Army. The delivering out of liquor two or three times a day on board our ships, will make the young seaman, however temperate when first entering the service, in a little time an habitual drinker, and induce a habit, which finally makes him a drunkard, and a sorrow to his family. Hundreds on hundreds of young men, who enter the service from the spirit of adventure, are thus finally ruined—disgracing themselves, dishonoring their kindred, and fearfully wrecking their hopes for this and the world to come. The broken hearts of a thousand mothers call loudly to those who legislate in the halls of Congress, to do away with this crying evil.[185]

Finally, the weight of public opinion inspired Congress to act on 14 July 1862. Much credit for the final enactment of this law has been given to Gustavus Vasa Fox, the Assistant Secretary of the Navy. Writing to Senator J. W. Grimes on 28 May 1862, Fox urged:

I beg of you for the enduring good of the service, which you have so much at heart, to add a proviso abolishing the spirit ration and forbidding any distilled liquors being placed on board any vessel belonging to, or chartered by the U. States, excepting of course, that in the Medical Dept. All insubordination, all misery, every deviltry on board ships can be traced to *rum*.[186]

The law read:

And be it further enacted, That from and after the first day of September, eighteen hundred and sixty-two, the spirit ration in the Navy of the United States shall forever cease, and thereafter no distilled spirituous liquors shall be admitted on board of vessels-of-war except as medical stores, and upon the order and under the control of the medical officers of such vessels, and to be used only for medical purposes. From and after the said first day of

---

[181] Nav. Rec. Coll., III:May 1843:87.
[182] *Ibid.*, MS. "Corporal Punishment and Spirit Rations."
[183] *Ibid.*
[184] Colton, *Deck and Port*, p. 19. See also pp. 303–4.
[185] Taylor, *The Broad Pennant*, p. 284.
[186] Thompson and Wainwright, *Confidential Correspondence of Gustavus Vasa Fox*, II:304.

September next there shall be allowed and paid to each person in the Navy now entitled to the spirit ration five cents per day in commutation and lieu thereof, which shall be in addition to their present pay.[187]

Thus a reform long recommended by chaplains became a reality. It is impossible after this lapse of time to evaluate accurately the chaplains' part in the reform, yet it is reasonable to believe that what they said and wrote on the subject helped to create that public opinion which reflected itself in the action of Congress.

IN REVIEW

The period 1841–1860 was marked by significant gains in the history of the Navy Chaplain Corps. The status of chaplains was now more clearly defined in *Navy Regulations*. Greater recognition was given to them by increasing their pay and regulating more carefully their uniform. Their usefulness was greatly facilitated by defining their duties in more detail, improving their working conditions aboard ship, and liberalizing the policy of supplying material equipment for their use. Divine Service was more generally observed throughout the Navy in this period than it had been in the previous twenty years. The regulation of 1860 permitted each chaplain to conduct Divine Service according to the manner and forms of his own church.

A more careful procedure was adopted in selecting chaplains and their number was increased from nine to twenty-four. After a heated debate in Congress relative to the wisdom of employing chaplains in the various branches of the United States Government, naval chaplains emerged from the discussion more firmly established than ever before.

During the period, chaplains contributed to the achievement of a number of reforms, the most notable being the abolition of flogging and the final elimination of the daily ration of grog. These were important gains which contributed directly and indirectly to the effectiveness and efficiency of the chaplains' ministry.

[187] 12 *Stat.* 565.

CHAPTER EIGHT

THE CIVIL WAR AND AFTERWARDS

1861–1880

Open hostilities in the Civil War began on 12 April 1861 when Fort Sumter was fired upon. On 4 March 1861, the United States Navy consisted of sixty-nine vessels. Forty-two of this number were in commission, including two frigates, eleven sloops, one screw frigate, five first-class steam sloops, three side-wheel steamers, and eight second-class steam sloops. Twenty-seven vessels were in ordinary and dismantled. When hostilities broke out, the North had but twelve vessels in commission in home waters, including three small steamers.[1] The other vessels in commission were in the various squadrons scattered in Asiatic and European waters.

According to the Secretary of the Navy, Gideon Welles, 322 naval officers "traitorously abandoned the service" at the outbreak of the war.[2] This was about one-fifth of the total officer personnel. However, the North did not lose a single ship in this shift of individual loyalties. The Naval Academy at Annapolis lost 106 of its midshipmen, which represented a much larger percentage of southern sympathizers than was found among the Regular Navy officer personnel.

In addition to such serious defections came the loss of the Navy Yard at Norfolk on 20 April, only eight days after the beginning of hostilities. The Confederates captured there a rich store of heavy ordnance, ammunition, and other war supplies. The steam frigate, *Merrimack,* was burned and scuttled before the yard was abandoned. Later the Confederates raised the hull, rebuilt her, reinforced the sides with sheets of iron, and sent her forth to spread devastation upon the squadron of Northern vessels in Hampton Roads.

Although the titanic struggles in this war were fought largely by armies, the United States Navy was called upon to perform the tremendously important task of blockading 3,000 miles of coast, of capturing important sea ports, and of gaining control of the Mississippi River. During the war years, the personnel of the Navy expanded nearly seven-fold —from 7,600 enlisted men at the beginning to 51,500 at its close.[3] It is estimated that during 1861–1865, more than 118,000 men served in the Navy for varying periods. By the time the war ended, the Navy had nearly 700 vessels in commission, most of these being small craft suitable only for coastal operations.

The United States emerged from the war one of the strongest naval powers in the world. Yet, within five years, most of the 700 vessels were sold or otherwise disposed of. The fleet of monitors rotted at their anchorages. By 1880, the United States had only forty-eight serviceable ships of which "not more than thirty wooden corvettes and sloops were fit for service abroad." During this period of stagnation, the United States dropped to twelfth place in naval strength, ranking below Denmark, Chile, and China.[4]

This, in brief, constituted the background for the services rendered by the Chaplain Corps during this period.

CHAPLAINCY IN THE CONFEDERATE NAVY

There were seven articles in the Confederate regulations governing the work of chaplains.[5] These articles were taken over from the proposed regulations which had been submitted to Congress by the Secretary of the Navy in 1858. These articles read as follows:

ARTICLE 1.

Respect to the Chaplain.

All officers are required to treat the chaplain with the respect due to his holy office, and to afford him every proper facility for the discharge of his sacred duties.

[1] *Report of Secretary of Navy,* 4 July 1861, p. 10.
[2] *Ibid.,* 1865.

[3] *Ibid.,* p. 130.
[4] Alden and Westcott, *The United States Navy,* p. 281.
[5] *Regulations for the Navy of the Confederate States,* pp. 139–40.

ARTICLE 2.

He will be attentive to the performance of his duties.

He is to be attentive to perform with due solemnity the duties of the Lord's Day, that the ship's company may be impressed with devotion, and he will carefully adapt his discourses to the capacity of the majority of his hearers, that his instructions may be intelligible and beneficial to them.

ARTICLE 3.

To perform divine service when required.

He will perform divine and funeral service when required to do so by his commanding officer.

ARTICLE 4.

He will be attentive to the requests of persons requiring his attendance.

He shall be very attentive to the requests of all sick persons who may desire his attendance, and shall, although not requested, visit all such as may be dangerously ill, and offer such consolations as they may require.

ARTICLE 5.

Religious instruction of boys, &c.

He is to instruct in the principles of the Christian religion the boys and such other persons as the captain may put under his care.

ARTICLE 6.

Instruction of the boys in the elementary branches of education.

He will apply to the captain to direct some intelligent and well-disposed person of the crew to instruct, under his directions, the boys of the ship in reading, writing, and the first rules of arithmetic, and will examine the boys often, and will report to the captain those whom he may find diligent and well-disposed, that they may be rewarded.

ARTICLE 7.

Chaplain's reports.

He will make to the commander of the ship a report on the first of January, April, July and October, of the duties performed by him in the previous three months, and a condensed report at the end of the cruise.

These Confederate regulations mention neither compulsory church attendance on Sunday nor daily prayers. The commanding officer determined when Divine Service should be held. Article 5 refers to the "Christian religion" which was not specificallly mentioned in any United States naval regulations before this date.

Even though provision was made for chaplains in the Confederate Navy, none were commissioned. The Confederates did not have the larger naval vessels which ordinarily rated chaplains.

Among the ex-naval chaplains who supported the Southern cause was Joseph Pere Bell Wilmer, who served in the United States Navy from 7 March 1839 to 23 July 1844. Wilmer went to England in 1863 to purchase medical supplies and Bibles for the Confederate Army and was taken prisoner on the return voyage. He with others was taken to Washington and confined for a time in the Old Capitol Prison. Claiming that he had been on an errand of mercy, he was soon released.[6]

CIVIL WAR CHAPLAINS

The year 1861 opened with the full complement of twenty-four chaplains—eight awaiting orders, ten at shore stations, and six at sea. Of those at sea, Chaplain Stewart was on the *Niagara* in Japanese waters, Chaplain Lewis was with the Pacific Squadron on the *Lancaster*, Chaplain Bartow was in the East Indies on the *Hartford*, Chaplain Blake was off the coast of Brazil on the *Congress*, Chaplain Davis was on the *Richmond* in the Mediterranean, and only Chaplain Lenhart on the *Cumberland* was serving aboard any vessel of the Home Squadron.[7]

By the law of 21 December 1861, any officer of the Navy was eligible for retirement "whose name had been borne on the Naval Register forty-five years" or who had attained the age of sixty-two.[8] This provision made it possible for a number of the older officers to turn over the responsibilities of their office to younger men during the critical days of the war. In the 1862 *Navy Register*, the following seven chaplains were listed as having been retired under this law: Charles Stewart, T. J. Harrison, George Jones, Moses Chase, J. W. Newton, John Watson, and Henry Wood. Actually, at least half of the chaplains listed as retired continued on duty during the war.

The first Navy chaplain to die as the result of enemy action was Chaplain John L. Lenhart, who was on board the *Cumberland* when she was rammed and sunk by the C. S. S. *Virginia* (formerly the *Merrimack*) at Hampton Roads on 8 March 1862, with a loss of 150 men. One account of the action stated that

> when it was seen that the Cumberland must go down all the officers in charge of the wounded were ordered to

[6] Hibben, *Navy-Yard, Washington*, p. 121. Dr. Wilmer was elected Bishop of Louisiana for the Episcopal Church in 1866.
[7] *Navy Register*, 1861.
[8] *Report of the Secretary of Navy*, 1862, p. 39.

John Lenhart, 1847–1862. First naval chaplain to be killed in action. (From painting by Clayton Braun)

John Beugless, 1864–1887. Appears to be wearing a sword. (From photograph)

TWO CIVIL WAR CHAPLAINS
Dates refer to naval service.

The first known reference to the wearing by chaplains of "a silver cross" on shoulder straps is found in the 1864 *Uniform Regulations.* Since Chaplain Lenhart lost his life in the sinking of the *Cumberland* 8 March 1862 before the wearing of the cross and before relative rank were authorized, it is evident that the insignia had been added to the picture.

come on deck and bring with them such of the wounded as there might be some hope of saving which order was obeyed by the surgeons and others. The Chaplain, instead of coming on deck, went into his room and shut the door when in a few minutes he met his fate, the ship going speedily down.[9]

It was thought that the door swung shut after the Chaplain entered his room and that he was unable to open it, due perhaps to damage it sustained. During World War II, a small chapel, the Lenhart Oratory, a part of the larger Royce Chapel at the Naval Training Station at Sampson, New York, honored the memory of this first naval chaplain to lose his life in the service of his country.[10]

At no time during the war, after 1861, was the active list of chaplains up to the authorized twenty-four. The average was nineteen, some of whom were not on active duty. One new chaplain entered in 1861, one in 1862, six in 1863, and four in 1864—a total of twelve. In 1861, there was on the average one chaplain for every 320 enlisted men; by 1865, the average had fallen to one chaplain for every 2,450 men.

During the first part of the war, the Navy Department felt that it had too many chaplains. Writing to G. S. Blake, Superintendent of the Naval Academy on 30 September 1862, Gustavus Vasa Fox, then Assistant Secretary of the Navy, asked:

> We have more chaplains in the Navy than can be used, and are about to appoint a few more. Should you like to have any of them as Asst. Prof's, temporarily attached to the Academy?[11]

The suggestion seems to have been well received, for four chaplains were instructing at Annapolis in 1864. Seven were serving in this capacity in 1865: Chaplains Noble, Hale, Wallace, McLaren, Smith, Henderson, and Hibben.

The following proffer reveals the spirit of patriotism so frequently noted in the letters of chaplains during this crisis.

> 6 March 1861
>
> I feel constrained to request, and do request, that if unfortunately it shall be found necessary to order any ship carrying a Chaplain or having accommodations for a Chaplain, to order to the discharge, on or off our coasts, of any duty of special peril, I may be ordered to the ship and share the duty and the peril.
>
> Chester Newell[12]

[9] From notes by Chaplain Hoes taken from 1862 *General Minutes of Newark Conference.*

[10] Names selected by Chaplain W. W. Edel, senior chaplain at Sampson at the time of the erection of the Chapel.

[11] Thompson and Wainwright, *op. cit.,* II:389.

[12] Nav. Rec. Coll., III:Mar. 1861:79. Newell received no answer to this letter. He wrote again offering his services on 17 June.

FOUR EARLY NINETEENTH CENTURY CHAPLAINS

All pictures except McLaren's from paintings by Clayton Braun. Dates refer to naval service.

Joseph Stockbridge, 1841–1873

Chester Newell, 1841–1865

FIRST CHAPLAINS TO BE MADE COMMODORES

Both promoted in 1883 while retired

Donald McLaren, 1863–1896
Commissioned Rear Admiral 1906
From photograph in Edel Collection

Thomas A. Gill, 1874–1902
Commissioned Rear Admiral 1902

FIRST CHAPLAINS TO BE MADE REAR ADMIRALS

Both promoted while retired

Among the outstanding chaplains of the war period was Joseph Stockbridge who was on the inactive list impatiently awaiting orders during the summer of 1861. He had served in the Navy since September 1841 and was most effective in his ministry to sailors. He had a sharp tongue and an equally pointed pen, which at times got him into trouble for he was never awed by rank when he felt he was defending truth. Writing to the Secretary of the Navy on 24 August 1861, Stockbridge appealed for orders:

> Sir,
>
> The U. S. Receiving Ship N. Carolina has been without a Chaplain for several months & at a time too when a large number of recruits needed the services of such an officer.
>
> If no other chaplain is ready for that duty, I am.
>
> . . . . .
>
> Where there is so much work to be done, I prefer not to be idle and am ready for anything.[13]

Stockbridge was ordered to the *Lancaster* in September 1861 and served aboard that vessel until October 1863. He was then stationed at the Naval Hospital at Norfolk. From this place, comes a story written years later by Admiral Robley D. Evans. It appears that Evans had been wounded in the attack on Fort Fisher and was taken to the hospital for treatment. Referring to Chaplain Stockbridge, Evans observed:

[13] Nav. Rec. Coll., III:3 of Aug. 1861:30.

Several times the doctors gave me up, and though they never told me so, I knew when they thought I was going to die by the appearance of the chaplain, who never hesitated to tell me I was dying and also just where I was going to bring up after I was dead.[14]

The Miltonian eschatological views of the old, frank chaplain evidently had a tonic effect upon the young naval officer who always rallied after such interviews.

THE CHRISTIAN COMMISSION

On 14 and 15 November 1861, a national convention of the YMCA was held in New York, which sponsored the organization of the United States Christian Commission. This Commission, the forerunner of the modern United Services Organization, grew out of the following resolution adopted at the YMCA convention:

Resolved: That it is the duty of the Young Men's Christian Association to take active measures to promote the spiritual and temporal welfare of the soldiers in the army and the sailors in the navy, in co-operation with the chaplains.[15]

With the exception of the Bible societies which had been supplying copies of the Scriptures to Navy personnel for years, this Christian Commission marks the beginning of organized inter-denominational effort on the part of the people of the United States to supplement the chaplains' work. Pioneering in a new field, the Commission rendered outstanding service. Quantities of religious literature were distributed at Navy Yards and aboard ships. Religious meetings were conducted for sailors. Perhaps, one of the best services rendered by the Commission was the awakening of the Christian conscience of the United States to the needs of the service men.

"ETERNAL FATHER STRONG TO SAVE"

Belonging, by virtue of chronology, to this period is the story of the familiar hymn, so widely used today in Navy Divine Services, "Eternal Father, Strong to Save." This hymn was written in 1860 by an Englishman, William Whiting, and was first published in England the following year in a volume compiled by William Monk entitled: *Hymns, Ancient and Modern*.[16] Whiting wrote several hymns all of which are almost forgotten except "Eternal Father, Strong to Save."

[14] Evans, *A Sailor's Log*, p. 104.
[15] Moss, *Annals of the United Station Christian Commission*, p. 105.
[16] Julian, *A Dictionary of Hymnology*, p. 356.

The poem consists of four stanzas and is modeled after a favorite device of ancient Latin hymns. Each of the first three verses is dedicated to a person of the Trinity, Father, Son, and Holy Spirit. The fourth stanza, as a summary, centers about the idea of the Trinity itself.

The hymn was printed in the United States in 1870 in *Hymns, Ancient and Modern by* William Henry Monk. The differences between this version and the modern rendering, as found in the 1941 edition of *Song and Service Book for Ship and Field*, is indicated below. The changes are indicated by italics.

1870 VERSION

Eternal Father, strong to save,
Whose arm hath bound the restless wave,
Who bidd'st the mighty ocean deep,
Its own appointed limits keep;

O Christ, whose voice the waters heard
And hushed their raging at Thy word,
Who walkest on the foaming deep,
And calm amidst its rage didst sleep;

Most Holy Spirit, who didst brood
Upon the chaos dark and rude,
And bid its angry tumult cease,
And give, for wild confusion, peace;

1941 VERSION

Eternal Father, strong to save,
Whose arm *doth bind* the restless wave,
Who bidd'st the mighty ocean deep,
Its own appointed limits keep;

O *Saviour,* whose *almighty word*
The winds and waves submissive heard,
Who walked'st on the foaming deep,
And calm amidst its rage didst sleep.

O Sacred Spirit, who didst brood
Upon the chaos dark and rude,
Who bad'st its angry tumult cease,
And *gavest light and life and peace;*

The chorus and the fourth stanza, with its chorus, as follows were unchanged:

O hear us when we cry to Thee
For those in peril on the sea
O Trinity of love and power,
Our brethren shield in danger's hour;
From rock and tempest, fire and foe,
Protect them where-soe'er they go,

Thus ever let there rise to Thee
Glad hymns of praise from land and sea.

The composer of the music was John Bacchus Dykes, also an Englishman. His contributions to *Hymns, Ancient and Modern,* which appeared in 1861, mark him as one of the great church composers of modern times. Among his compositions are the tunes for the well known hymns, "Holy, Holy, Holy," "The King of Love My Shepherd is," "Lead Kindly Light," and "Jesus the Very Thought of Thee." Dykes named the tune composed for "Eternal Father" "Melita" after the island of Malta.

The lack of good hymns dealing with the sea, together with the merits of the poem and the restrained melody of the tune, insured the popularity of this hymn from its first publication. It was soon as familiar to the British tars as "Rule Britannia." Today the English Navy uses the last stanza only.

As with many English naval customs and traditions, the hymn was freely appropriated by the United States Navy. Undoubtedly, the use of this hymn at the United States Naval Academy has accounted for its wide-spread adoption throughout the Navy today. Chaplain W. N. Thomas, formerly attached to the United States Naval Academy, states:

> The tradition of singing this hymn at the Naval Academy goes back to 1879 when a young officer, Charles J. Train, who later became an Admiral in the Navy, was in charge of the midshipman choir and initiated the singing of the hymn in the Chapel. The custom has continued from that time.[17]

The present custom calls for the singing of the first stanza only at the close of each chapel service, including the Baccalaureate Service, with the congregation kneeling.

The hymn has been translated into French and now belongs to the French naval tradition as well. Its popularity has spread, moreover, into civilian circles. It is now included in most church hymnals used by the larger denominations throughout the United States.

THE CHAPLAIN'S UNIFORM

During the years under review, the matter of the chaplain's uniform was again under consideration. The regulations of 1853, which substituted black covered buttons for the regulation buttons, removed the only distinctive naval insignia from the chaplain's uniform. In all other respects such as cut of cloth, wearing of sword, etc. the chaplain's uniform was similar to that worn by other naval officers. It is to be noted, however, that the chaplain wore no insignia of rank or any badge of his office.

One of the chaplains who took a leading part in agitating changes in the uniform was Chaplain Joseph Stockbridge, whose letters to the Department reflect an independent spirit.[18] When he was ordered back to duty on 30 August 1861, he was troubled about the uniform he would be obliged to wear, and wrote to the Secretary of the Navy on 2 September as follows:

> During my last cruise, I often experienced inconvenience & sometimes rudeness in foreign ports from the fact that a Chaplain's uniform does not distinguish him from a citizen; the Navy button does not appear on it. It seems to me but just to give us the protection which the button affords. I propose that, *in addition* to the present uniform, Chaplains be permitted to wear the following described coat:—
>
> A double-breasted blue frock-coat with rolling collar, having six Navy buttons on each breast & the usual number on the sleeves & skirt. Chaplains who have been twelve years, or less, will wear one strip of gold lace on the cuffs, those over twelve years two strips, the same as for Surgeons. The lace is necessary to distinguish the wearer from a forward officer; & the absence of the shoulder-strap will distinguish him from all other commissioned officers.
>
> To the Hon. Secretary of the Navy, this may seem a matter of little importance, but, to us who are left by the present regulations to appear from our dress like citizens, whether on ship or shore, it is a matter of grave importance. There is almost universal dissatisfaction in the corps with the present uniform, & if the Department has not heard of it, it is because many Chaplains wear a coat which was once in the book of regulations, but, has long since disappeared.
>
> The dress proposed is a plain, inexpensive service coat which, being *in addition* to the black coat, will be worn by only those who prefer it. Whether adopted or not, I beg permission to wear such a coat "during the war."[19]

Stockbridge followed up his letter of 2 September with another on the 4th to the Assistant Secretary of

[17] CoC., Thomas File, letter of 17 Mar. 1945.

[18] Stockbridge was a Baptist and loyal to the historical tradition of that church. He had a total disregard for clericalism and had taken an active part in the prayer-book controversy earlier. Concerning his inactive status, he wrote on 20 March 1861 to the Honorable Hannibal Hamlin: "You probably remember the long & fierce struggle I had with Episcopalians on the use of the Prayer Book in the Navy. That controversy brought me into disfavor at the Navy Department, & as soon as Mr. Toucey could find an excuse for it, he put me off duty." (Nav. Rec. Coll., III: Mar. 1861:264).

[19] Nav. Rec. Coll., III: 1 of Sept. 1861: 15. A reference to the chaplain "corps" appears in Stockbridge's letter. The Chaplain Corps as such was not organized with a Chief or Director until October 1917, yet there was a growing tendency among naval chaplains during the Civil War and the years following to refer to their number as comprising a Corps.

the Navy, on the subject of the uniform. He declared:

At present, the only regulation dress is such a coat as is worn by Catholic Priests & Episcopal ministers, a dress which in no manner identifies us with the Navy. Unless my face is known, I am nowhere, even on duty such as visiting other ships of the Squadron, recognized by sailors as an officer. I wear a cap, but, anybody can buy one for a few shillings, & may wear it where he pleases. The older Chaplains, when they go abroad, extemporize a uniform with the Navy button on it; but, I am a "law & order" man & prefer to have the authority of the Department.[20]

The Secretary informed Stockbridge that the whole subject of proposed uniform changes would be investigated when "more time is afforded the Department."

Stockbdidge discussed the matter of uniform with his friend and colleague, Chaplain Chester Newell, who wrote to the Navy on 18 September 1861. Newell confessed that he had given little thought to the subject, but, felt that some kind of insignia would be helpful. He was not altogether in accord with Stockbridge's proposals as the following reveals:

Mr. Stockbridge proposes a blue frock coat with Navy buttons, stripes of gold lace, etc. The cut & style of a coat may be of small importance, but I should prefer a blue frock coat, single breasted, with black velvet rolling collar, ordinary silk-covered buttons and some sufficiently conspicuous cord, same color & material, on the cuffs. I am opposed to the Navy buttons & gold lace, as unclerical.[21]

The desired changes were not made until relative rank was granted to chaplains in 1863, at which time they were authorized to wear the Latin cross on their shoulder straps and caps together with other insignia of rank. The cross measured seven-eighths of an inch in length and one-half inch in width. It was placed in a sloping position on both the shoulder straps and cap, with the foot of the cross forward. Chaplains with less than twelve years service wore the insignia of a lieutenant commander and those with twelve years or more that of commander. The cross was silver in both cases. The regulations also provided: "Chaplains, when performing divine service, may wear either the vestments of the church to which they belong or the uniform prescribed in the regulations."[22]

[20] Nav. Rec. Coll., III:1 of Sept. 1861:88.
[21] *Ibid.*, III:2 of Sept. 1861:185.
[22] *Uniform Regulations for the United States Navy*, 1866.

832159—49——8

SHOULDER STRAP AND CAP INSIGNIA FOR CHAPLAINS, 1863–1898.

From *Uniform Regulations for 1869.* Chaplains with the relative rank of lieutenant commander wore the gold leaf with the gold cross. Those who held the relative rank of commander wore the silver leaf with the silver cross. The original illustrations are in color.

The first illustrations of the chaplain's insignia appear in the 1863 *Uniform Regulations* in black and white. The cross was to be ⅞" in length and ½" in width. It was to be placed "obliquely in centre."

NAVY REGULATIONS

No less than five editions of Navy regulations were issued during the period under review, 1861–1880. The action of the Attorney General, in declaring the regulations formulated by the Secretary of the Navy in 1853 invalid, brought about action on the part of Congress. The following became law on 14 July 1862:

And be it further enacted, That the orders, regulations, and instructions heretofore issued by the Secretary of the Navy be, and they are hereby, recognized as the regulations of the Navy Department, subject, however, to such alterations as the Secretary of the Navy may adopt, with the approbation of the President of the United States.[23]

Many new provisions relating to chaplains and their duties appear during this period, and such subjects as "Divine Service," "Sunday Observance," "Gambling," and "Funeral Ceremonies" are covered in some detail. A comparative study of these changing regulations throws considerable light upon the status of the chaplain and his responsibilities. Most of the present regulations bearing on these subjects

[23] 12 *Stat.* 565.

may be traced to the editions issued in the sixties or the seventies.

NAVY REGULATIONS GOVERNING DIVINE SERVICE

It should be recalled that since 1799 attendance at Divine Services on board ship in the Navy was compulsory and on shore stations entirely voluntary. Chaplain E. C. Bittinger, in his letter of 10 March 1862 to the Navy Department, contrasted the problem faced by chaplains at sea with those on shore. Of his experience on board ship he wrote:

> The Religious Services on board the Princeton have been invariably well attended by every man so that I have been greatly encouraged to labor for their improvement in morals, their instructions in righteousness & their Salvation from the power & the pollution on Sin. *Sailors like soldiers will only do what is required of them.*[24]

Quite different was Bittinger's experience at the Philadelphia Navy Yard. Of the Marines stationed there, he said he had so "signally failed" to persuade them to attend church that he found it necessary to suspend Sunday services for the three preceding months. His letter continued:

> I deem it my solemn duty to inform the Department in regard to these Marines that seldom now attend divine service on the Sabbath while connected with this Yard; as soon as they are transferred to one of our men-of-war, having a Chaplain on board, they are required to attend divine service after the customary Sunday morning inspection, & I have never known a solitary Marine refusing to obey the regulations of the Service, but as soon as they are transferred from a man-of-war where they have attended divine service during the cruise to this Yard, they are no longer required to attend divine service at this Yard, provided for them by the Government, & therefore you must see how very discouraging it must be to us as well as how demoralising it must be to them when instead of attending upon the means of grace in the Chapel after the customary Sunday morning inspection, they are allowed to do just as they please.[25]

The rules governing church attendance were radically changed on 17 July 1862 when Congress passed "An Act for the better Government of the Navy of the United States." Article 2 of this Act, referring to Divine Service, made no mention of compulsory attendance or to daily morning and evening prayers aboard ship. The new article read:

> The commanders of vessels and naval stations, to which chaplains are attached, shall cause divine service to be performed on Sunday, whenever the weather and other circumstances will allow it to be done; and it is earnestly recommended to all officers, seamen, and others in the naval service diligently to attend to every performance of the worship of Almighty God. Any irreverent or unbecoming behavior during divine service shall be punished as a general or summary court-martial shall direct.[26]

The effects of this ruling were so radical that this date may well be taken as the end of one era in the history of the United States Navy chaplains and the beginning of another.

The burden of maintaining attendance at Divine Service was thrown upon the chaplain. Men who entered the Navy from non-church homes, who did not have the church-going habit, were no longer ordered to attend Divine Service. It became the chaplain's duty to reach them by the persuasive power of religion projecting itself through a friendly personality. Under the earlier regulations, a chaplain could, and some undoubtedly did, take their official duties lightly. Under the new regulation, he was obliged to use all the initiative and ingenuity he possessed in order to reach a fourth or even a tenth of the ship's company. These conditions demanded a more vigorous chaplain with a more aggressive program.

The change in the regulation was in keeping with the constitutional provision which guaranteed freedom of religion to citizens. At the time of the Revolutionary War and in the early days of our national history, the Roman Catholic proportion of the population was very small. According to a Congressional document dated 27 March 1854, the Roman Catholics then had 1,081 priests in the United States and 620,000 "worshippers." The Protestants listed 27,122 ministers and 13,228,946 "worshippers." Beginning with the Civil War period, there are increasing references to the presence of Roman Catholic personnel in the Navy. It was not until 1888, however, that the United States Navy had a Roman Catholic chaplain. It is quite possible that the growing number of Catholic personnel in the service was one of the reasons the regulation requiring attendance by all available hands at Divine Service was changed to voluntary attendance.[27]

Chaplain George Dorrance, writing his annual report on 31 December 1866, from the *Hartford,* then at Hong Kong, stated that the average attendance at

[24] Nav. Rec. Coll., III:1 of Mar. 1862:208.
[25] *Ibid.*

[26] 12 *Stat.* 601. The article is unchanged and still in effect (1946), with this exception—the last sentence is now article 3 of present day regulations.
[27] 33rd Cong., 1st sess.; *House Report,* No. 124.

his Divine Services had been small. Part of the reason for this was that "a considerable number of the crew profess to be Roman Catholics." However, the chaplain added: "The simple recommendation for men to attend on the worship of Almighty God, where all other duties on board ship are enforced by an order, has little or no influence upon them."[28]

On the other hand, Chaplain Stockbridge pointed out that more attended the daily evening prayer meeting which he organized in the fall of 1862 aboard the *Lancaster* than had attended Sunday services when attendance was compulsory.[29] Beginning with three "professed Christians," Stockbridge soon had so many attending that the quarters became too crowded. "Many sinners were awakened," he wrote, "and seven of the number converted, five of whom united with a mariners church in California." Interest in the prayer meetings was so great that they were held every night for ten months.

Among the stories told about the famous Admiral Louis M. Goldborough is that of an incident which occurred aboard his Flagship *Colorado* when it was in the Mediterranean shortly after the war. On a bright Sunday morning, the quarter-deck was rigged for church and all hands were called "up and aft." Some 600 officers and men assembled for Divine Service, the men being arranged by divisions and the officers seated immediately in front of the pulpit. Directly in the center was a big arm chair reserved for the Admiral.

When all were present, an orderly was sent to notify the Admiral that Divine Service was about to start. The organist played through a hymn on the portable organ. Minutes passed. Chaplain Charles R. Hale showed signs of restlessness. The orderly was sent a second time and the organist pumped out another hymn. Still no Admiral. The Commanding Officer, thinking that the Admiral was not planning to attend, nodded to Chaplain Hale to commence.

With reverential voice and ministerial solemnity, the Chaplain bowed his head and intoned: "The Lord is his Holy Temple . . ." and came to a full stop, for at that precise moment the Admiral suddenly appeared and shouted:

"Hold on, Chaplain, hold on. I'll have you to understand that the Lord is not in his Holy Temple until I get there!"[30]

[28] Nav. Rec. Coll., III: Dec. 1866: 378.
[29] *Ibid.*, IV: Oct. 1862: 157.
[30] *Ibid.*, Chaplain's File, MS. "The Chaplain's Mistake."

REGULATIONS REGARDING SUNDAY OBSERVANCE

The series of Navy regulations under review contain several references to the importance of observing Sunday aboard vessels and at shore stations in an orderly and reverential manner. It may be that repeated emphasis was necessary as a substitute for the old regulation requiring the attendance of all available hands at Divine Service aboard ship.

On 15 November 1862 President Lincoln issued the following General Order:

> The President, Commander-in-Chief of the Army and Navy, desires and enjoins the orderly observance of the Sabbath by the officers and men in the military and naval service. The importance for man and beast of the prescribed weekly rest, the sacred rights of Christian soldiers and sailors, a becoming deference to the best sentiments of a Christian people, and a due regard for the Divine will, demand that Sunday labor in the Army and Navy be reduced to the measure of strict necessity.
>
> The discipline and character of the national forces should not suffer, nor the cause they defend be imperiled, by the profanation of the day or name of the Most High. "At this time of public distress," adopting the words of Washington in 1776, "men may find enough to do in the service of God and their country, without abandoning themselves to vice and immorality." The first General Order, issued by the Father of his Country after the Declaration of Independence, indicates the spirit in which our institutions were founded and should ever be defended: *"The General hopes and trusts that every officer and man will endeavor to live and act as becomes a Christian soldier defending the dearest rights and liberties of his country."*
>
> Abraham Lincoln[31]

The *Regulations* of 1865 contained the following statement under article IV entitled "General Instructions":

> 210. Sunday must be observed on board of all vessels of the navy, and at all stations and navy yards, in an orderly manner, by officers and men. All labor or duty will be reduced to the measure of strict necessity. The religious tendencies of officers and men are to be encouraged, and suitable times and places will be assigned for Divine worship. It is hoped that the religious rights of Christian sailors will at all times, receive due regard from all commanding officers and others in authority.[32]

The next edition of regulations, in 1869, assigned the responsibility for Sunday observance to the Commander-in-Chief and to individual officers commanding vessels. They were not to permit unneces-

[31] *General Orders and Circulars issued by the Navy Dept.*, 1863–1887, p. 6.
[32] This section was repeated in subsequent revisions of the *Regulations* except for the deletion of the last sentence.

sary work on board their ships on Sunday and were to "see that divine service is performed every Sunday." Commanding officers were instructed to "require that all persons on board, passengers included, . . . conduct themselves with propriety, and that silence and order be observed in every part of the vessel, by those who do not attend the service."[33]

Deference to Sunday at the Naval Academy was provided by law on 15 July 1870.

That the Secretaries of War and the Navy be, and they are hereby, authorized and directed so to arrange the course of studies and the order of regulations at the Military and Naval Academies that the students in said institutions will not be required to pursue their studies on Sunday.

The proper observance of Sunday was assured in still another manner, and again by regulation in 1876. Punishments had continued to be inflicted on Sunday, even after flogging was outlawed. The 1876 *Regulations*, however, forbade this practice.

All minor punishments, except to prisoners confined in the cells, or to whom punishments have been awarded by a general or summary court-martial, are to be discontinued during Sunday, and punishments are not to be inflicted on Sunday except in cases where immediate action is necessary.

REGULATIONS OUTLINING THE CHAPLAIN'S DUTIES

The *Navy Regulations* of 1865 outlined the duties of the chaplain in section 10 of article VII as follows:

573. The Chaplain is to perform divine service and to offer prayers when duly requested by the Commanding Officer.

574. He is, with the consent of the senior Medical Officer, to visit the sick and afford them consolation.

575. He is to instruct in the principles of the Christian religion the boys and such other persons as the Commander of the vessel may commit to his care.

576. Should there be no schoolmaster on board, he is to apply to the Commanding Officer to detail an intelligent and well-disposed person of the crew to instruct, under his direction, the boys of the vessel, in reading, writing, and the elementary rules of arithmetic; and he is frequently to examine the boys, and report from time to time to the Commanding Officer those whom he may find to be diligent and well-disposed, in order that they may be suitably encouraged and rewarded.

577. He is to make to his Commanding Officer, on the first of January, April, July, and October, a report of the duties performed by him during the previous three months, and also a condensed report of the kind at the end of the cruise.

[33] *Op. cit.*, article 3, sections 66, 67, 404, 406.

These regulations are similar to regulations proposed in 1853, submitted to Congress in 1858, and adopted by the Confederate Navy Department. The *Regulations* of 1869 and 1870 substituted "ordered" for "requested" in No. 573. The *Regulations* of 1876 read: "He will perform Divine Service and offer prayers at such times as the Commanding Officer may designate."

Under these regulations chaplains were to "instruct in the principles of the Christian religion." No Jewish chaplain was commissioned until the First World War when such a regulation was no longer in effect.

The *Regulations* of 1876 added the following paragraph regarding the chaplain:

A clergyman appointed Chaplain in the Navy of the United States must consider it his duty that the morality of his conduct and the propriety of his manners are such as become his sacred office, and such as shall inspire officers and crews with reverence and respect toward him.

MISCELLANEOUS PROVISIONS

Several miscellaneous provisions, which bear upon the chaplain's work are to be found in the various editions of *Navy Regulations* issued during the years under review. Article X of the 1869 *Regulations* strictly prohibited gambling "on board vessels of the Navy and in navy yards, and at all places and stations belonging to, or under the control of the Navy Department."

This rule was repeated in subsequent *Regulations*, and is still (1946) in effect. More than one sailor has made an involuntary contribution to Navy Relief or some other charitable fund at the time a gambling game was raided.

Article VI of the 1870 *Regulations* prescribed the order of appearance of staff officers in the *Navy Register*. Chaplains were tenth in line, following surgeons, paymasters, and engineers, and the assistants in those respective corps. The engineers, who were organized with the advent of steam, although later in point of time, took precedence over chaplains.

Article VII of the 1870 edition ruled that the same military honors, rendered at the funeral of a line officer including the three volleys of musketry, were to be given at that of a staff officer. Sections 741 and 743 of this article gave directions for funerals.

Funeral processions will shove off in the following order: 1st, Music and firing party; 2d, boat with chaplain and surgeon; 3d, boat carrying the corpse, with body bearers; 4th, boats with pall-bearers; 5th, boats with officers of the ship to which the deceased was attached; 6th, boats from other vessels of the United States, in the inverse order of the rank of Commanding Officers: 7th, boats from foreign ships, arranged from van to rear in the inverse order of the rank of their several senior officers, and when such seniors are of the same grade, then length of service on the station will decide relative positions. If the deceased be a Commander of a squadron or of a single vessel, his flag or pennant will be carried at half-mast in the bow of the boat containing his coffin.

On reaching the shore, the procession should be formed under the command of an officer, senior to the officers commanding the firing party and the details of men from different vessels who are to form a part of the procession. The order of formation will be as follows: see plan.

Music.
Firing party.
Chaplain and surgeon.
Pall or hearse.
Men from different vessels, in squads, commanded by their own officers.
Officers of the vessel to which the deceased was attached, juniors leading.
Officers from the fleet or squadron, juniors leading.
Foreign officers, arranged as directed for procession in boats.

By an Act of Congress, 3 March 1873, all honorably discharged soldiers, sailors, and marines who served during the Civil War and died subsequent to the passage of the Act could be buried in any national cemetery of the United States free of cost. Among those who lie buried at Arlington and other national cemeteries are several who served as naval chaplains. A special section called Chaplain's Hill, has been set aside in the Arlington Cemetery for Army and Navy chaplains.

THE APPOINTMENT OF CHAPLAINS

A law of 1 June 1860 provided that a naval chaplain would not be less than twenty-one or more than thirty-five at the time of his appointment. (The *Regulations* of 1841, while not strictly observed, required that a chaplain be not more than thirty years of age at the time of appointment.) This same regulation was repeated in an amendment to a Naval Appropriation Act of 17 July 1862 and in the *Regulations* of 1865, 1869, and 1870. However, the *Regulations* of 1876 lowered the upper age to thirty.

Section 1395 of an Act by Congress on 15 July 1870 repeated the limitation of twenty-four naval chaplains "in actual service."

An examination of the known ages of sixteen chaplains appointed during the years 1862–1869 shows that the requirements of the law were met. The average age was twenty-nine. Eleven of the chaplains appointed during the years 1870–1880 also averaged twenty-nine years of age when they were appointed.[34]

In addition to the age limitation, the Navy Department during this period under review determined the health requirements for officers. Gideon Welles, Secretary of the Navy in Lincoln's cabinet, issued a General Order on 15 October 1862 which stipulated:

> Hereafter, no person will be appointed to any Commission, or Warrant Office in the Navy, until a Medical Board shall have declared the Candidate to be free from physical defects; or any obvious tendency to any form of disease which would be likely to interfere with the prompt and efficient discharge of duty.[35]

This was an important ruling, for in previous years, many clergymen, whose health did not permit them to continue the strenuous duties of a parish, sought appointments as naval chaplains. The following letter of application from the Reverend Flavel S. Mines from New York dated 30 October 1840 is typical:

> I have, in time past, frequently thought of a *chaplaincy* in the Navy as promising to be congenial to my taste & feelings; & now the additional consideration of my health has more definitely turned my attention to the subject. . . . I profess to be something of a sailor having been a number of times to sea. If, however, I might be permitted to express a preference, it would be to be sent immediately to some quite southern port or the Mediterranean or some similar *climate* for the present.[36]

With this letter was one from a friend who stated that Mines had been compelled "to abandon his parochial duties owing to his ill health." Flavel Mines did not receive the appointment he sought.

There is much evidence to show that many chaplains during the earlier period were old, ill, or infirm. Chaplain Crawford, writing to the Secretary of the Navy on 26 March 1878, stated:

> Appointments, orders, leave, etc., have been secured on purely political grounds, without any reference to worth or fitness. Chaplains have been appointed without physical examination, making the Navy an asylum for invalids who are incapable of sharing disagreeable duty with other officers.[37]

---

[34] This number did not include Chaplain Elijah W. Hagar, who is believed to have been born in 1819. If this date is correct, he was fifty-four when appointed in 1873. It is quite possible that a mistake was made in recording his age.
[35] Nav. Rec. Coll., V:2:395.
[36] *Ibid.*, File "NN"–1859, "Chaplains and Professors."
[37] *Ibid.*, III:Mar.–Apr. 1878:180.

The substance of the circular order issued by Welles was included in the 1865 edition of *Navy Regulations*:

No person will be appointed to any commissioned or warranted office in the Navy until he shall have passed a physical and a professional examination, except Chaplains and Professors of Mathematics, who are not required to undergo the latter. . . .

Although this edition of *Regulations* did not require a candidate for the chaplaincy to pass a professional examination, it did specify that he "must be a regularly ordained minister of some religious denomination, and of unimpeached character." The three following editions of the *Regulations* repeated this condition with the exception of the two latter clauses.

The Navy Department was beginning to insist on younger and stronger men for the chaplaincy. By 1862, the day had passed when an old or sick clergyman could turn to the Navy for an easy berth.

RELATIVE RANK

The growth of the Navy during the war made necessary a clarification of the status and a decision on the order of precedence of the various grades of naval officers. On 13 March 1863, Secretary of the Navy Welles issued a circular which permitted chaplains and professors of mathematics of more than twelve years' service to rank with commanders. Those with less than twelve years' service were to rank with lieutenant commanders.[38]

The *Navy Regulations* of 1865 listed the various grades of officers, including chaplains, who were classified as "staff officers" as distinguished from the "line officers." The *Regulations* then defined "the relative rank" between the officers of the two classes, using the basis set forth by Welles in 1863 concerning chaplains. The establishment of relative rank meant that staff officers wore the insignia, and had other prerogatives pertaining to their respective positions, but did not necessarily draw the pay of that rank nor did they hold a commission as such. Relative rank merely fixed precedence at official and social functions.

The first chaplains to hold the relative rank of commander, received by right of seniority, were the following ten listed in the 1863 *Navy Register:* Rodman Lewis, F. W. Taylor, M. R. Talbot, Chester Newell, T. B. Bartow, Joseph Stockbridge, Photius Fisk, Nathaniel Frost, John Blake, and E. C. Bittinger. In the *Register* of the following year, six chaplains on the retired list, including C. S. Stewart, George Jones, and Moses B. Chase, were also listed as commanders.

The revised *Regulations* for 1869 contained nothing about relative rank for chaplains with the result that the *Navy Registers* of 1 July 1869, 1 January 1870, and 1 January 1871 listed the chaplains simply under two categories: "Active" and "Retired." On the basis of subsequent events, it may be assumed that having once tasted the privileges of rank, the chaplains were unhappy with the reversion to their former ungraded status. The fact that Congress, on 3 March 1871, passed the following Act implies that pressure was brought to bear by an influential chaplain or group of chaplains: "Chaplains shall have relative rank as follows: Four, the relative rank of captain, seven, that of commander; and not more than seven that of lieutenant commander or lieutenant." This accounted for eighteen of the authorized twenty-four. The failure of this Act to define the status of the remaining six, when the full quota of twenty-four was in service, caused some embarrassment to those left unclassified.

The first four chaplains to attain the "relative rank" of captain, listed in the 1872 *Navy Register,* were Joseph Stockbridge, John Blake, E. C. Bittinger, and Robert Givin. All attained the rank on the basis of seniority. As vacancies occurred in the higher grades, names were moved up, with the result that some chaplains were given the relative rank of commander with less than five years service. Such was the case with Chaplain W. O. Holway in 1873. Some held the relative rank of lieutenant commander, with less than two years service, as did Chaplain J. B. Van Meter in 1873. This may have brought criticism from other officers, for, beginning with the *Register* of 1878, the grade of lieutenant was substituted for that of lieutenant commander, which was entirely in accord with the law. This explains the demotion of Chaplains Rose, Crawford, Rawson, Van Meter, Tribou, Clark, and Hager from lieutenant commander in 1877 to lieutenant in 1878.

It is to be noted that the Act of 1871 removed the requirement of twelve years' service before a chaplain could be ranked with commanders. For more than twenty years following 1878, chaplains were listed in the *Navy Registers* under the four categories: Captains, Commanders, Lieutenants, and Chaplains.

The first chaplains to be given the relative rank

[38] V:2:406.

of commodore were Chester Newell and Joseph Stockbridge. These chaplains, both retired at the time they were privileged to wear the broad stripe, were the only chaplains to rise above the relative rank of captain during the nineteenth century.[39]

While most chaplains of that generation, undoubtedly, welcomed the official recognition which came with relative rank, there was at least one who registered a protest to the innovation. He was Chaplain Van Meter, who in a letter to the Secretary of the Navy dated 11 April 1878, objected:

> Perfect freedom of intercourse with all grades of the service are essential to a proper performance of a Chaplain's work. There should be no artificial barriers between him and the men on the one hand, or the officers upon the other. He should have the same access to them as is had by a civilian, and they should have such access to him. *No rank should be attached to such an office as this,*—it is utterly incongruous with its nature. The conferring of what is termed "relative rank" upon the Chaplaincy is a comparatively recent thing, and I cannot but regard it as a mistake. Rank is a device for effecting and maintaining organization and regulating subordination and command. But the Chaplain is no part of a military organization. He does not exist for the sake of anything which that organization contemplates. His relations are with men as men, not as officers and seamen, commanders and subordinates, and his office relates to matters altogether outside of, and above, military discipline. He commands no one,—in his duties no one can command him. There is no reason, in the nature of his office and its relations to the service, why any one, even the third class boy or scullion should regard him as a superior,—no reason why an Admiral should consider him a subordinate. Rank digs a gulf between him and those to whom he should minister as the servant of all, separating him on the one hand from the men, who see in him a superior to whom they must defer,—on the other from the outranking officers, not necessarily by any fault or incivility of theirs, but by the feeling of estrangement on both sides which these distinctions engender. . . .[40]

Van Meter's colleagues felt, not that "the chaplain is no part of a military organization" but that, since he must live and work within the naval organization, even the nebulous relative rank gave him a certain official standing. It enabled the chaplain to be more effective in his work, for it gave him protection and an allotted place among his brother officers, and imposed limits upon the discriminatory acts of the occasional unsympathetic commanding officer. Also, there were certain financial considerations that went with rank, including longevity and retirement pay, which were important factors in attracting the right kind of men to the chaplaincy.

Rank in the military service is an issue that has brought criticism from civilian clergymen, who claim that the office of a Christian or Jewish chaplain is incompatible with the gradations which rank imposes. Most chaplains connected with the service, however, while admitting that there are limitations in accepting rank, consider it indispensable in the effectiveness of their work.

During the seventies, a number of incidents occurred which involved the assignment of chaplains to cabins on board ship. A difference of opinion existed as to whether the chaplain should occupy the fourth or the fifth cabin on the port side off the wardroom.[41] While the whole matter may seem rather trivial, yet, it involved the prestige of the chaplain aboard ship.

When Chaplain John D. Beugless reported for duty aboard the *Susquehanna* in August 1864, he found the marine officer occupying the fourth room. "As a matter of accommodation to Mr. Wallace," wrote Beugless to the Secretary of the Navy on 17 March 1865, "I took, temporarily, the after room." A little later, Lieutenant Wallace took up quarters on shore and the Chaplain occupied his room. The marine officer soon returned and demanded the fourth room. "We have agreed," wrote the Chaplain, "to refer the matter to the Department, and most respectfully beg leave to solicit your decision as to who is entitled to the said fourth (4th) room on the Port side of the ship."[42]

On 3 April the Secretary sent back his decision:

> "All rooms abaft the one assigned to the Surgeon or Assistant Surgeon in charge of the Medical Department to be occupied by the officers entitled to rooms on port side in the order of their rank of seniority, whether designated as Fleet Officers, Marine Officers, or otherwise."[43]

As Chaplain Beugless did not join the Navy until just before reporting for duty on the *Susquehanna,* it is probable that he was the junior officer and, therefore, the one who moved to the fifth room.

[39] Promotion to the relative rank of commodore was made possible by the Act of 3 March 1871 which provided that "Officers of the Medical, Pay, and Engineer Corps, chaplains, professors of mathematics, and constructors, who shall have served faithfully for forty-five years, shall, when retired, have the relative rank of commodore." The Act also provided that officers with forty years service who retired at the age of sixty-two could be promoted to this rank. Both Newell and Stockbridge came under this latter provision for both were commissioned on 8 September 1841.

[40] Nav. Rec. Coll., III: Mar.-Apr., 1878:327.

[41] Original plans of the frigate *United States* in Nav. Rec. Coll. show that the fifth cabin was allotted to the chaplain. Several chaplains, on the other hand, testify to having occupied the fourth.

[42] Nav. Rec. Coll., III:2 of Mar. 1865:344.

[43] *Ibid.,* I:Apr. 1865:280.

Article 2 of the *Navy Regulations* of 1865 prescribed stateroom occupancy as follows:

> The state-rooms opening into the wardroom country will be occupied, on the port side, by the Staff Officers, as follows: the forward room shall be occupied by the Senior Engineer, in charge of the engines, and if there be no such officer on board, then by the Paymaster, or Assistant Paymaster in charge of the Pay Department; the next room by the Surgeon, or Assistant in charge of the Medical Department; the next room by the Marine Officer in charge of the guard; and all the rooms abaft this by Staff Officers, in the order of their rank. In flag-ships the Engineer, Paymaster, and Surgeon, in charge of their respective departments, shall occupy rooms comformable to the above rule, and all other officers entitled to rooms on the port side, according to their rank. All other rooms, not assigned by the Department, shall be occupied as the Commander-in-Chief may direct.

The chaplain's position aboard ship was not entirely overlooked, however, for a section of article 12 of the same *Regulations* included the chaplain and the professor of mathematics as mess-room officers and provided that each was "to have the state-room connected therewith prescribed for his own accommodation, if any such there be."

Another stateroom problem involving a chaplain arose on the *Piscataqua*, the Flagship of the Asiatic Squadron, in October 1867 when Chaplain E. C. Bittinger reported. So many officers were already on board that the junior medical officer was forced to vacate his room in order to accommodate the Chaplain. The Captain wrote to the Secretary of the Navy requesting that the Chaplain's orders be revoked because of these crowded conditions. He added:

> The vessel will be cruising in a region where missionaries are numerous and who are gratified at an opportunity of holding service on board ship. I write with the concurrence of Admiral Rowan that the orders of the Chaplain may be revoked, and, thus, all the staff officers be provided with rooms, which seems almost necessary on so long a cruise and in that climate.[44]

The Secretary of the Navy replied on 18 October. While the settlement of the room controversy is not disclosed, he did insist that the Chaplain remain aboard. "In reply to your letter of the 15th Inst.," the Secretary wrote, "you are informed that the Department declines to withdraw the Chaplain, considering it advisable to have one at least of that grade in each Squadron."[45]

On 2 October 1871, Chaplain John Blake, who had shortly before been given the relative rank of captain, wrote to the Secretary of the Navy regarding his claim to the fourth stateroom on the port side of the *California*. This vessel, then at Mare Island, was about to become the Flagship of the Pacific Squadron. When the Chaplain reported aboard, he was told that the fourth room belonged to the marine officer and that the fifth room was assigned to the chaplain.

Troubled by this reversal of an old tradition, the Chaplain argued his case with the Secretary of the Navy in a letter dated 9 September 1871.

> As this room has, for many years, on all Flag Ships, been the Chaplain's room, and as Regulation 976 explicitly requires that it be assigned to officers according to rank, it is the opinion of many officers on this station that the directions of the Department have been misunderstood and will not bear the construction here placed upon them.
>
> As the rooms of the Chaplain and Marine Officer adjoin each other, it is evident that this change was not required as a matter of convenience, and as the last Congress saw fit to confer on me the relative rank of Captain, it seems impossible that any other room than the fourth could be assigned me in accordance with Regulations. If the law which confers this rank is regarded as defective or inoperative, then of course I am entitled to no room—no rights—and no protection—on board ship, as the discretion formerly given Commanding Officers, whereby they were enabled to assign certain rooms to persons without rank has been superseded by the present Regulations.
>
> In conclusion, I respectfully remark that having been for about thirty years in the public service, nearly twenty-five of which have been in the Navy, and having been ordered south on the Blockade during the War, where I remained for nearly three years, it was not without humiliation that I received the Order of the Admiral, requring me, thus, to give precedence to a young man with only the assimulated rank of Lieutenant and less than ten years in the service.[46]

The Secretary replied: "The ruling of the department is that the senior marine officer always occupies the fourth room on the port side of the wardroom."[46]

There are two instances of chaplains who, being without relative rank, were not assigned cabins, but, were obliged to make themselves as comfortable as possible in the steerage. The first case involved Chaplain S. D. Boorom, who was appointed in January 1876, and the second Chaplain Richard Hayward, who entered the Navy in October of that year. Since the quota of eighteen chaplains with relative rank was filled, they were listed simply as "chaplains." The difficulty came when each reported aboard his respec-

---

[44] *Ibid.*, II: July-Dec. 1867: 166.

[45] *Ibid.*, I: 83: 460.

[46] *Ibid.*, III: Oct. 1871: 11.

tive ship for duty. In both cases, the commanding officers refused to recognize their claims to a room.

Writing from Yokohama on 22 June 1877, Rear Admiral Reynolds on board the *Tennessee* laid the following difficulty before the Secretary of the Navy:

> Chaplain Hayward, on reporting for duty on this ship, wore the uniform of a Lieutenant, and supposed himself to have the relative rank of a Lieutenant. Immediately arose a question as to assignment of a State-room in the Ward Room.
>
> On referring to the law, I find that the relative rank for Eighteen Chaplains only, out of twenty-four is prescribed. On examining the Navy Register for 1877 it appears that the rank of Lt. Comdr. is assigned to the seven Chaplains who are to have the rank of Lieut-Comdr. or Lieutenant, thus, exhausting the quota to whom relative rank is allowed by law. The remaining Chaplains apparently have no rank at all, either by law or regulation.
>
> Under these circumstances, I have not assigned a room to him in the Ward Room as if he had the relative rank of a Lieutenant, and beg leave to request a decision of the Department in this case[47]

The Secretary of the Navy submitted the problem of what to do with these unclassified chaplains to the Navy Solicitor, John A. Bolles, who in his reply referred to the earlier case of Chaplain Boorom. Bolles briefly summarized the problem and made his recommendation:

> that the Secretary by genl order, or regulation circular, direct that all Chaplains whose relative rank is not fixed by law shall, in the assignment of quarters, be treated as having the relative rank of Lieutenants.
>
> As like doubts may arise in regard to the uniform of this class of officers, & other matters affecting their official status, the Solicitor respectfully submits a form of order that will provide for this whole subject, as follows:—
>
> "Genl. O. No...... *Chaplains whose relative rank is not fixed by law will, in the assignment of quarters,—in the matter of uniform,—& in all cases affecting their official status, be regarded & treated as having the relative rank of Lieutenants.*"[48]

The general order recommended was officially issued on 1 August 1877.[49] However, the *Navy Registers* continued to list all chaplains over the allotted number of eighteen as "chaplains" until 1899 when full rank was granted by law.

On 14 August 1878, the Secretary of the Navy issued another order to clarify the vexing problem of precedence in regard to the respective claims of the chaplain and the marine officer to the fourth cabin on the port side of the wardroom. The circular read as follows:

> On board of all vessels of the Navy, the marine officer in command of the marine guard will occupy the fourth room from forward, on the port side of the ward-room.
>
> All other marine officers will be assigned quarters in accordance with their rank.[50]

Thus, an issue raised by Chaplain Blake in 1871 was officially settled—in favor of the marine officer.

THE CHURCH PENNANT

Instead of the blue Greek cross which the United States Navy used, the Confederate Navy adopted a red Latin cross on a white background, with the long arm of the cross vertical. An illustration of the Southern church pennant appears in the *Signal Book* of 1861 which was compiled for the use of the Confederate Navy.

The earliest discovered order regarding the display of the church pennant is that issued by Admiral Farragut from his Flagship *Hartford,* off New Orleans on 26 April 1862:

> Eleven o'clock this morning is the hour appointed for all the officers and crews of the fleet to return thanks to Almighty God for His great goodness and mercy in permitting us to pass through the events of the last two days with so little loss of life and blood.
>
> At that hour, the church pennant will be hoisted on every vessel of the fleet, and their crews assembled will, in humiliation and prayer, make their acknowledgment therefore to the great Dispenser of all human events.[51]

The *United States Navy Signal Code* for 1867 gives the first known official directions for the display of the church pennant. "The Church Pendant will be hoisted immediately above the ensign at the peak or flagstaff at the time of commencing, and kept hoisted during the continuance of divine service on board all vessels of the Navy."[52] This regulation merely puts into words the traditional usage of the church pennant.

A book by Captain G. H. Preble, published in 1872, entitled *Our Flag,* contains the following description of the church pennant and its usage. It is a "white pendant, without swallow tails charged

[47] Nav. Rec. Coll., Letters from Asiatic Squadron for 1877:61.

[48] *Ibid.*

[49] Nav. Rec. Coll., V:8.

[50] *Ibid.*, V:352.

[51] *Official Records of the Union and Confederate Navies in the War of The Rebellion,* Series 1, XVIII:152. The *Hartford* was the Flagship Admiral Farragut took into Mobile Bay during the Civil War when he is reported to have ordered: "Damn the torpedoes. Full steam ahead." The old steam frigate, without its masts, remained tied up at the dock for years at the Washington Navy Yard. In 1945, the vessel was taken to Norfolk to be restored.

[52] *Op. cit.*, article 45, p. 11.

with a blue Latin cross, to be hoisted at the peak, during divine service, *over the ensign.*" It is the "only flag to which the national ensign shows such submission."[53] The pennant is described as it is today with a Latin cross while early signal books indicate it as having a Greek cross. Indeed, even in Preble's book, the illustration of the church pennant shows the Greek cross while the text refers to the Latin cross.

BUGLE CALL FOR HAMMOCKS

Conclusive evidence is not available as to when the bugle call for hammocks was introduced into the Navy. Tradition has it that in the "old days" when hammocks was sounded in the evening, all men not on watch stood by their hammocks, faced aft, and maintained silence while the chaplain read the evening prayer. It is quite possible that this custom began during the years, 1861-1880, under review.

HAMMOCK BUGLE CALL

Ship and Gunnery Drills, 1927, p. 327.

The call closely resembles taps. The custom of a few moments of silence at hammocks has continued on some ships down to the present time, but the fact of its origin in the chaplains' practice of reading evening prayers has almost been forgotten.

ANNUAL REPORTS

According to an old yarn, which must have originated around the scuttlebutt (*i.e.* the drinking fountain), one sailor asked another: "What does the chaplain do?" The other replied: "He does nothing six days a week and disturbs the peace on Sunday." According to another anecdote, one bluejacket said to another: "Who has less to do, the chaplain or the marine officer?" The answer was: "The marine officer because he has a second lieutenant to help him."

Chaplains carried on a great number of activities, however, and their annual reports for these years indicate that the Navy chaplaincy was composed of capable and conscientious men whose duties occupied them seven days of the week. These reports are submitted to the Secretary of the Navy in accordance with the Act of Congress dated 1 June 1860, and today furnish a good index to the varied activities of chaplains.[54]

This directive appears to have had faulty promulgation in the first years after its adoption. Chaplain William H. Stewart, who had been commissioned in 1863, in a letter to the Secretary on 10 June four years later, apologetically confessed that he had failed to get the word: "I beg leave to state that my attention not having been called to his law, I was ignorant that annual reports to the Secretary of the Navy were required until the present time, and hence my failure to report heretofore." Once the making of annual reports became established as a routine duty, they provided comprehensive and statistical data on the work of Navy chaplains through the years.

Chaplain T. B. Bartow, writing from the *Hartford* at Shanghai, China, on 29 May 1861, before the regulation requiring attendance at Divine Service had been changed, revealed a conscientious attention to his duties:

> The law of Congress enjoining the worship of Almighty God has been carefully observed on board this ship. In obedience to that law "daily morning and evening prayers have been offered and sermons preached on Sundays, whenever bad weather did not prevent." The funerals of six sailors have been solemnized. I have preached on board the Adams, Saginaw and Dacotah when our ships have been together. Regular services have been held for the sick, and on suitable occasions the Holy Communion has been administered to the religious men of the ship. Good books & tracts, abundantly supplied from home, have been distributed and when we were in the interior I have given books in their own language to the amiable and friendly Chinese, who seemed to receive them as most acceptable.[55]

Chaplains covered in these reports such activities as conducting Sunday Schools, weekly prayer meetings, visiting the sick, teaching classes, distributing Bibles and tracts, and operating libraries. Chaplain George Williamson Smith, who was commissioned in July 1864, writing from the *Franklin,* gave the following description of her library:

> As there is much leisure time on board a ship a library is very desirable, that the men may have instruction oc-

---

[53] *Op. cit.,* p. 469.

[54] The 1865 edition of *Navy Regulations,* and the three subsequent editions, directed chaplains to report to their commanding officers. Some chaplains assumed that this superseded the Act of Congress and reported only to their respective commanding officers, with the result that such reports are not on file in Naval Archives. A sufficient number, however, sent direct to the Secretary remain to paint the picture.

[55] Nav. Rec. Coll., III:2 of May 1861:235.

cupation and that there may be less temptation to peruse the unseemly literature which is always found among them.

Accordingly, by the consent of Captn. C. R. P. Rogers and the hearty cooperation of the officers under his command, a proposition was laid before the crew to raise by subscription money to start a library. The appeal was liberally responded to and the books purchased.

Without exaggeration, the library has been a success. The number of volumes from first to last is upwards of 1400, of which there are now on hand 536 volumes. Many of the books first purchased have been worn out, so that the discrepancy between these numbers is accounted for.

The number of members of the library is 325. The total amount of money raised and expended is $622.47. The highest cost to any member for 30 months has been $1.80.

Newspapers and periodicals were subscribed for, according to the money on hand; lamps and also stationery were purchased, that means of letter writing might be within reach.

From 4 P. M. to 8 or 9 P. M. whenever circumstances permitted, tables were placed on the starboard side of the half-deck, and such persons as desired have had opportunity for reading and writing. Under the tuition of the Schoolmaster, a number of the crew have learned to read and write during the hours above indicated.[56]

In their annual reports, chaplains frequently mentioned their work of distributing the Scriptures to naval personnel. During the Civil War, the American Bible Society was active in providing copies of the Scriptures to members of both the Army and the Navy. The records do not distinguish between the two branches of the service, but, indicate that during the war a grand total of 5,297,732 Bibles or portions were distributed on both sides of the fighting line. The Society reported that many naval vessels were supplied, including gunboats on the Ohio and Mississippi Rivers.

Chaplain Stewart's report covering the war years, reveals his own activity in this work:

By order of Rear Admiral Porter, dated Jan. 14th, 1864, I made a journey down the Mississippi River, visiting the vessels for the purpose of distributing Bibles and other religious publications, and resumed my duties on the Receiving Ship on my return.

On the 1st of March I made another journey up the Tennessee River for the same purpose.[57]

That the chaplains' services to naval personnel were somewhat similar all over the world is apparent from the statement of Chaplain Stockbridge. Writing from the Bay of Panama on 27 October 1862, he described his duties as a Squadron Chaplain:

I have during the last twelve months delivered on board this ship one hundred & twenty sermons & lectures, and besides, have preached on the Isthmus at Aspinwall & Panama, & in California at San Francisco, Vallejo & Healdsburg. I have visited all the ships of the Squadron and distributed among them the Sacred Scripture, tracts & other religious reading.[58]

One tradition associated with certain branches of the Protestant church is the offering for charity or the "deacon's fund" taken at the communion services. An early instance of the taking of a collection at a Divine Service conducted by a naval chaplain appears in the 5 June 1868 annual report of Chaplain G. W. Smith. Smith had been stationed at the Naval Academy since 20 September 1864. He stated:

The Sacrament of the Lord's Supper has been administered nine (9) times during the Academic year. The average number of communicants has been thirty-three (33). The alms collected on these occasions and distributed amount is $135.06. It may be proper to add in this connection that the number of Midshipmen who are in full membership with some Christian body is eighty-five (85), being rather more than 25 perc't of the whole number of Midshipmen.[59]

While only twenty-five percent of the midshipmen were communicant members of a Christian church, this ratio corresponds favorably with the national church membership of that day.

THE CHAPLAIN'S PAY

Chaplains received two pay advances during the years 1861-1880. The first came by an opinion of the Attorney General dated 4 September 1862 which allowed chaplains the same pay as that specified for lieutenants in the Act of 16 July 1862.[60] According to this pay scale, chaplains received $1,875 per annum while on sea duty and $1,500 on shore. They were to receive $1,200 while on leave or awaiting orders and $1,000 if retired.

An act approved 21 April 1864 provided that the retired pay of staff officers, including chaplains, be the same as that of retired line officers with whom they had relative rank. Hence, while chaplains on active duty were paid as lieutenants, on the retired lists they might receive the retired pay of a commander.

The Naval Appropriations Act of 15 July 1870 provided the following pay scale for chaplains:

	At sea	*On shore*	*On leave or awaiting orders*
First five years	$2,500	$2,000	$1,600
After five years	$2,800	$2,300	$1,900

[56] *Ibid.*, III: Nov. and Dec. 1871:66.
[57] *Ibid.*, III: June 1867:88.
[58] *Ibid.*, IV:3:157.
[59] *Ibid.*, III: June 1868:36.
[60] See *Regulations*, 1869, Article XXXIII.

While this increase of pay was acceptable, the law still discriminated against chaplains for it allowed the following compensation to surgeons, paymasters, and chief engineers:

	At sea	On shore	On leave or awaiting orders
1st five years	$2,800	$2,400	$2,000
2nd five years	$3,200	$2,800	$2,400
3rd five years	$3,500	$3,200	$2,600
4th five years	$3,700	$3,600	$2,800
After 20 years	$4,200	$4,000	$3,000[61]

According to this law, chaplains had but one increase of pay while their brother staff officers had four. An effort was made in 1878 to have Congress place the chaplain's pay on an equality with the surgeon, paymaster, and chief engineer. The friends of the proposed law included a section in the bill to reduce the number of chaplains from twenty-four to twenty so as to provide for the increase of pay without placing a heavier burden on the government. This was to silence critics who might have opposed the measure on economy grounds.

The Naval Affairs Committee of the Senate reported this bill favorably on 24 January 1879, and commented on the section calling for the increase of pay as follows:

This section increases the pay of chaplains, placing them on an equality, in this respect, with surgeons or paymasters, and chief engineers. No professional man can be expected to leave civil life and subject himself to long absences from his family, unless he has the inducement of such preferment and increased pay as his talents and constantly enlarging experience would give him in civil life. This is recognized in every corps except that of chaplains. All other staff officers, after certain grades, have their pay increased up to 15 and 20 yrs. of service. Chaplains have now only one increase, at the end of 5 yrs., so that a chaplain who has served 5 yrs. gets as much pay as one who has served 40 yrs.

The natural tendency of this is to drive out the men we need to keep and keep the men we could well spare. Such has been the fact, except in the case of a few worthy men who have remained from conscientious principles. Of the 24 chaps. now in the service, only 2 have served 16 years, and 15 have not served 10 years. This shows a rapidity of change which is not healthy

Other things being equal, the longer a man is in the service the more efficient he can be, especially in dealing religiously with sailors, who form a class so distinctly *sui generis*. Chaplains simply ask to be put on the same footing with officers whose equals they are expected to be in every respect, and with whom they share equally all expenses, including heavy mess bills and fees for official entertainments, which last are provided for in other Navies at the national expense.

Chaplains have the entire charge of the religious and educational instruction of men and boys, including the hundreds of apprentices under the new system. They have no subordinates or clerks to assist them as other officers have. Abundant testimony can be brought to the fact that no other officer can be so efficient in securing comfort, good order, and good discipline among the men. The question then comes down to this: "How shall we secure an efficient worthy corps?" Only by making it an object for good men to come in and stay in. If chaplains are to accomplish any good, it must be evident that the power which creates them intends them to be respectable and respected.[62]

This bill passed the Senate, but failed in the House. Another effort was made in 1880 which also failed. Twenty replacements entered the Corps between 1861 and 1870 and twelve in the following decade, a turnover which suggests discontent.

SUGGESTIONS FOR IMPROVEMENT

During the seventies, a number of excellent suggestions were sent to the Secretary of the Navy by individual chaplains and by groups of chaplains offering ideas for the improvement of their work. Some, however, were not adopted until World War I and others were delayed until World War II.

Chaplain G. W. Smith, on 6 November 1871, submitted a paper to the Secretary entitled "More Chaplains Desirable." He had just returned from a cruise aboard the *Franklin* and wrote from the Boston Navy Yard.

It seems also desirable that there should be Chaplains on more of our Naval vessels than there are now. . . . Under the present system the great majority of sailors are unused to the presence of such an officer, and when they happen to be sent to a ship carrying a Chaplain, he does not seem to them a part of the ordinary ship's complement with ordinary duties to perform, but rather an *ornamental appendage* to give dignity to the ship or superior officer. He is not, in their minds, on the same footing as other officers whose usefulness they see exemplified every day, and hence he is an object of indifference or perhaps of dislike; and unless favorable circumstances enable him to overcome this feeling, he is tempted to become in reality, what he is considered by the great majority of people on the ship to be, viz., a respectable Nonentity.

Smith found, as has the chaplain in the naval service of this generation, that much of his time was spent in counselling with the men. He requested that space be provided where the chaplain could interview privately those coming to him. He wrote:

The work done by a Chaplain is not simply the holding

[61] 16 *Stat.* 330–1.

[62] 45th Cong., 3rd sess., *Senate Report*, No. 626; 46th Cong., 1st sess., *Senate Report*, No. 209.

of services on Sunday as required by Regulations. That which tells most is the intercourse with individuals one by one. At present, the Chaplain is quartered in the Ward-room with the other commissioned officers and has no place to see those of the crew who desire to consult him. It is awkward to have to talk about delicate matters at the Wardroom door, or on the Gun-deck, or forward between the chains, and it is hardly proper to take one of the crew into the Officers' quarters. When preparing a small number to receive the Holy Communion the only place obtainable was the Yeoman's Storeroom.[63]

One of the most penetrating analyses of the weaknesses of the chaplains' status in general was made by Chaplain W. O. Holway on the invitation of the Secretary of the Navy. In his report of 20 March 1878, he pointed out the need of a unified corps with an experienced chaplain at its head to watch over and direct the affairs of chaplains.

Holway pronounced the system of appointing and detailing chaplains "radically defective." Candidates were not selected with care, he wrote, and once they had been commissioned, no rules were laid down to guide them and no supervision was exercised over their movements. Good men, he further pointed out, were "frequently chilled by the coldness and unconcern which everywhere confront them, and soon learn that it is a matter of indifference whether they perform their appropriate duties or not." Holway continued:

It is not to be wondered at that such results follow a system, so incomplete and faulty. With no responsible head, no concert of action, no coherence as a body, mere units without union, no regular system of detail, no circulars incorporating those methods of usefulness which have proved effective by actual experience, no sympathy either afloat or ashore, it is no wonder that the corps is signally inefficient, and that many sincere and thoughtful officers believe that the time has come for its excision.

Holway believed that "the first and most important step toward a reform" would be the appointment of some older chaplain who would act as the head of the Chaplain Corps:

First, to exercise supervision over Chaplains. None but a Chaplain is really competent for the work, as none but a Surgeon is competent to supervise Surgeons. Communications relating to Chaplains should be referred to him. He should invite correspondence with Chaplains, learn their methods of usefulness, stir them up to their duty, devise systems of reports, prepare circulars, and perform such other duties as may be needful for a thorough, intelligent and systematic supervision.

Second, watch over the detail of Chaplains—see that every Chaplain is regularly and appropriately employed. The time has been that there were three Chaplains at one time in the European fleet and none in the North Pacific or South Atlantic. A strong desire has been expressed by some earnest Chaplains that they should be regularly represented on board every first and second rate ship in commission, ironclads excepted. A proper system of detail would set every Chaplain at work, and, further, give each his appropriate work. . . .

Third, to guard the entrance to the Chaplaincy, the Chaplain in charge should carefully investigate the claims of applicants, ascertain by correspondence the fitness of each for the work, and be prepared at the first vacancy to nominate to the Secretary the right man. . . .

Fourth, to assist in the scheme of training boys for the naval service. . . .

Fifth, to represent the Chaplaincy at the Department. There must, occasionally, arise questions which only a Chaplain could answer. It is only right and just that a Chaplain should be within call at the Department to explain or defend the work of his office.

Chaplain Holway was serving aboard the *Minnesota* when he submitted his report. Captain S. B. Luce of that vessel added the following endorsement:

The careful examination by a competent board into the character, qualification, and special fitness of candidiates for appointments to Chaplaincies in the Navy is shown by experience to be very necessary. It is believed that this class of officers form the single exception to the practice of holding such examinations previous to appointment.[64]

Writing from the Boston Navy Yard on 26 March 1878, also on the invitation of the Secretary of the Navy, Chaplain G. A. Crawford likewise stressed the imperative need for reform. His frankness and sincerity are convincing and many of his suggestions were similar to those advocated by Holway.

On 2 November 1878, a self-appointed committee of chaplains consisting of Holway and Crawford, together with Chaplains W. H. Stewart, J. D. Beugless, J. M. R. Matthews, E. K. Rawson, and D. H. Tribou, submitted to the Secretary of the Navy these "suggestions" for the "greater efficiency" of the Corps:

I. That a roster for sea duty be kept, and that all Chaplains, except the first four, be sent to sea in regular order, due respect being had to rotation in stations.

II. That each Fleet have at least one Chaplain. Three of our six Fleets are without Chaplains at the present time. In other Corps from one third to one half of the officers are at sea.

III. That Chaplains be required to make, through the usual channels, Quarterly Reports to the *Department*, and that some one there be authorized to examine and act upon the same.

[63] Nav. Rec. Coll., III:Nov. and Dec. 1871:66.

[64] *Ibid.*, III:Mar. and Apr. 1878:142.

George A. Crawford, 1870–1889
Showing summer uniform.

David H. Tribou, 1872–1922
With more than forty years of active service.

Alfred L. Royce, 1881–1902
A Spanish-American War Chaplain.

Roswell R. Hoes, 1882–1912
Historian of the Chaplain Corps.

Charles H. Parks, 1888–1890
First Roman Catholic Navy Chaplain.

William H. I. Reaney, 1892–1915
Prominent Catholic Chaplain.

Dates refer to naval service. All from paintings by Clayton Braun except pictures of Chaplains Hoes and Reaney which are taken from photographs.

IV. That all Chaplains who have not on record the certificate of a Medical Examining Board be ordered before such a board, and all not found fit for sea duty be retired. A large number of Chaplains have *never been subjected to a physical examination.*

V. That the U. S. Statute requiring a physical examination before appointment, and also before each promotion, be enforced in the case of Chaplains.

VI. That Chaplains be allowed stationery. They are now required to make all communications upon official paper, but, are not allowed the means of complying with the regulation.

VII. That each ship to which a Chaplain is attached be allowed a School-master, who shall be competent to play an organ and lead ordinary singing.[65]

The efforts which were made in 1878 and 1880 to have Congress pass a law to give the chaplains more pay included other reforms. Section two of a proposed bill submitted to the House of Representatives in 1878 referred to the appointment and qualifications of applicants for the naval chaplaincy.

No person shall be appointed as chaplain, until he shall furnish proof that he is a regularly ordained minister of some religious denomination in good standing at the time of his appointment, together with a recommendation for such appointment from some authorized ecclesiastical body, or, from not less than five accredited ministers of said denomination; and, further, no person shall be appointed as chaplain, unless he has taken the degree of bachelor of arts or bachelor of divinity in regular course, and has been pastor or assistant pastor of some church at least one year.[66]

The House opposed the bill for the selfsame clause that won the Senate's approval; namely, the condition that all appointees have either a B. A. or a B. D. degree. The House took the stand that such a provision eliminated many candidates from those denominations with low educational standards for the ministry. The attitude of the Senate was expressed on 29 January 1879 by Senator Sargent who argued that naval chaplains especially should have at least an M.A. degree "because when they go abroad they are invited to preach before foreign officers and on other occasions and ought to display to a certain degree the culture of their profession among the American people."[67]

The provision is now an accepted part of the Navy chaplain's qualifications, but was too extreme for that generation and the bill was defeated.

The bill of 1880 omitted the requirement for a B. A. or a B. D. degree but retained the provision requiring a year's practical experience. However, the effort made in 1880 to get Congress to take action for the improvement of the Chaplain Corps of the Navy failed. Thus, these unsettled issues remained as focal points of agitation and restlessness among the Navy chaplains for twenty years or more until a satisfactory solution was reached.

DATA ABOUT CHAPLAINS

In 1861, the full quota of twenty-four chaplains, including those awaiting orders, was carried on the rolls. Beginning with the *Navy Register* of 1862, chaplains were divided into two main classifications, active and retired. During the years 1862 to 1865 inclusive, the number of chaplains on the active list totaled for the respective years 16, 18, 20, and 21. The full quota of twenty-four chaplains on active duty was not reached again until 1875.

Among the new chaplains commissioned were the following five, who had served as Army chaplains during the Civil War: E. D. Winslow, J. D. Beugless, H. B. Hibben, G. D. Henderson, and J. B. Rose. Samuel Kramer also served as an Army chaplain during the war, and, although not commissioned, acted as a chaplain at the Washington Navy Yard beginning in 1880. J. J. Kane served in the Navy as a mate and ensign during the war and was commissioned a chaplain three years after he left the service. W. R. Cobb, who had been wounded and captured while serving in the Confederate Army, became a naval chaplain in 1869. Two, who served as privates in the Union Army, T. A. Gill and A. L. Royce, were commissioned naval chaplains in 1874 and 1881 respectively. In all, ten who had military or naval experience during the war became naval chaplains.

Among the best known and most capable chaplains of this period was the saintly and scholarly George Williamson Smith, who served from 1864, to 1876, the years 1865 to 1868 being spent at the Naval Academy. He was responsible, while at Annapolis, for the building of a new chapel. Since the old one built like a Greek temple was inadequate, Smith interested Vice Admiral D. D. Porter in the project. Porter in turn presented the need to Congress, and in July 1866 an appropriation of $25,000 was voted for the erection of a new chapel. An impressive Gothic structure with a high thin spire was built, and dedicated on 24 May 1868

[65] *Ibid.*, III: Nov. and Dec. 1879: 18.
[66] 45th Cong., 2nd sess., *House Report*, No. 4477.
[67] *Congressional Record* for 29 Jan. 1879.

Smith was elected a bishop of the Protestant Episcopal Church three times and each time declined the office. He became President of Trinity College at Hartford in 1883.

A fine commendation of the conscientious ministry rendered by Chaplain Smith is found in a letter written to the Secretary of the Navy by Commodore C. R. P. Rodgers on 14 March 1872. The Commodore wrote:

> I desire to place on file in the Navy Department my opinion of the services and character of Chaplain G. Williamson Smith, who served under my orders during the two years in which I commanded the "Franklin." Chaplain Smith by his devotion to his duty and his benevolence to the crew commanded their respect and won their affection in a remarkable degree, contributing largely to their contentment on board ship. When sixty of the Franklin's men were ill with smallpox at Flushing, (England), their Chaplain went with them to the X ship's hospital on shore, where he remained with them through the most loathsome stages of the disease, rendering the services of his sacred office to the dying with the most tender assiduity,—nursing and comforting the sick by night and day, and evincing a sense of duty and an indifference to danger that secured for him respect and honor from every officer and man on board. It is in such a battle that a Chaplain best exhibits his courage and zeal, and I desire that Chaplain Smith receive the credit to which he is entitled.

The original letter in the Navy Records Collection bears the pencilled notation "File and send copy to Chaplain Smith." Today, an official letter of commendation would approximate this gesture of the Navy Department in recognition of the Chaplain's distinguished service.

A number of important names appear on the list of the chaplains for the first time during these years. Special mention should be made of the following:

Charles R. Hale, widely known as a scholar and author, spoke Russian and Greek fluently, travelled extensively throughout the near East, later became Episcopal bishop of Cairo, Illinois;

Edward Kirk Rawson, later a professor of mathematics in the Navy, Superintendent of Navy War Records of the Navy Department, author of well known naval histories;

David Howard Tribou, "one of the rarest and sweetest souls ever to have worn the uniform of the Corps of Chaplains,[68]" an active chaplain for a record of more than forty-three years at the time of his death in 1922, historian of the Chaplain Corps; and

Henry Howard Clark, for nineteen years Chaplain at the Naval Academy, author of a number of books dealing with boy life in the Navy, loved by generations of midshipmen who passed under his influence while at Annapolis.

Mention should also be made of the Reverend Samuel Kramer who served as chaplain at the Washington Navy Yard for almost four years, beginning this duty at the age of seventy-two. Kramer's interest in the Navy undoubtedly went back to the years 1822-1830 when he served as a seaman. His name is the last on the list of non-commissioned chaplains. Yet, he performed a chaplain's duty at the request of the Navy Department and received compensation at the rate of $75 a month for the first two years of his service. He served without pay from 1 July 1882 to 26 March 1884. Kramer began his work at the yard as the relief of Chaplain E. W. Hager, who died a few days after Kramer began his duties. Friends of Kramer urged his claims upon Congress which granted him the retirement pay of a chaplain on 5 July 1884. President Arthur, however, did not sign the bill.[69]

It should not be forgotten that through the years chaplains of the United States Navy have enjoyed the hearty cooperation of many consecrated Christian officers. During the war, Stephen Decatur Trenchard, later rear admiral, commanded the *Rhode Island* and conducted Divine Service himself so faithfully that by the close of the war his ship was known throughout the Navy as the "Gospel Gunboat."

E. S. Maclay in his *Reminiscences of the Old Navy* gave the following account of Trenchard's religious activities:

> Sunday always was a sacred day with him. In the four years he commanded the *Rhode Island*, he held service with uniform regularity, much to the disgust of some of the wicked tars who preferred to lounge away the time reading cheap literature, or teasing the ship's cat. It has been estimated that during the time he commanded this vessel, Trenchard failed to hold service on Sunday only twice, and then only because the weather made church service out of the question.[70]

Chaplain Henderson, writing to the Secretary of the Navy on 18 May 1871, commented on the efforts many line officers had made for the religious welfare of the men under their command:

> . . . we are glad to know that there have always been Naval officers of all grades who have engaged in religious work on board of their respective ships. The honored

[68] Edel, "Navy Chaplains from 1775–1917," pp. 13-4.

[69] Hibben, *Navy-Yard, Washington*, p. 178.

[70] *Op. cit.*, p. 26.

names of Stockton, Dupont, and Foot are familiar to us all in this connection, and they are but examples out of many who, ever since the unfurling of our flag, have thus hallowed it on every sea.[71]

The period under review closed with twenty-three names on the roll of active chaplains. Four of these had the relative rank of captain. Bittinger, age seventy-two, and Given, age sixty-two, retired in March 1881, leaving Stewart and Wallace on the active list, each being fifty years of age. The average age of the seven who had the relative rank of commander was forty-six; that of the seven lieutenants was thirty-nine; and the five "chaplains" averaged thirty-eight. Never before was there such a well-trained and stable group of men serving as naval chaplains. The fact that most of these men joined the naval chaplaincy in the early years of their ministry with the intention of making it their life work brought in a stabilizing element which augured well for the spiritual welfare of naval personnel.

IN SUMMARY

The events of the years 1861–1880 were epoch-making in the history of the naval chaplaincy. For the Navy itself, the great expansion during the Civil War was followed by a period of contraction and stagnation.

[71] *Nav. Rec. Coll.*, III:May 1871:138.

832159—49——9

Navy chaplains during the period were developing a sense of unity in a common task. They were tending more and more to refer to their group as a Corps. The provision for relative rank and the permission to wear the sign of their ministry, the cross, on their uniform, helped to create a fine *espirit de corps*. Stricter rules had been adopted regarding the age and health of new appointees, yet, great and beneficial as the changes were, many chaplains were eager for still more improvement. They proposed good and timely suggestions for the betterment of their work, many of which, however, were not immediately adopted. This agitation for reform was a healthy sign.

The old tradition of the church pennant flying above the national emblem during Divine Service aboard ship was now officially recognized. Another tradition taking root was the use in Divine Services of the hymn appropriated from the English, "Eternal Father, Strong to Save." During the Civil War, came the Christian Commission, an organization notable for its service to naval and military personnel.

Some radical changes, such as permitting voluntary attendance instead of enforcing attendance at Divine Services, were made. This was a period when the whole subject of naval regulations was in flux, and out of the series of improved regulations came those guiding principles which, for the most part, are still in effect.

CHAPTER NINE

THE NEW NAVY STEAMS FORTH

1881–1900

Agitation for a new Navy began about 1881. In March 1883 Congress authorized the 1,485 ton dispatch vessel *Dolphin;* the unarmored cruisers of 3,000 tons, *Atlanta* and *Boston;* and the 4,000 ton cruiser *Chicago.* The commissioning of these vessels marked the beginning of what was called the "New Navy." During the years 1887–1893, some thirty-six vessels aggregating more than 130,000 tons were added to the Navy, including thirteen armored cruisers. Also added were the *Maine* and *New York,* listed at first as heavy cruisers and afterwards classified as battleships, and the battleships *Texas, Indiana, Massachusetts, Oregon,* and *Iowa.* Other vessels were built and commissioned during the years 1894 to 1896, including the battleship *Kearsarge.*

The names of large ships are as important in the history of the Navy Chaplain Corps as are the names of dioceses and parishes in the history of the Christian church. These larger ships were the floating parishes of naval chaplains. While some clergyman ashore might take great pride in being pastor of a certain church in a large city, his brother in naval uniform would be equally proud, if not more so, to be chaplain of a cruiser or battleship. The chaplain had the advantage over his brother on land, in that the farthest any member of his parish could be removed from him at sea was the distance of only several hundred feet.

The period of naval decay from the war peak extended to 1881, a span of sixteen years, and this was paralleled by a like period, 1883–1898, in the ascent again to naval power. A new and vastly stronger Navy than ever before came into being in the years 1881–1900. Not only was the usefulness of the old sailing vessels as potential members of the fleet ended, but also consigned to the limbo of obsoletism were the hybrid combinations of sail and steam. Such ships as the *Constitution, Constellation,* and *Hartford,* were kept in commission merely as training ships and because of their historic significance.

THE CHAPLAINS OF THE NEW NAVY

Twenty-three chaplains were on duty in 1881. With the addition of new and more powerful ships came the inevitable increase in naval personnel. The number of officers and men increased from 9,840 in 1881 to 18,515 in 1900. In addition, the number of officers and men in the Marine Corps grew from 1902 to 5,414 during the same period. Since the quota of chaplains remained unchanged at twenty-four, there was, in 1881, a ration of one chaplain to 487 naval personnel and in 1900 the ratio was one to 997. In other words, considered numerically, the chaplain's responsibility had doubled. During the peak enrollment of the Spanish-American War, the ratio was one chaplain to 1,075 men.

The full quota of twenty-four chaplains was usually on duty during the years 1881 to 1900. Never before was the turnover in the chaplaincy so small. This is evidenced by the fact that only nine new chaplains were commissioned during the decade, 1881–1890, and fifteen in the ten-year period following. In the previous period, 1861–1880, thirty-two chaplains had been appointed.

Another indication of stability is found in the increased average length of service of chaplains. The nine men who entered the chaplaincy from 1881 to 1890 averaged 18.5 years of service each. Such men as Frank Thompson and Roswell R. Hoes each served about thirty-nine years. Only three out of the nine served less than ten years. During the next decade, the average length of service of the fifteen who entered the Navy as chaplains was about thirteen years. The men of this group who remained in the longest were W. E. Edmonson, twenty-five years; John Brown Frazier, thirty; W. G. Cassard, twenty-two; Curtis H. Dickins, thirty-one; and W. G. Isaacs, thirty-six years.

There is no instrument that can measure the increase of usefulness and effectiveness which accompanied this new stability. Even though the chaplain were not being paid on an equal basis with their fellow staff officers, yet for the most part the men who entered the chaplaincy during these years found such joy and satisfaction in their work that they remained in the Navy in spite of the discrimination. Of the twenty-four chaplains who were commissioned during these years, five died while on active duty, seven remained sufficiently long enough to be retired, eight resigned, three were asked to resign, and one was court-martialed.

During these years, the following procedure was followed by a clergyman desiring to obtain a commission as a naval chaplain. He applied directly to the President of the United States or to the Secretary of the Navy. Often the applicant asked an influential member of Congress to endorse his application. For instance, Senator Angus Cameron recommended Alfred L. Royce to the Secretary when the death of Chaplain Elijah W. Hager created a vacancy. Sometimes ecclesiastical superiors wrote either to the President or to the Secretary of the Navy. Such a recommendation was made in behalf of Father Louis Paul Rennolds. James Cardinal Gibbons wrote on 13 January 1900 to the President, requesting: "Rev. Father Parks one of the Catholic chaplains in the Navy intends to resign. It is the chaplaincy thus made vacant which Father Rennolds desires to obtain.[1]" Beyond any such letters from ecclesiastical superiors or brother clergymen, no endorsement by the denomination an applicant represented was required.

These letters were kept on file in the Secretary's office, and, when a vacancy occurred, the most promising candidate was nominated by the President. The Senate, before the man could be commissioned, had to give official confirmation. The successful candidate was then required to pass a physical examination.

An effort was frequently made to select new chaplains of the same denomination as those who were being replaced. The Navy Department, however, does not appear to have pursued a policy of a fixed quota for the separate denominations.

FIRST CHAPLAIN FROM THE DISCIPLES CHURCH

The number of men who are known to have served in the United States Navy as chaplains down to 1881 total 159, of which 58 were not ordained. The classification of the ordained men by denominations is as follows: Baptist, 12; Congregational, 7; Methodist, 22; Protestant Episcopal, 42; Presbyterian, 14; Unitarian, 3; and Universalist, 1. Since the total is 101, the figures given show the approximate percentage. Two denominations, the Methodist and the Protestant Episcopal, furnished about two-thirds of all ordained naval chaplains for the first eighty years of the Navy's history.

As was to be expected, the time came when other churches requested the privilege of having some of their clergymen commissioned as chaplains. In this twenty year period under review, two new denominations were represented among the chaplains for the first time, the Disciples of Christ and the Roman Catholic.

The first Disciples chaplain in the Navy was Carroll Quinn Wright, who was appointed in 1885. In applying for a commission, Wright in a letter to the Secretary of the Navy dated 7 January 1885, stated: "The Denomination which I represent ranks about 3rd in the U. S. numerically. None of our ministry has ever held a chaplaincy in the U. S. Navy." A letter of recommendation from Senator John B. Mitchell to President Arthur read: "He is endorsed by some of the prominent Republicans of Philadelphia very strongly, and I have no doubt his appointment would give very great satisfaction in that city." This suggests that a would-be chaplain's political affiliation was influential in his appointment.

Chaplain Wright rendered notable service in the Navy from 1885 to the time of his retirement in 1920. Chaplain Edel appraised Chaplain Wright as "one of the most effective and convincing preachers the Corps ever had, and one of the most lovable of men." Chaplain Frazier, the first Chief of the Chaplains Division, referred to Chaplain Wright as "a Captain who never commanded a ship, but whose sphere of command was the hearts of all his shipmates, and who won, not the glories of war, but the love of all with whom he came in touch.[2]"

FIRST ROMAN CATHOLIC CHAPLAINS

The Reverend Charles Henry Parks, who was commissioned 30 April 1888, was the first Roman Catholic priest to become a chaplain in the United States Navy. An effort, however, had been made in 1846 to obtain a Catholic chaplain when President Polk had appealed to Bishop Hughes of New York,

[1] CoC., Correspondence File.

[2] Edel, "Navy Chaplains from 1775 to 1917," pp. 21–2.

but was informed that no priest was available at that time.[3]

With the increase of Catholic personnel in the service after the Civil War, a rapid expansion in the number of chaplains of this faith was logical. Three more Catholics were commissioned before the end of the century. They were: William Henry Ironsides Reaney, 1892; John Patrick Sylvester Chidwick, 1895; and Louis Paul Rennolds, 1900. Reaney, surprisingly, was born on the old frigate *Constitution* in New York harbor and, thus, received the unusual third name of Ironsides. Chaplain Edel in his appraisal of Chaplain Reaney said that he

> became one of the outstanding characters in the whole history of the Navy. . . . As a boy, he looked forward to service in the Navy and when his vocation opened before him, his dearest wish became that he might be a Catholic Chaplain and serve in that capacity. He was appointed to the Corps in 1892 and served twenty-three years ashore and afloat, becoming easily the best known Chaplain of his day. . . . Chaplain Reaney was a big man, erect of carriage, with handsome, leonine features and the trained body of a fine athlete. His ready Irish wit and instant friendly affection for those around him won an equally instant friendliness for him in turn. He loved all athletic sports and threw himself with unexampled vigor into the promotion of an athletic program for the men of the Navy. . . . He promoted baseball, football, rowing and boxing, especially boxing. He, himself, was a boxer of skill and resource, winner of many bouts, sparring partner of President Theodore Roosevelt. . . . He fitted the Chaplaincy as the hand fits the glove, and in the battle for righteousness he used his office with the same skill and zest that he used the gloves when he fought in the squared circle. . . . And neither before his day nor since has there been one like him.[4]

During the first decade of the twentieth century, six of the allotted twenty-four chaplains were Catholics. No definite quota of denominational representation has ever been set for the Navy Chaplain Corps, although an effort has been made to select men from the larger denominations more or less on the basis of the national religious census returns.

OUTSTANDING CHAPLAINS

There were certain outstanding chaplains whose interest in and activity for the men under their care set them apart as worthy of special mention.

First in order of seniority is the name of Chaplain John S. Wallace, who was commissioned in 1863. Perhaps, his most outstanding service was achieved during his tour of duty on the receiving ship *Wabash* from 1891 to 1893. At that time, a number of naval prisoners were confined aboard this ship, a common practice after the middle of the century, when flogging was abolished. Wallace was deeply interested in the spiritual and moral welfare of these men and zealously ministered to their needs.

Shortly before Chaplain Wallace retired in 1893, a former naval prisoner who described himself as "one of those that stand at the lowest rung of the social ladder," wrote to the Chaplain in appreciation, saying,

> allow me at this time an opportunity to accord you my humble, but heartfelt and sincere thanks for your kind words and actions towards myself while I was undergoing a sentence at the naval prison. You brought with your presence into that prison some of the brightest sunbeams that ever straggled over the pathway of a then dark and dismal portion of my life.
>
> You may never, perhaps, in this life fully comprehend how precious and anxiously looked for those Sunday afternoon visits were to my comrades in misery.[5]

It is said that the agitation for drastic reforms in the treatment of naval prisoners, in which Chaplain Wallace took a part, led to the construction of the new naval prison at Portsmouth, New Hampshire, in 1904.

Among the reasons Chaplain Wesley O. Holway should be remembered is that he introduced physical drill into the Navy. In the old days, sailors kept physically fit by climbing the rigging and handling the canvas. When steam was introduced and the donkey engine provided power for the hoists, the lack of hard labor had an unfavorable effect upon the physique of the enlisted personnel. Sailors tended to become soft-muscled and round-shouldered. Captain Bunce of the Newport Training Station requested Chaplain Holway to set up some form of compulsory drill. Holway prepared a booklet which set forth some light dumbbell drills based upon a pamphlet issued by the YMCA. The success of these drills was such as to attract the attention of the Bureau of Medicine and Surgery and of the Navy Department. In 1898, the Navy Department issued *A Manual of Physical Drill* which had been prepared by Chaplain Holway.[6]

One of the youngest chaplains ever to be commissioned in the Navy was George A. Crawford, born in Maine, 29 April 1849 and commissioned 23 May 1870, when he was a few days more than twenty-one years old. Ten years later, he was ordered to the

---

[3] Germain, *Catholic Military and Naval Chaplains*, p. 143.
[4] Edel, *op. cit.*, pp. 17–8.
[5] Wallace File, Alumni Records, Harvard University.
[6] Copy in Edel Coll.

Richmond, then on the Asiatic station, where he remained for about four years. The story of the religious revival which occurred on the *Richmond* is told by Chaplain Edel:

> Chaplain Crawford had already served ten years before he joined the RICHMOND in 1880, but he found laid out for him on that vessel the work for which he was to become famous. When he arrived on the Asiatic Station, he found the RICHMOND a most unpromising field for Christian effort, with a greatly lowered state of morale. The crew are said to have indulged freely in the revolting vices of the foreign waterfronts, they were addicted to drugs and drink, and they scoffed loudly at the idea that any Chaplain could have any influence over them. Only a handful of men greeted the Chaplain with friendliness, but the Chaplain using these men as the nucleus of his Christian activities, preached with boldness the need for instant conversion. To the surprise of officers and men, there broke out almost at once a great revival of religion, which continued until the majority of the crew and many of the officers professed conversion and banded themselves into prayer groups, led by the valiant Chaplain. The use of intoxicants and drugs nearly disappeared, liberty parties returned to the ship clean and sober, and officers and men began to refer to the RICHMOND as a "happy ship."[7]

Chaplain Crawford often referred to his temperance work aboard the *Richmond.* Writing on 4 January 1884, he commented on the "Lodge of Good Templars," a temperance organization, which had been organized aboard his ship. He said it "continued in a flourishing condition, and nearly all of its members have been true to their pledge.[8]"

Among the new names which appear on the roll of Navy chaplains, during these years, is that of Roswell Randall Hoes, who was commissioned in 1882. His usual self-introduction was "My name is Hoes, it rhymes with goose." His hobby was history and nothing delighted him more than to dig into the records of the past. Through the years, he built a notable collection of early and rare Americana. Sections of this collection are now in such libraries as the Library of Congress, the Henry E. Huntington Library at San Marino, California, and the New York Historical Society. By request of the Queen of the Hawaiian Islands, Chaplain Hoes spent some time in arranging archives there and in preparing a bibliography of the Kingdom of Hawaii. He also was interested in several local histories of New York State and in the early story of the Dutch settlers in New Amsterdam. Chaplain Henry H. Clark referring to his colleague once remarked that an "eminent antiquary was spoiled when Chaplain Hoes entered the corps."

With such predilections, it was expected that Chaplain Hoes would become interested in the history of the naval chaplains. He was the first to begin collecting pictures, records, and other chaplain data from the beginning of the Navy down to and including his own day. His successors who have carried on researches in the history of the Chaplain Corps have built on the foundations which Chaplain Hoes so well laid.

In addition to those mentioned, certain chaplains who entered the service before 1881, as William H. Stewart, Donald McLaren, David Tribou, Henry Howard Clark, Adam A. McAlister, Thomas A. Gill, and Sylvester Boorom, were also in their prime during this period.

THE BEGINNING OF THE NAVY YMCA

One of the most effective and efficient civilian agencies organized to minister to the physical, mental, and spiritual needs of naval personnel, especially when off duty, is the Navy Young Men's Christian Association. Naval chaplains whose duty brought them within reach of the Navy Y activities are quick to praise the splendid service rendered by this branch of the National Council YMCA.[9] It is interesting to note that two naval chaplains, David H. Tribou and Robert Edward Steele, played an important role in the organization of the first Navy YMCA.

The YMCA was first organized in London by George Williams in 1844. Transplanted to the United States, the organization quickly took root and reached its fullest development. As has been mentioned, the YMCA was concerned with the welfare of the men in service during the Civil War and took a leading part in the organization of the United States Christian Commission. When the Spanish-American War broke out in 1898, the Y again became active in ministering to the men in uniform. Out of this conflict came the Army and Navy YMCA, an organization which during the following forty years established branches in China, the Phillipines, the Canal Zone, at military posts and in the leading naval port cities of the United States. Wherever the United States Navy went, the Y followed if circumstances permitted.

[7] Edel, *op. cit.*, p. 13.

[8] Nav. Rec. Coll., III: Jan. 1884:55.

[9] The Army and Navy Department of the International Committee of the YMCA of the United States and Canada was transferred to the National Council of the YMCA of the United States upon the establishment of the latter body in 1924.

The first Navy Y was organized in 1898 at Brooklyn, New York. The following historical item is found on the souvenir program printed for the dedication of the "Brooklyn Branch of the Naval Young Men's Christian Association 15 May 1902."

> As early as January, 1898, before there was any real belief in the possibility of war, Chaplain D. H. Tribou, U. S. N., in his annual report to the Department at Washington, suggested that in his judgment the Young Men's Christian Association could minister to the needs of the young men in the Navy, and offered to take up the question with the International Committee. This he was instructed to do by the Department, and as a result the Committee received a communication from him dated January 27, 1898. In response to this a secretary of the Committee went to Boston to confer with Chaplain Tribou, and a feasible plan of procedure was outlined. Thus, even before the war began, forces were set in motion which prepared the way for the prompt action needed when it was actually declared.

In the meantime, the Reverend Robert E. Steele was also studying the possible application of the techniques developed by the Y to the needs of naval personnel. Steele, a Presbyterian minister, became secretary of a Mission for Sailors organized at Brunswick, Georgia, in 1886. After two years, he went to Savannah to study the work of the YMCA there. Steele later became the chaplain of the New Orleans Seaman's Friend Society.

Steele saw the need for a building in the city which could be used for the entertainment of naval personnel while on liberty. Through the generosity of a number of citizens whose interest in the project had been cultivated by Steele, a temporary "Sailor's Rest" or "Bethel" was secured on Chartes Street, just off Canal. Hundreds of naval men used the facilities of this place while their ships remained in port. This was the first successful attempt in the United States to deal with the problem of entertaining naval personnel while on liberty.

Early in 1897, the battleships *Maine* and *Texas* visited New Orleans to permit their crews to take part in the famed Mardi Gras festivities. The mayor of New Orleans gave Steele full police authority over all intoxicated and disorderly bluejackets while the ships were in port. Steele did not permit a single sailor to remain in jail. As soon as he was notified of an arrest, he secured the man's release and after sobering him up, saw that he was returned to his ship. The success of this unusual experiment attracted the attention of the commanding officers of the ships, who extended their thanks to Steele. Writing to the Board of Directors of the Bethel on 26 March 1897, Captain A. S. Crowninshield of the *Maine* gave this word of testimony;

> Through the untiring efforts of your Chaplain, the Reverend R. E. Steele, many men were sheltered in comfort where the alternative might have been jail and the police courts. On the day of sailing, two men who would, undoubtedly, have been left behind, were brought off to this vessel by your Chaplain, and on other occasions men who would have broken their liberty were, undoubtedly, induced by him to return to their vessel.[10]

In May, following the declaration of war on Spain, Steele offered his services to the Army Christian Commission, under which name the YMCA was seeking to carry on its activities with Army personnel. Nothing had been done as yet for the Navy men. Steele offered to go to Key West and inaugurate work there for the men of the fleet. The Y accepted Steele's offer. He arrived at Key West on 4 June and rented a large tobacco factory which he called the "Naval Headquarters of the Army and Navy Christian Commission." Here he carried on a multiplicity of duties. He visited the sick, provided reading matter for both soldiers and sailors, and held religious meetings almost every evening in the tobacco factory. His excellent work inspired a letter of commendation from Commander George C. Remey to the Navy Department.

On 4 July 1898 Steele was commissioned a Chaplain in the Navy and given duty on the *Lancaster* then at Key West. The fact that Steele was chosen out of more than seventy applicants to fill one of the three vacancies, speaks well for the high regard the Navy Department had for the work he had performed. The outbreak of yellow fever at Key West in August compelled the withdrawal of the naval forces. Chaplain Steele accompanied his ship to Portsmouth where he was detached and ordered to the Navy Yard at Brooklyn. Here, he was given the special assignment of investigating the problems to be met in handling naval personnel on liberty.

On 19 September 1898, Steele submitted his report to Rear Admiral F. M. Bunce. In pointing out the necessity for a Naval Young Men's Christian Association, Steele wrote:

> No society or other instrumentality is, at present, maintaining resorts entirely suited to meet the real demands of our sailors, nor can I find that any comprehensive wide-reaching plan such as that of the Young Men's Christian Association, has ever been formulated for this purpose.
>
> Perhaps, one of the best proofs of the practical demand for this proposed organization is the fact that from all

[10] Edel Coll., Steele File.

parts of the country, and from different sources in the Navy, come suggestions very similar in detail and all looking to the same results.[11]

Steele envisaged a work that would be navy-wide in scope and not limited to any particular port. He recommended the erection of buildings at Brooklyn, Boston, Norfolk, and San Francisco. He suggested that the building at Brooklyn be erected first because the largest Navy Yard was located there and because it was near the headquarters of the International Committee of the YMCA. Steele felt that this latter fact would make it possible to secure able supervision for the new project.

Admiral Bunce gave his hearty approval to the plan and forwarded it to the Department. On 8 October 1898, Secretary of the Navy John D. Long wrote to Chaplain Steele approving the plan. Secretary Long suggested, however, that a change be made in the rules of the new organization to permit all members the right to vote and hold office. Steele, true to his own convictions and the generally accepted policy of the Y, had divided membership into two categories, active and associate. Those who became active members had to be Christians, and were called upon to sign a credal statement of faith in Jesus Christ. This excuded such faiths as the Unitarians and it so happened that Secretary Long was of this persuasion.

Steele replied on 18 October, pointing out that the suggested change did not exist "in any other branch of the YMCA" and that the change would have the effect "of emasculating the whole movement of its strongest element." He stated that, in his mind, the main purpose of this rule was "to organize and identify the Christians of the service, both officers and men" for the evangelizing of the Navy. On 21 October, Long notified him that since "this proposed branch organization is not a part of the Naval Service, the Department is at present of the opinion that it may be left to establish its own rules."

The Secretary wrote a strong letter of endorsement for the proposed "Naval Branch of the Young Men's Christian Association in the City of New York," which was evidently to be used in raising funds. "The plan," wrote the Secretary, "provides for the proper entertainment of the sailors of our Navy when on shore, their protection from the temptations which beset them, and the general care and good influence which such an institution will exert."

A number of wealthy persons became interested in this work. Miss Helen Gould provided funds to purchase property at 167 Sands Street, Brooklyn, within three blocks of the main entrance of the Navy Yard. The building there was opened on 1 March 1899. Men who were critical of anything resembling a mission were at first prejudiced against the institution. It was not long, however, before the Navy Y had their full confidence and respect.

Steele was ordered to the *Hartford* in September 1899, and on the 25th of that month submitted a report to Rear Admiral J. W. Philip at the Navy Yard, New York, which summed up the part he had played in the organization of this branch of the YMCA. Steele first referred to the fact that he had been assigned by Admiral Bunce "to special duty of organizing the Naval Young Men's Christian Association, and the establishment of a Home for enlisted men on liberty, near the Yard." Steele reported that the Navy Y had sixty-four charter members, and that "A Naval General Committee has been appointed, a constitution prepared and adopted, and arrangements perfected for the establishment of branches on ship board and at stations."

After referring to the building which had been opened for the accommodation of the men at 167 Sands St., Brooklyn, Steele wrote of the work in other places as follows:

> At Newport, R. I., Boston, Mass., Portland, Me., and at Bar Harbor, Me., buildings were fitted up by the local Young Men's Christian Association for temporary sleeping quarters during the visit of the N. A. Squadron. This was in response to my requests, excepting at Newport, where the local Association responded to the request of Chaplain Royce of the Flagship New York. At Philadelphia, Pa., the Association extended all the privileges to our men but was unable to secure a building for sleeping quarters. At Tompkinsville, L. I., the Naval Association fitted up, under my direction, a large Hotel building near the landing, with cots, a piano, lunch room, etc. This was used by the men of the N. A. Squadron during and just preceding the Reception of Admiral Dewey.
>
> I have visited Washington, D. C., Boston, Mass., Newport, R. I., Philadelphia, Pa., Albany, N. Y., West Point, N. Y., Grand Rapids, Michigan, and Tarrytown, N. Y., in the interest of the Association. Have held divine service on board the Detroit, New Orleans, Newark, Texas, and Massachusetts, and on the Cob Dock, and at the Naval Hospital.

The Navy Y owes much to the indefatigable zeal and vision of Chaplain Robert E. Steele. Basic principles, which have guided the Army and Navy YMCA in its activities from the very beginning of its history, were first outlined and put into practice by Chaplain Steele. His plan of providing clean and

[11] *Ibid.*

comfortable sleeping accommodations, reading and recreational rooms, and eating facilities are still included in the arrangements of every Navy Y. Many a United States sailor has been the beneficiary of this little known Navy chaplain.

EVOLVING NAVY REGULATIONS

Navy regulations were still in flux during the years 1881 to 1900. New editions appeared in 1893, 1896, and 1900. Gradually, the articles governing the duties of the chaplain evolved into their present form.

Chapter XVII of the 1893 edition was entitled "The Chaplain." The *Regulations* of 1876 stated that the chaplain "will perform Divine Service and offer prayers at such times as the Commanding Officer may designate." In the 1893 edition, the following article appeared:

> 734. The chaplain shall perform divine service and offer prayers on board of the ship to which he is attached at such times as the captain may prescribe; also on board other ships to which chaplains are not attached, or at shore stations and naval hospitals, when so directed by the senior officer present.

The "will perform" of the 1876 *Regulations* was supplanted by the more forceful expression, "shall perform." It is interesting to note that the chaplain was also instructed to conduct services on other ships, shore stations, and at hospitals where no chaplains were attached "when so directed by the senior officer present." This provision greatly increased the chaplain's usefulness. The 1896 and 1900 editions of the *Regulations* repeated the provisions of article 734.

The following articles from the 1893 *Regulations* were also repeated in the 1896 and 1900 editions:

> 735. He shall be permitted to conduct public worship according to the manner and forms of the church of which he is a minister.
>
> 736. He shall facilitate, so far as possible, the performance of divine service by clergymen of churches other than his own, who may be permitted by the captain to visit the ship for that purpose.
>
> 737. He may, with the sanction of the captain, form voluntary classes for religious instruction.
>
> 738. He shall visit the sick frequently, unless the condition of the sick renders such visits unadvisable.
>
> 739. Under the direction of the captain, he shall supervise the instruction of boys, and of such other persons in the Navy as may need to be taught the elementary principles of reading, writing, arithmetic, and geography. He shall report in writing to the captain at the end of each week the character of instruction given, the number of hours under instruction, and the progress made by each person.

Article 736, which appears for the first time in 1893, suggests that some chaplains were not as diligent as they might have been in providing religious services for those of the crew who differed from them in faith. In the early years of the Navy, it was assumed that all were Protestants. With the commissioning of Roman Catholic chaplains in 1888 and 1892, a policy had to be evolved to care for Protestant men on ships served by the Catholics and to make Catholic services available for those ships and stations which had Protestant chaplains.

Article 737 modified the earlier provisions which authorized the chaplain "to instruct in the principles of the Christian religion the boys and such other persons as the Commander of the vessel may commit to his care." All reference to the "Christian religion" was stricken out. Thus, the way was prepared for the services of Jewish chaplains.

The 1893 edition contained the following two new articles:

> 740. He shall always report at quarters for inspection, when on board. His station at quarters for battle and for inspection, shall be as the captain may direct.
>
> 741. The chaplain's duty in battle is to aid the wounded.

The 1896 *Regulations* combined the two articles by inserting article 741 after the first sentence in article 740. The 1900 edition eliminated article 741 of the earlier edition.

Article 1685 of the 1893 *Regulations* stated that newly appointed chaplains were to be not less than twenty-one or more than thirty-five years of age. The 1876 *Regulations* had placed the upper age at thirty. This article also ruled the applicant for the chaplaincy had to pass a physical examination and "be a regularly ordained minister of good standing in his denomination." No previous ruling on this subject had been so specific. The *Regulations* of 1896 and 1900 repeated this article without change.

The law of 1860 which required chaplains to report annually to the Secretary of the Navy was frequently disregarded in the years that followed. The *Regulations* of 1896, still in force, reaffirmed the earlier law: "Chaplains shall report annually to the Secretary of the Navy the official services performed by them."

It took the Civil War to give relative rank to the staff officers of the Navy and the Spanish-American War to give them actual rank. Congress, on 3 March 1899, passed "An Act to reorganize and increase the efficiency of the personnel of the Navy and Marine Corps of the United States." Section 7 stated that "all sections of the Revised Statutes, which in defining the rank of officers or positions in the Navy, contain the words 'relative rank of' are hereby amended so as to read 'the rank of'." All staff officers, including chaplains, and those on the retired list were in this way placed on an equal footing with line officers in respect to rank. Nevertheless, discrimination still existed, for the maximum pay chaplains could receive, regardless of rank, was that of lieutenant commander. Section 13 provided further: "That naval chaplains, who do not possess relative rank, shall have the rank of lieutenant in the Navy."[12] Newly commissioned chaplains received the rank of senior grade lieutenant. The rank of lieutenant (junior grade) was not given to chaplains until 1905 and the grade of acting chaplain with status of lieutenant (junior grade), did not apply until 1915.

In accordance with this 1899 law, the *Regulations* of 1900 provided the rank of lieutenant for "the remaining chaplains" whose rank had not been determined by the 3 March 1871 Act of Congress. The distribution for the other ranks remained the same as in the earlier Act, and promotion was still on the basis of seniority.

Several regulations, dealing with the observance of Sunday, are of interest. The *Regulations* of 1876 stated: "When a national anniversary, to be celebrated, occurs on Sunday, all the ceremonies are to be deferred until following day; and in no case shall a salute be fired on Sunday, unless the failure to do so would give offense to foreign authorities; but salutes may be returned on that day." Even earlier editions of the *Regulations* carried similar provisions.

Article 239 of the 1893 edition reads as follows:

> 1. Officers commanding squadrons or ships shall not permit women to reside on board of, or take passage in, any ship of the Navy in commission for sea service.
>
> 2. Sunday shall be observed on board of all ships and at naval stations in an orderly manner. All labor shall be reduced to the requirements of necessary duty. The religious tendencies of officers and men shall be recognized and encouraged. The religious institutions and customs of foreign countries visited by ships of the Navy must be respected.

The *Regulations* of 1896 added a section which relieved somewhat the pressure of work on Sunday. It stated that Saturday afternoon "shall in general be regarded on board ships as a half holiday," thus permitting the men to perform incidental chores which had been attended to on Sunday. It also provided: "All punishments consisting of extra duties shall be discontinued on Sunday." More and more Sunday was being protected and set apart as a day for rest and worship. All this was a great help to the Navy chaplain in his work.

"THE SMOKING LAMP IS OUT"

Before the invention of matches, and cigarette lighters, a smoking lamp hung from a cord near the cook's stove. During certain hours of the day, the wick was lighted and all who wished to light their pipes or cigars had to go there to do so. Later a lantern was substituted for the old smoking lamp, but the name itself remained.

Mention of smoking is found in the section dealing with the duties of the commanding officer in the 1865 edition of *Navy Regulations*. It is as follows:

> He will permit a lighted lantern to be hung up in a suitable place during meal hours, and after evening quarters until tattoo, or the setting of the watch, from which pipes or cigars may be lighted. No pipes or cigars shall be lighted at the galley or on the berth deck.

In the 1876 edition of *Regulations* the following restrictions were laid down:

> He will not permit smoking in the wardroom, steerage, cockpit, or on any part of the berth-deck, and will himself designate the places where smoking will be permitted.

A further restriction on smoking was included in the 1893 and subsequent *Regulations* which read:

> 378. (4). Smoking during divine service is forbidden.

Here, then, is the origin of the traditional passing of the word by the boatswain's mate: "The smoking lamp is out. Knock off all card games. Keep silence about the deck during Divine Service." The smoking lamp and the smoking lantern have long since disappeared. Yet the call "The smoking lamp is out" is still used ashore and afloat to mean, "knock off smoking."[13]

At the conclusion of Divine Service, it was customary on some ships and stations to have the boatswain's mate notify the ship's company by singing out: "The smoking lamp is lighted." This was a signal for the resumption of all activities which were suspended during the church hour.

---

[12] 30 *Stat.* 1006–7. It is to be noted that the granting of full rank to chaplains did not permit them to wear the same uniform as that worn by other naval officers. This was provided in subsequent legislation, as was equalization in pay.

[13] The traditional church call has been amended today to include "Turn off all radios." The installation of a public address system on the larger vessels permits the sound of the boatswain's pipe and voice to be carried to every corner of the ship or station with ease. In line with the tradition that all unnecessary activities cease during Divine Service, the ship's service store and the library are usually closed at this time. Upon the suggestion of the chaplain, the call can be given sufficiently in advance of the opening of the service to permit all hands to assemble.

THE CHAPLAIN'S UNIFORM

When chaplains were given the relative rank of commander and of lieutenant commander in 1863, they were authorized to wear the regular naval officer's uniform, including full-dress, with the insignia of their respective ranks. The cross was to be worn on the shoulder straps and on the cap as the Corps device. Some chaplains, however, preferred to wear their civilian clerical clothes. On 25 August 1881, Chaplain W. F. Morrision wrote to the Secretary of the Navy requesting permission "to wear the ordinary black clerical suit when on ship." The Secretary, however, ruled: "Chaplains must wear their uniform as well as other officers."[14]

The 1883 edition of *Uniform Regulations* made optional the wearing of the authorized uniform by chaplains: "In place of the prescribed uniform, chaplains may wear the single-breasted coat, waistcoat, and trousers commonly worn by clergymen, made of black or dark navy-blue cloth." These *Regulations* also permitted the wearing of a summer uniform, a white duck coat with white braid and without distinctive marks on the collar.

The edition of *Uniform Regulations*, approved 1 July 1885, contains only the following regulation relative to chaplains:

> Chaplains shall wear the dress commonly worn by clergymen, consisting of a single-breasted coat, with standing collar, waistcoat, and trousers of black or dark navy-blue cloth, and black, low-crowned soft felt hat.

Since no mention was made of the full-dress uniform, chaplains were obliged to wear civilian dress on formal occasions.

This often caused embarrassment. A typical illustration of this is found in an experience of Chaplain H. H. Clark when he was ordered by Admiral Kirkland from the bridge of the Flagship *New York* at the opening ceremonics of the Kiel Canal, 27 June 1895.

Writing to the Secretary of the Navy, Chaplain Clark stated that as chaplain of that ship he had received "a typewritten order . . . to be present on the ship's bridge during the Emperor's review of the fleet." Chaplain Clark appeared in the authorized full dress uniform of a chaplain, the frock coat of a civilian clergyman. He was then ordered below, the Admiral explaining that his "dress was unsuitable for such distinctively military assemblage as that on the bridge." Chaplain Clark then added:

> I did not think, nor could I think, that there was the slightest intention to discriminate against me; but the conclusion was obvious that in the judgment of the Commander-in-Chief my dress afforded sufficient reason for debarring me from participation in an occasion in a manner that had been made a common privilege and duty for all other commissioned officers of the ship. I did not experience any sense of personal or professional injury, but I could not resist the inferences, that by my exclusion from the bridge my uniform had deprived me of a statutory right, and that from a military point of view I had suffered humiliation.

Inasmuch as the prescribed chaplain's uniform for dress occasions set them apart from all other naval officers, Chaplain Clark respectfully requested that "a uniform be devised for Chaplains which shall express their relative relations to the naval establishment, and which shall raise no question as to their availability for, or presentability in, any military formation or parade, or shall seem unfitting to any occasion, of which, by virtue of law and custom, they are a legitimate and organic part."[15]

On 3 September 1895, the Secretary of the Navy replied that the Admiral had "been advised of the Department's disapproval of his action in directing that you be ordered from the bridge of the U. S. Flagship *New York* upon the occasion of the review of the naval fleet, at Kiel, by the German Emperor." Regarding the request for an appropriate full-dress uniform, the Secretary wrote that "the Department does not deem it desirable, at the present time, to make any change in the existing dress for officers of this corps."[16]

The new *Uniform Regulations* of 1 July 1897 contained the following article:

> Chaplains shall wear the dress commonly worn by clergymen consisting of a single-breasted frock coat, with standing collar, waistcoat, and trousers, of black or navy-blue material, and a black hat. A navy cap with black buttons and strap, and without ornaments, may be worn.

An Addenda appeared on 23 December 1898 which specifically prohibited chaplains from wearing the full-dress and the special full-dress uniforms authorized for other Navy officers. Chaplains were authorized to wear for special full-dress a "Frock coat, plain blue trousers, blue cap." For less formal occasions, white trousers were permitted and a white cover with the cap if such were ordered. The chaplain's service dress consisted of "Blue or white service coat, plain blue or white trousers, and blue caps (with white cover if ordered), and his evening dress

[14] Nav. Rec. Coll., III:Aug. 1881:131.

[15] Original in Clark File, Chaplain's Office, Naval Academy.

[16] From copy in Clark File, Naval Academy.

consisted of "Evening dress coat and waistcoat, plain blue trousers and blue cap. Chaplains who desire may wear the frock coat instead of the evening dress coat, with this uniform." In other words chaplains were not permitted to wear the cocked hat, full-dress frock coat, laced trousers, shoulder knots, sword, and leather belt.

Drastic changes were also made in the service uniform. In general, the chaplains' dress was to be separate and distinct from that of their brother officers. All brass buttons, which were the regular Navy eagle buttons, were removed from the chaplain's uniform. The result was that the white jacket of the summer uniforms closely resembled that of the mess attendant.

While other officers had a double-breasted frock coat with two rows of the official buttons, the frock coat worn by the chaplain was to be as follows:

> For chaplains—The frock coat shall be of dark navy-blue cloth, faced with the same and lined with black silk serge, single-breasted, made to button to the neck, with one row of navy size plain flat silk buttons on the breast, seven in number; plain standing collar; skirts to be full, and to descend four-fifths of the distance from the hip bone to the knee, with two buttons at the waist behind; cuffs to be closed, without buttons, and from 2½ to 3 inches deep. On each side of the collar shall be embroidered in high relief the corps badge and grade device.

Sleeve ornaments, too, were ruled out: "The sleeve ornaments for chaplains shall be of lustrous black braid, and shall correspond in width and disposition with those of line officers of the same relative rank." Insignia of rank and corps appeared on the collar, the cross being worn at a 15° angle, as it had previously been worn on the cap and shoulder straps. Some have theorized that the cross was first placed at an angle on the collar to correct the natural slope of the collar and make the cross appear vertical. It is, however, to be noted that the cross was worn at an angle on the cap and on the shoulders before it appeared on the collar.

A distinctive change was also made in the cap. Chaplains were authorized to wear the usual blue cap, with the white cover when ordered, but with "the sliding chin strap and buttons" of black mohair. "Chaplains with the ranks of captain and commander shall have on the front edge of the visor a band of lustrous black mohair, one-half inch wide." Thus, all gold braid was removed from the chaplain's uniform along with the brass buttons.

The earlier provision which permitted chaplains to wear during Divine Services the vestments of the church to which they belonged was repeated in the *Regulations.*

THE CHAPLAIN AND HIS WORK

During the years before the building of the White Squadron, when the fortunes of the Navy were on the decline, when ships were rotting away at their anchorages, and when the total number of naval personnel dropped below 10,000, the opportunities for an effective ministry on the part of naval chaplains were naturally becoming more and more restricted. Undoubtedly, during these years, some chaplains were having an easier time and a larger income than were some of their brother ministers in civilian parishes. Since there was no organized Chaplain Corps with a Director or Chief empowered to supervise chaplains' activities and take remedial action where necessary, it is possible that among the chaplains on duty were a few who were indolent or inefficient.

Whatever the circumstances in the background may be, a barbed-pen article, from which the following extracts are taken, appeared in a prominent New York newspaper in 1885.

UNCLE SAM'S NAVAL CHAPLAINS

Tweny-one Fortunate Clergymen who are Blessed with Ease and Plenty

> Out of the many thousand gentlemen who find snug shelter beneath the Government's fostering wing, none enjoy the peace, prosperity, and general happiness in equal measure with the twenty-one who are chaplains in the navy. They toil not and seldom pray, but draw their salaries with elegance and precision. These salaries are large, or would be for an ununiformed worldling. For the first five years of service each chaplain culls $2500 per year and his rations, if at sea, from a benevolent national Treasury, $2000 if on shore duty, and $1600 if on waiting orders, the last named being a condition of complete inertia. Even after the five years mentioned, the remuneration is $300 more, in all stages of service. Rations consist in the main of hardtack and pork, with a smack of sugar and coffee thrown in, and can be traded off for canned chicken or anything palatable to wardroom appetites.
>
> With the exception of giving the ship schoolmaster an occasional lift in his duty of driving simple educational facts through saline skulls, the chaplain does nothing. He holds no services, except now and then at a stray funeral. The only Sunday features of a man-of-war are extra clean decks and officers in full dress uniforms, including buttons. The chaplain wears just as many buttons as any of them, but he does no preaching. Except for a provision in the naval regulations it would be hard to tell why any ships carry a chaplain. This says they must, and they do. There is no dodging naval rules with salaries attached to them. No chaplain on shipboard wearies himself with stir-

ring up religious sentiment among the men, or worries about their tarry souls.

Not every ship carries a chaplain. Only flagships are thus adorned. Chaplains are too expensive.

Not a few of the shore chaplains are prosperously quartered on denominational parishes, thus, securing pleasing additions to their several incomes. Those stationed at navy yards make the most show of professional usefulness. Local churches or missions usually afford services of some sort on the receiving ships in which the chaplain cooperates. But, altogether the lot of a chaplain is a most happy one, and that of chaplains at sea especially so. Without irksome labor of any sort or responsibility, he can join the Captain of marines, another flagship luxury, in a perennial siesta on the wardroom sofas.[17]

The evident prejudice, misstatement, and overstatement which colors this editorial is so extreme as to make it humorous. The writer does not take into account the fact that naval officers often spent as much as $600 for a complete uniform outfit. They had no manse ashore for their families, but, had to keep up not only the usual wardroom expenses on ship, but, also a separate establishment for their families on shore. These were extra expenses which civilian clergymen did not have. Moreover, the chaplain was the lowest paid of all naval officers of the same relative rank.

It should be remembered that the critic wrote in 1885 just as the White Squadron was beginning to take form. With the growth of the Navy and the increase of naval personnel, without any corresponding increase in chaplains, the work of the latter, naturally, became more important.

The official reports of contemporary chaplains, moreover, furnished a highly conclusive rebuttal to the newspaper criticism. Chaplain David H. Tribou, who served in the *Powhatan,* reported that he provided Sunday morning services forty-nine out of the fifty-two Sundays in 1882, and evening services on forty-six of those Sundays. While the ship was in the West Indies and the crew were unable to procure good reading matter, he was able, "through the kindness of some of the Exchange Editors of the daily papers in Boston and the Secretary of the Young Men's Christian Association of New York . . . to distribute hundreds of newspapers among the men . . ."[18]

Chaplain Tribou's report indicated his interest in the spiritual, mental, and physical welfare of his men. In days when the benevolence of the Navy Department did not include such church gear as organs and hymn books, Chaplain Tribou raised by popular subscription the money necessary to secure them. His report for 1883 is more specific in regard to the men's response to Divine Services.

During the year, I have been attached to, and serving on board this ship, this being my second full year on board. I have held divine service ninety-three times, forty-seven times in the morning and forty-six times in the evening. The attendance has been excellent, averaging seventy-seven at the morning service and ninety-seven at the evening service,—or one hundred and seventy-four for each Sunday.[19]

Since the *Powhatan* had a complement of about 300 men, it would appear that Chaplain Tribou reached an unusually high percentage of the personnel of the ship.

Chaplain Tribou's 1883 report delineated a score of collateral duties as well. He lectured to the men on the "history, government and resources of the islands" visited by the ship, and arranged for "friends from on shore" to give several concerts. The ship's library under his supervision, was

much enlarged during the year,—more than two hundred volumes having been added, and nearly one hundred volumes having been rebound, at an expense of more than two hundred dollars. The library now contains about four hundred and fifty volumes, and has cost for additions, books replaced and books rebound, about four hundred and twenty dollars. There was no library on board when I joined the ship two years ago, except some books belonging to the Seaman's Friend Society. It furnished almost all the reading matter for the men, and is very popular and well patronized.[20]

Tribou tried to induce men "to abstain from the use of intoxicating liquors." He summarized his activities with this observation:

The work of a Chaplain which can be reported is but a small part, and by far the least important part of his duty. While I have tried to attend to the more public duties, I have never lost sight of the fact that, in personal work among the men, lies the Chaplain's most promising opportunity, and while there is much to discourage one in such work yet it is by far the most satisfactory part of the work which I am sent to do.[21]

Tribou was one of the younger and more aggressive naval chaplains then on duty. His popularity with the men was based upon a genuine interest in their welfare.

The available reports of other chaplains indicate that Divine Services were held regularly unless inclement weather prevented. Sunday Schools were

[17] New York *Sun,* 28 Oct. 1885.
[18] Nav. Rec. Coll., III: Jan. 1883:6.

[19] *Ibid.,* III: Jan. 1884:6.
[20] *Ibid.*
[21] *Idem.*

usually conducted at shore stations for the children of naval personnel. All chaplains had the usual duties which included officiating at funerals and weddings, and administering the sacraments. They continued to look upon the promotion of temperance among the men as part of their duty.

Chaplain Hoes, writing from the receiving ship *Wabash* at the Boston Navy Yard on 23 January 1900, reported how he had increased church attendance by rigging church on the forward part of the quarter-deck. He secured a choir from a local church ashore whenever possible for his services. "This plan," wrote the Chaplain, "has met with great success, as in my opinion at least 80 percent of the men on the ship attend each Sunday the worship of Almighty God."[22]

Several chaplains commented on their activities in securing books and papers for the men. Chaplain Crawford writing on 2 January 1883 stated: "I raised money by subscription and purchased a large library for the use of officers and men—also a Cabinet Organ for use at service."[23] Chaplain Rawson wrote on 17 November 1884: "Owing to the kindness of the Directors of the Boston Public Library, a plentiful supply of books has been furnished under my personal supervision to the Marines at the barracks to the number of sixty to one hundred volumes per month . . . it is to their credit that not a single book has been defaced or lost."[24]

Chaplain A. A. McAlister, writing from the Naval Academy on 1 January 1883, reported a Sunday School for cadets which met after Divine Service on Sunday morning and a YMCA which met in the afternoon. "I have said prayers in the mess-hall daily immediately after breakfast," wrote Chaplain McAlister, "except on Sundays, Thanksgiving and Christmas, when Divine Service was held in the chapel."[25] This practice is still continued at the Academy. Duty at Annapolis has always been con-

[22] NRSO: 10608.

[23] Nav. Rec. Coll., III: Jan. 1883:35.

[24] Edel Coll., Rawson File.

[25] *Ibid.*, III: Jan. 1883:1.

Second Naval Academy Chapel, 1869–1903.

sidered a choice assignment by naval chaplains and letters of application from members of the Corps for this duty were often received by the Secretary of the Navy.

Chaplain H. H. Clark, who reported for duty at the Naval Academy in September 1890, noted in volume two of the Chapel Log that on his arrival: "About seventy of the cadets belonged to the several church squads attending the city churches, making the number attending the chapel about one hundred and eighty." This custom of sending ashore church parties composed of midshipmen who desire to worship in civilian churches of their choice has been long observed at Annapolis.

Chaplain Clark wrote in the Chapel Log about the activities of the YMCA at the Academy.

> The Y. M. C. A. began on the first Sunday in October in the chapel at 5:15. Chaplain Rawson had been very successful with the Association, & it was in a flourishing condition. It numbered about seventy-five members, & the attendance at the meetings was from thirty to fifty.

During the 1893 June week exercises at the Academy, Bibles were presented to the members of the graduating class by the American Seaman's Friend Society in accordance with a custom which was then long established.

Chaplain H. W. Jones, who was commissioned in June 1896, tells in his book, *The Experiences of a Chaplain Ashore and Afloat,* of his activities at the Naval Training Station at Newport, Rhode Island. Among other duties, he served as "a sort of overseeing schoolmaster" for 387 boys, ranging from fourteen to seventeen years of age. Jones had three assistants who took care of most of the week-day work.

> Sunday came; the boys were mustered as the bugle sounded, and all marched in a very orderly manner into the gymnasium, where divine services were to be conducted. The Stars and Stripes were lowered, and above them went the Church pennant. . . . I took my place behind the desk and commenced my first service as a Chaplain in the Navy. All the Protestant boys were compelled to attend this service. The Roman Catholic boys were sent ashore to attend Mass.[26]

As though he were answering the critical article in the *New York Sun* of 1885, Chaplain Jones compared the life and work of a naval chaplain with that of a minister in civilian life. He wrote:

> Many people had said to me that a Chaplain's life was an exceedingly easy one. I beg to differ with them, however, and say, not with the spirit of egotism, but because I think it is an absolute fact, a minister on shore has twice as easy an occupation as a Chaplain aboard a ship. A minister is known best by his own immediate family. In fact, no one knows him as well as his wife and children, and no one sees the minister as he really is, but those of his family circle. His congregation sees him in his pulpit at different times, and he is with them on their prayer-meeting nights during the week, but they don't see that minister as a man. They simply see him as a pastor and preacher; and I don't believe there is a minister in the world who would be willing to live with his whole congregation seven days in the week, that is, to have each member of it become one of his immediate family—but that is what the Chaplain has to do.[27]

Chaplain Jones then commented on the fact, noticed sooner or later by all Navy chaplains, that naval personnel, especially aboard ship where all have to live in such close proximity, are quick to detect the slightest inconsistency in the life of the chaplain. Jones added:

> In fact, I think a Chaplain's life is the hardest one a Christian man can live. He has no encouragements to help him in his work, no loved ones to stimulate him, nothing of the sweet discipleship of noble Christian women that the minister is blessed with at home; he has to stand there alone, single handed, to fight against the tremendous army of Satan. Many times, however, he is privileged to find sweet Christian men among the officers, and, oftentimes, in the crew are noble Christian fellows, striving to live for the Lord Jesus Christ; and where he is fortunate enough to find such men, it is of great assistance to him in reaching others, as example counts for much in the Navy. If the Commanding Officer absents himself from the divine services, it does tremendous injury in the matter of attendance; but if the Commanding Officer feels that it is his duty to recognize the services for the sake of the men, the effect is wonderful in influencing many of them to attend.[28]

An interesting anecdote handed down through the years concerns Chaplain Frederick Sherman, an Episcopalian, and his Commanding Officer, Captain Alfred T. Mahan. The incident occurred aboard the *Chicago* in the early nineties. It appears that the famous Captain objected to the continued discourses by the Chaplain on the subject of socialism. Reaching the end of his patience, Captain Mahan handed the Chaplain a list of subjects on which he was not to preach, a list which included all aspects of socialism. Chaplain Sherman resented what he considered to be an infringement of his clerical rights and expressed his displeasure by refusing to serve communion to

[26] *Op. cit.*, p. 110. Compulsory church attendance at recruit training activities known as "boot camps" was a common practice until it was pointed out by the Secretary of the Navy in 1944 that "no coercion or discrimination of any kind is to be sanctioned to compel such attendance." (SecNav. ltr 44-1302)

[27] Jones *op. cit.*, p. 122.
[28] *Ibid.*

Captain Mahan. This, in effect, meant excommunication. Suffice it to say, Chaplain Sherman was transferred to another ship.

This was not the only occasion on which Mahan had friction with a naval chaplain. In fairness to Mahan, however, it should be noted that he was a sincerely religious man and early in his career had considered entering the ministry. Throughout his service in the Navy, he was always interested in the religious welfare of his men. During the monotonous blockade duty in the Civil War, he had ample time for reflection and expressed in his correspondence his concern over this question: How could he influence his men for their spiritual benefit?[29] Still later, Mahan's deeply religious tendency culminated in the book, *The Harvest Within*, a spiritual and devout treatise on personal religion.

Evidence of Navy chaplains' interest in moral problems is found in a rare little sixteen page pamphlet printed at Nice, France, by Chaplain W. F. Morrision while he was serving as chaplain aboard the *Lancaster*, 20 August 1881—19 June 1884. The pamphlet contains the substance of two discourses delivered aboard the ship. Chaplain Morrison had some pointed remarks to make about "that infamous institution at Monte Carlo."[30]

One of the first if not the first chaplain to use the stereopticon lantern in his work was Chaplain Holway. Writing to the Secretary of the Navy in October 1897, Chaplain Holway requested permission to introduce the lanterns on sea-going vessels. He wrote:

> It has been demonstrated that lectures and slides can be prepared that may be used on board our ships by any intelligent petty officer or sailor. . . . Twenty-two of such lectures, illustrated by 571 slides, were prepared by the writer at the Newport Training Station in 1892–1895. They were intended to furnish the bulk of English instruction for the apprentices. . . . They proved to be exceedingly popular and instructive. . . . Some of the sailor men on the Station have heard these lectures scores of times and never seem to tire of them.
>
> Up to within a recent date, the introduction of lanterns on board sea-going ships has been impracticable. Both kerosene oil and compressed gases were prohibited by regulation. An electric arc light stereopticon has been invented which can be used on an incandescent circuit, and which works (at Newport) successfully. . . . One of these lanterns can be installed on board any ship provided with an electric plant. . . . A good lecture will stand repetition three or four times at least during a cruise. . . . There are over 10,000 slides already available from which selections could be made.

Chaplain Holway suggested that a special set of slides on Bible subjects be provided with other slides carrying the words and music of hymns. "Sailors love to sing," he wrote, ". . . and a cornetist or violinist can always be found on board ship who could 'lead' these hymns." Holway felt that this scheme "would prove particularly helpful to the *unchaplained many*." He offered to help arrange the lectures. His proposal, however, was rejected, since the Chief of the Bureau of Navigation did not "consider it practicable to carry the lecture outfit recommended by Chaplain Holway in any but the largest ships, on account of the lack of storage room." The plan, however, was approved for training stations.[31]

One of the first chaplains to use a stereopticon machine aboard ship was Chaplain J. P. S. Chidwick who purchased a machine with slides for use aboard the *Maine*. Writing to Admiral A. S. Crowninshield from the *New York* on 26 February 1901, Chidwick stated:

> The Bureau of Equipment has refused to furnish us with a stereopticon and lecture-slides which have been granted to some of our ships, and, as we will be away for a few years, I believe that lectures of this kind will be greatly beneficial to the crew. I have expended all my money on a piano, a phonograph and church effects and have not sufficient to spare for the lectures. . . .
>
> You are aware, perhaps, that I did furnish machine and slides from my own money on the "Maine" and was not reimbursed for it when lost on the ship. I lost about one hundred and twenty-eight dollars. . . .[32]

Chaplain Wright was one of the first chaplains to conduct sightseeing parties of sailors ashore. Out of this practice of guiding bluejackets on visits to ports of interest in the nineties, grew his two pamphlets. These, published in 1901, were entitled: *Notes on New York & Brooklyn* and *Guide to New York & Brooklyn Compiled for my Shipmates in the U. S. Navy.*

SUGGESTED CHANGES

Suggestions, which chaplains of this period made for the betterment of the Corps and of conditions affecting the men, throw light on the problems and difficulties these chaplains faced.

On 12 August 1881, a committee of chaplains including Beugless, Holway, and Rawson sent to the Secretary of the Navy a letter containing suggestions

[29] Puleston, *Mahan*, pp. 23–4, 37.
[30] CoC., Morrison File.

[31] NRSO:5330.
[32] NRSO:12299.

for the improvement of the Chaplain Corps. The tone of this petition was similar to the one submitted by six chaplains on 2 November 1878.

The petition opened with a statement that the chaplains felt that every possible safeguard should be "placed at the door of entrance to the Corps—at least as strong safeguards as those which protect admission to other Corps in the Navy." The report claimed that chaplains had been commissioned "without any kind of personal examination of the applicants, either physical, mental, moral, or professional." The chaplains earnestly recommended that

> in every case of future nomination to the Naval Chaplaincy, applicants shall be required by the Department to undergo a like physical examination with applicants for appointments to other Corps; and, further, that a Board of Chaplains shall be designated by the Department to investigate the claims and credentials of applicants as to their professional qualifications and fitness for the work.

The committee of chaplains made further recommendations concerning the assigment of members of their Corps to duty, as well as to quarters on shore and on board ship. They suggested that every ship to which a chaplain was attached be provided with an organ and "singing books" and pointed out that if the chaplain were permitted to select the ship's schoolmaster, he could "choose for the purpose a man capable of playing on the organ, and of conducting the singing." Finally, the committee urged the appointment of a chaplain "not below the grade of Commander" to the Washington Navy Yard, with "a desk at the Department, that through him the Department may be brought into closer communication with the Corps, and learn more of its need in order to its greater usefullness."[33]

The matter of the assignment of the fourth or fifth cabin continued to be discussed. Chaplain Hoes wrote to the Secretary on 24 November 1882, suggesting that a definite room be assigned the chaplain regardless of his rank. He wrote:

> Until recent years, the rooms upon the port side of the ward room were so arranged upon all Flag ships that the *fourth* room was assigned to the Chaplain. That room was subsequently assigned to the Senior Marine Officer and the fifth room was given to the Chaplain. Under the present arrangement the senior officer in each of the other corps has a room assigned to him, while the Chaplain has to take his chance of the rooms remaining, according to his rank; and as many of the Junior Chaplains have no rank, they sometimes have a very inferior room. As the Chaplains' duties involve the necessity of constant retirement and study, it seems but reasonable that a light and airy room should be secured to them. This could be accomplished by an order giving to Chaplains fixed rooms, irrespective of rank. Is there any good reason why the *fourth* or *fifth* room should not be assigned to him?[34]

Another attempt was made by friends of Navy chaplains to secure legislative action on certain desired reforms. Senate Bill No. 1551 introduced into the first session of the 47th Congress, raised the lower age limit of the applicant to twenty-five, called for the denominational approval of applicants, and authorized a longevity increase of pay after ten years service. While the bill had the approval of the Naval Affairs Committee of the Senate, it did not become a law.

In 1884, a unique suggestion was presented to the Secretary of the Navy by Chaplain Richard Hayward. He requested that he be sent to Europe for the purpose of studying the "moral and religious work" of foreign armies and navies.[35] He pointed out that chaplains in our services were without guides or precedents and that they worked in widely separated areas where they could not easily communicate with one another. It was his purpose, therefore to prepare a manual of instructions for chaplains which could be used as a guide. Hayward's suggestion was filed without action. In 1918, Chaplain Frazier brought out the first Navy chaplains' manual.

Another chaplain, J. H. H. Brown, stated that conditions under which he and his colleagues worked were so intolerable that a number of chaplains were leaving the service. He pointed out that there was no "proper provision made for the decorous celebration of Divine Services." He said that chaplains were not provided with suitable offices and places of study and added that the "wardroom is not the place for a student and a clergyman who wishes faithfully to discharge his duties." Totally discouraged, Chaplain Brown, after serving ten years, resigned.[36]

A little later, Chaplain Hoes, writing from the receiving ship *Wabash* at the Boston Navy Yard, confirmed Chaplain Brown's comment on the inadequate office quarters for chaplains aboard ship. Referring in particular to his own type of ship, he said:

> It is my belief that on this and every other receiving-

[33] Nav. Rec. Coll., III:Aug. 1881:111–2. The suggestion concerning the schoolmaster, no doubt, reminds Navy chaplains who served in World War II of the newly established naval rating for chaplain's assistants known as Specialist (W). The last suggestion was later put into effect by the Navy Department. For a number of years before the Chaplain Corps was officially organized with a Chief, the chaplain on duty at the Washington Yard served as an advisor to the Navy Department in many matters pertaining to the Corps.

[34] Nav. Rec. Coll., III:Nov. 1882:171.

[35] *Ibid.*, III:Sept. 1884:143.

[36] *Ibid.*, III:May 1884:146.

ship, and, as far as practicable, on every sea-going ship, a suitable place should be provided where the Chaplain may see the men in private. I believe that the reasons for this are too obvious to require explanation. I beg leave to suggest to the Department the importance of making provision for such office or other place on this ship, and I respectfully ask that it be done.[37]

The Acting Chief of the Bureau added the following comment to Chaplain Hoes' suggestion:

The Bureau would recommend any accommodation for a Chaplain which their interests on a vessel permit, but it does not regard it as advisable that offices for chaplains be permitted on board modern vessels where every square foot of space is needed for other purposes. Space can be canvassed off where necessary.

The report with this notation was then returned to the Chaplain's Commanding Officer who added: "In my opinion, an office for the Chaplain on board this ship is not necessary or desirable, and is not practicable, without using space which is required for other purposes." The Bureau sent the suggestion to the Board of Construction for its comments. This Board reported that it concurred in the endorsement of the Commanding Officer of the *Wabash*. There the matter rested.

Chaplains frequently offered suggestions for the improvement of conditions which affected the welfare of the men. Chaplain McAlister was concerned with the improvidence of the sailor and his habit of squandering his money a few days after he had received it. He presented the problem:

Alluring schemes are presented to him by designing men on shore, and especially when he is under the influence of liquor, the temptation to take what money he can get and desert, is too strong for so weak a mortal to resist. At the end of a week or month after deserting, he may wish himself back in the service, but he can return only in disgrace as a deserter.[38]

McAlister suggested that a plan, such as an "ordinary savings bank" which would protect the sailors' funds and also pay them interest, be adopted by the Navy. Years passed, however, before his suggestion was put into practice.

One of the first projects undertaken by the newly organized Navy YMCA at Brooklyn was that of arranging with a bank for the care of money deposited by honorably discharged enlisted men. Chaplain Steele stated on 25 September 1899 that the Navy Y had handled over $13,000 since it opened in Brooklyn on 1 March of that year. Through the years, the Navy Y has encouraged thrift by providing to naval personnel means whereby money could be sent home or deposited in a bank.

Another excellent suggestion was contributed to the Department by Chaplain Tribou in his letter of 1 January 1883. He wrote:

On several occasions, there have been entertainments given on board, consisting of reading and singing, without expense to the men. Such exercises have a good influence on the crew, and were it not for the expense attending them they might be well increased in numbers. If some regular fund could be provided for this purpose, I am sure good would result. The men as a rule are better off on board ship than roaming about the streets and whatever helps to keep them contentedly on board should be encouraged.[39]

The subject of compulsory attendance at Divine Service aboard ship was reopened in 1882 by Chaplain T. A. Gill in his letter from the *Tennessee*. He reminded the Department that attendance was required of the boys when they were on training ships, and that certain commanding officers of ships at sea demanded it. On the *Tennessee*, however, church attendance was entirely voluntary even for the young boys. "As a consequence," commented the Chaplain, "very few of them are ever seen at a Religious Service, and these attend with a not very exemplary regularity." The Chaplain had taken up the matter with his Commanding Officer who refused to take the responsibility of compelling the boys to attend because he felt "that such a course would not be sustained by the Department." Gill then laid his case before the Secretary of the Navy.

The position maintained by the Chaplain has been, in these several instances, that the Department sustains toward these Boys, till their majority, a relation substantially that of "In Loco Parentis"; and although it may not *compel* the attendance of the crew at large at Religious Services, as in the British Navy, yet it may properly do so with these Boys, other than those of the Catholic Faith. . . .[40]

The Department, however, refused to change the regulations which permitted officers only to "recommend" church attendance.

THE SPANISH-AMERICAN WAR

The New Navy was initiated in the early eighties and the fleet was in relatively good condition when war with Spain broke out in the spring of 1898. While friction with that country had been of long standing, it was the sinking of the battleship *Maine*

[37] NRSO: 10608.
[38] Nav. Rec. Coll., III: Nov. and Dec. 1879: 319.
[39] *Ibid.*, III: Jan. 1883: 6.
[40] *Ibid.*, III: Apr. 1882: 261.

832159—49——10

in the Havana harbor on 15 February 1898 that precipitated hostilities. Among the eighty-nine survivors in the doomed ship's complement of 355 was the first and only chaplain to serve aboard the *Maine*—John P. S. Chidwick, the third Roman Catholic to be appointed to the chaplaincy.

When the explosion occurred, Chaplain Chidwick was in his cabin and had just finished reading his office of the day. Uninjured, he rushed to the deck where he joined with others in rescue work. His zealous efforts in behalf of the wounded and dying inspired a letter of commendation from Captain N. Sigsbee of the *Maine* to the Secretary of the Navy. On 9 April 1898, Secretary John D. Long sent the following letter to Chaplain Chidwick:

> My dear Chaplain:
>
> The Department is in receipt of a letter from Captain Sigsbee, dated yesterday, calling attention to the manner in which you performed the duties which devolved upon you in connection with the disaster to the Maine.
>
> In transmitting to you a copy of this letter, I cannot refrain speaking for the Department, and myself, from adding this further expression of commendation. Your heroic devotion to duty; your tender sympathy with suffering; your care for the dead; your fearless fidelity to your post, mark you as a true servant to the Master. You have set an example for the emulation of every chaplain in the Navy, and are entitled to the gratitude of the Department and of every American citizen.[41]

The only naval chaplain known to have been wounded in action during the war was Harry W. Jones, who served aboard the *Texas*.[42] Jones wrote two books on his war experiences: *A Chaplain's Experience Ashore and Afloat, The Texas under Fire*, 1901, and *The Battle of Santiago on board the U. S. Battleship Texas*, 1913.

Chaplain Jones gives the following description of a Divine Service held aboard the *Texas* shortly before the battle of Santiago:

> Sunday morning the sun was shining as bright as ever; it was a perfect day; the bugle sounded the church call, and about two hundred of our men mustered to attend divine service. Our Church pennant went up, together with the other Church pennants of the ships of the Squadron.
>
> The contrast was most striking: here we were telling the men of Jesus and His love; and, yet, at any moment we might have to cease our discourse and enter an engagement. Our men were very attentive all through this service, and they seemed to have far greater interest in spiritual matters every time I preached to them during these days, when we were standing by ready for action.[43]

Many chaplains, especially in the subsequent World Wars, have confirmed Jones's observations. In their reports, they note that fear and danger often make the mind attentive and the heart receptive to spiritual truth.

During the Spanish-American War, the Navy Department was not equipped to handle casualty notices promptly and efficiently. After the battle of Santiago, Jones found that one of his duties was to notify the next of kin of battle casualties. Of this he wrote:

> I immediately took the names of the wounded and the addresses of their people, taking down messages for them that they wanted me to wire to their homes, so that their friends would not be too much broken up when reading the possible exaggerated newspaper reports regarding their injuries.[44]

Two other naval chaplains, W. G. Cassard and Roswell R. Hoes, published works on their war experiences. Cassard, who served aboard the *Indiana*, wrote *The Battleship Indiana and her part in the Spanish-American War*. Chaplain Hoes published *God's Hand at Santiago, A Sermon preached on board the U. S. Battleship Iowa in Guantanamo Bay, Cuba, July 10, 1898, the Sunday following the Naval Battle of Santiago*. This outburst of literary activity was practically the first endeavor of the kind since the days of the scholarly Walter Colton and George Jones.

The name of the Spanish-American war hero, Admiral William T. Sampson, has been perpetuated in the great naval training station established on Seneca Lake in World War II. On the recommendation of the senior chaplain on duty at the time, W. W. Edel, the two chapels which were erected were named in memory of two outstanding Spanish-American War chaplains—John P. S. Chidwick and Alfred Lee Royce. Royce was the chaplain of Admiral Sampson's Flagship, *New York*.

Chaplain Joseph P. McIntire was on board the *Oregon* when she made her famous 14,700 mile cruise from San Francisco around South America to Cuba in the record time of sixty-six days.[45] Other chaplains who saw sea duty during the war were McAlister on the *Brooklyn*, Isaacs on the *Massachu-*

[41] Chidwick, John P. S., *Remember the Maine*. This pamphlet was published by Harry T. Cook to whom the Right Reverend Monsignor Chidwick dictated his memories of the *Maine* disaster shortly before his death in 1935.

[42] Jones, *Battle of Santiago*, p. 11, refers to having been wounded 2 July 1898.

[43] Jones, *A Chaplain's Experience*, pp. 215–6.

[44] *Ibid.*, p. 222.

[45] Chaplain McIntire stated that he received $75 as his share of the prize money awarded the crew of the *Oregon*. (NRSO:5670–78.)

setts, Parks on the *Essex,* Reaney on the *Columbia,* Freeman on the *Baltimore,* and Frazier on the *Olympia.*

Frazier and Freeman served under Commodore George Dewey at the Battle of Manila Bay. Chaplain Edel wrote of Frazier:

> Chaplain Frazier had been appointed in 1895 and was Dewey's Chaplain at Manila Bay in 1898. During the battle the young Chaplain thrust his head up out of a hatch on the OLYMPIA, was seen by the Commodore and called to the bridge. For part of the battle he stood at Dewey's side on the bridge, and he stood by Dewey's side in friendship for the years that followed, and in 1917 when Dewey's body lay in state in the rotunda of the Capitol at Washington it was in response to the Admiral's own dying request that Chaplain Frazier conducted the final rites.[46]

There were twenty-three chaplains on the active list on 1 January 1898.[47] Four new chaplains were appointed in 1898 to maintain the quota of twenty-four. Even though the country was involved in war, the Government made no move to increase the allotted number of chaplains.

With the outbreak of hostilities, patriotic clergymen of all of the larger denominations offered their services to the Government as naval chaplains. Some seventy applications, from twenty-four different states and the District of Columbia, were received by the Navy Department. The average age of the applicants was thirty-one. Many of those seeking appointments sent letters of endorsement from senators, congressmen, and others prominent in political affairs.

A Presbyterian missionary from San Louis Potosi, Mexico, volunteered to serve as a chaplain. "It might be that my knowledge of the Spanish language," he wrote, "would be of use in the coming conflict." A Christian Science reader of Dallas, Texas, offered his services. No chaplain from this denomination, however, was commissioned in the Navy until 1918. That nearly twenty times as many applications for commissions were received as could be granted reflected the patriotic spirit of the clergy.

[46] Edel, "Navy Chaplains from 1775 to 1917," pp. 24–5.
[47] *Navy Register,* 1 Jan. 1898.

IN CONCLUSION

The years under review saw an increase in naval personnel brought about by the building of the "New Navy." While naval personnel doubled in these years, the number of chaplains remained unchanged. Twenty-three chaplains were on duty in 1881; twenty-four carried the work of the Corps over into the new century. Among these twenty-four were some of the best loved men who ever wore the chaplain's insignia on the naval uniform.

During the period, Navy chaplains were still more closely integrated into the life of the Navy. They were given actual rank; their duties were better defined. In addition to such recognized duties as conducting Divine Services and religious instruction, establishing and maintaining libraries, ministering to the sick and imprisoned, and supervising educational activities of members of the crew, chaplains were also taking a leading part in providing wholesome entertainment aboard ship, conducting sightseeing tours on shore, and promoting other activities designed to interest the sailor in his off-duty hours. Chaplains were among the first to recognize the fact that good morals contribute to high morale.

The morale of the chaplains themselves, however, was undermined by continued discrimination in both congressional legislation and Navy regulations. Deferentially worded petitions, requesting such reforms as prerogatives of rank and increase in pay, were submitted to the Secretary of the Navy with little success. There was no organized Chaplain Corps to build up an *esprit de corps;* no Director or Chief in Washington to intercede in their behalf; no medium through which they could exchange helpful ideas and methods of work; and little if any interest in their work on the part of the denominations from which they came. With conditions so discouraging, it is understandable that several able chaplains resigned their commissions. The marvel is that so many good and capable men remained to carry on the work.

CHAPTER TEN

PRELUDE TO WORLD WAR I

1901–1916

After all previous wars, the United States Navy had experienced a decline in men and ships. This was not true after the Spanish-American War. The events of this conflict made the Navy popular and the American public navy-minded. The acquisition of new possessions in the Far East and in the West Indies called for adequate naval protection and the canal at Panama which was soon under construction presented new strategic problems. In Theodore Roosevelt, who became President in 1901, the Navy found a warm supporter whose policies demanded a strong modern Navy. The new President was a disciple of Admiral Mahan, the leading authority on sea power, who taught that capital ships were the strength of a fighting force. It was but natural then that the United States Navy experienced a vigorous expansion and that it devoted its main attention to battleships which were showing the most marked development in the early years of the nineteenth century.[1]

AGITATION FOR REFORM

When the twentieth century opened, in addition to the inequalities of pay, Navy chaplains were dissatisfied with the arbitrary quota and certain uniform regulations. They had expected much of the Act of 1899, which had placed chaplains on the same basis as other staff officers. The law, however, in its practical application gave chaplains actual rank with their brother staff officers but kept them on the discriminatory pay scale adopted in 1870.

Another provision of this Act which contributed to the discontent among chaplains was the provision in section 13 which stated that

> all officers, including warrant officers, who have been or will be appointed to the Navy from civil life shall, on the date of appointment, be credited, for computing their pay, with five years service.

The intent of this provision was to give men who received their professional training at their own expense equality to some extent with those officers who were trained at Government expense at Annapolis. Since no exception was made in this section for chaplains, the newest appointee to the Chaplain Corps did not wait five years for the first advance in pay but started with exactly the same pay as the oldest chaplain in the service. In other words, while the new chaplain had the actual rank of lieutenant, he received a lieutenant commander's base pay. Although this was quite acceptable to new chaplains, it made the older ones feel that the Government had failed to recognize long and faithful service.

A full realization of the discriminatory sections of the Act as they affected the pay of the older chaplains was not appreciated at first. A year or so elapsed before active agitation began for modification of these provisions. Civilian clergymen took up the cause of the Navy chaplains, who by regulation were not permitted to engage in activities which were intended to influence legislation.

The most outspoken civilian friend of the Chaplain Corps was the Reverend G. E. Strobridge, the author of a pamphlet which put the chaplains' case before the public. Strobridge began with the statement that "the chaplains of the United States Navy are the victims of invidious discrimination as compared with the other officers of the service." Concerning the inequality in pay, Strobridge wrote:

> Out of this, insignificant and jealously curtailed salary, the chaplain must pay an average of \$30 per month mess-bill, support the establishment of his family on shore, and meet the expensive outlay necessary to his station on shipboard. This latter item is no fiction. The officers, as a rule live high, and although the chaplain may be quiet in his tastes and moderate in his habits, he is constantly taxed to contribute his share to meet the demands of his extravagant environment.
>
> . . . . .
>
> In the Personnel Bill, passed at the last session of Congress, while every branch of the naval service was

[1] Clark and others, *A Short History of the United States Navy*, pp. 477–8; Sprout, H. and M., *The Rise of American Naval Power*, pp. 251ff.

advanced in pay, the salary of the chaplains was not increased. The omission is specially and painfully conspicuous, not only because such an increase would have been eminently just, but because this discrimination against the chaplains is despicable and degrading. . . . One can readily see how this neglect puts the chaplains in a pillory of mortification before their fellow-officers and even before the men.

Pertinent comments on pay were also contributed by an anonymous twenty-four page pamphlet attributed to Chaplain Tribou:

> Chaplains are the only sea-going officers who have no allowances for quarters when on shore duty.
>
> Except at the Naval Academy, Chaplains are not provided with quarters, while other officers of rank, both line and staff, are furnished houses easily worth a rental of $1000 a year.
>
> Chaplains are the lowest paid of all commissioned officers of the Navy; and . . . the pay of a warrant officer, in certain grades, exceeds that of a Chaplain.
>
> Every officer and every enlisted man is given longevity pay but chaplains are not.

This writer, too, claimed that such "invidious discrimination" put the chaplain "at a disadvantage before the men to whom he is sent to minister."

THE CHAPLAINS AND SECRETARY LONG

Another source of information regarding the feelings and attitudes of chaplains toward pay is found in several House and Senate documents of the second session of the 57th Congress. The most informative of these, *Senate Document* No. 138, contained "all correspondence relating to the pay and status of naval chaplains between the secretary and naval chaplains from November 1, 1901, to the present time [1903]." In brief the story is as follows:

Chaplain W. G. Cassard wrote, on 16 December 1901, from the Naval Training Station at Newport, Rhode Island, respectfully requesting permission to submit suggestions to the Secretary of the Navy for legislation in behalf of Navy chaplains. Admiral A. S. Crowninshield, Chief of the Bureau of Navigation, replied on 23 December granting permission. Cassard submitted his proposals on 4 January 1902 and stated that he was voicing the sentiment of twelve chaplains. He addressed his appeal to the Secretary, he said, because he believed that the discrimination against chaplains was neither deserved nor intended, "but has come to pass because the corps has always been small and never self-assertive and has simply been overlooked in legislation in behalf of the personnel of the Navy."

Most of the Secretary's 19 January reply was taken up with a defense of the old pay scale. The following are extracts from his letter:

> As to pay, there are many points in that regard which have escaped your attention. The question of pay must depend somewhat upon the usual consideration of supply and demand, and upon the rates paid for like service outside of the Navy.
>
> With regard to chaplains, for instance, I should be glad to have you advise me what is the average pay of clergymen at large throughout the United States, and also what was the highest salary paid any present naval chaplain prior to his appointment. The pay of a chaplain is now, as you say, $2,300 on shore and $2,800 at sea after his first five years. Am I wrong in assuming that this is two or three times the average salary of a clergyman at large?

Regarding the discrimination in longevity pay, the Secretary claimed that the services of other staff officers increased in value to the Navy with years of experience "while the duties of a chaplain are the same upon his entrance into the service as they are afterwards." The Secretary added:

> I regret, therefore, that you should speak of "the present pay table as a reflection" upon naval chaplains and "a blow at their self-respect." It seems to me that this is a frame of mind which is not to be commended. The character of the chaplain and his self-respect certainly do not depend upon the question whether his salary is somewhat more or less than that of other officers. . . .

Long pointed out that there could not be any reflection upon the self-respect of the chaplains, since the Department was overwhelmed with applications for appointments to the Corps while the quota for the Medical Corps had not been filled for years.

Cassard sent a long reply on 21 February in which he gave rebuttal arguments and submitted additional evidence to strengthen former statements. His letter contained this information:

> In conclusion, sir, I should be untrue to myself if I should fail to tell you that I was deeply hurt to see your answer to my letter published in numerous daily papers and also in the Army and Navy Gazette. In the Baltimore Sun, the letter was published under these headlines: "Rebukes a chaplain—Secretary Long writes minister a stinging letter—He complains about pay."
>
> The publication of your letter with the construction placed upon it by the papers, without any word to represent the position of the chaplains, is calculated to estrange interest from our position and to make impossible any favorable action in Congress. . . .

THE NEWSPAPER CAMPAIGN FOR PAY REFORM

The pamphlet by Strobridge, the booklet credited to Tribou, the release for publication of Long's letter to Cassard, and the discussion of this letter on the floor of the Senate together with its appearance as a Senate document, all combined to furnish interested editors with verbal ammunition for a barrage of explosive editorials. Among the many papers that published editorials favorable to the chaplains were: *Chicago Advance,* 30 October 1902; *Interior,* Chicago, 11 December (written by Dr. Henry van Dyke of Princeton); *Christian Intelligencer,* 25 December: *The Churchman,* 27 December, 10 January 1903, 31 January; *Christian Register,* 1 January (written by Edward Everett Hale); *New York Examiner,* 8 January; *Boston Herald,* 10 January; 11 January; *New York Sun,* 10 January; *Army and Navy Register,* 10 January; *New York Christian Advocate,* 15 January; and *The Christian Advocate,* 29 January. Such a sudden outburst of editorial opinion in so many papers scattered throughout the country could not have been accidental. This is evidence of well laid plans and weeks of careful preparation.

These editorials were collected and published by authority of the United States Senate in its *Document No. 129,* second session, 57th Congress, under date of 2 February 1903.

The Personnel Bill of 1899, with its discriminatory pay scale for chaplains, was denounced in these editorials as being "unjust," "a cruel wrong," "a stigma upon the chaplain corps," and a "brand of dishonor." The influential *Army and Navy Register* stated:

> The day is long since past when a minister of the gospel, whether in or out of the service of his country, can be treated with conspicuous injustice without invoking the vigorous protest of all patriotic American citizens.
>
> . . . .
>
> The wholly gratuitous statement that has been presented by a few men of contracted intellectual horizon . . . that the national government should enter the theological labor market and bid in its chaplain at the lowest merchantable rate, as it would buy its coal or its pork, is an argument that perhaps might appeal to the strictly utilitarian mind, but will never commend itself or its wisdom in our day to the conservative judgment of the American people. That was an argument that was not applied to Navy paymasters or surgeons, and nothing short of the most cruel injustice can by any possibility make it apply to navy chaplains.

An editorial in *The Churchman* was contributed by Dr. George Williamson Smith, President of Trinity College, Hartford, who had served nearly twelve years as a Navy chaplain. He wrote:

> The injustice of this discrimination against the representatives of the Christian Church in this branch of the national service is manifest . . . .
>
> To brand his office and work as inferior in comparison with that of the others, is to put him at a disadvantage and render him a nonentity or an object of contempt to the carelessly minded.

THE CHAPLAINS ARGUE THEIR CASE

It appears that this outburst of public opinion had an immediate effect upon the Committee of Naval Affairs of the House, for a meeting was arranged with representative naval chaplains on 12 January 1903. For years the majority sentiment in the Senate had been favorable to naval chaplains. The House, on the other hand, contained influential members who blocked proposed changes. For the first time, Navy chaplains had the opportunity to present directly and officially their point of view to members of Congress. Chaplains Hoes, Cassard, and Clark were called to testify.

The report of the committee meeting appeared as *House Document 42,* second session, 57th Congress. Although matters of pay were discussed at length, the chaplains were insistent that the real issues troubling them went much deeper. Hoes declared:

> We feel most deeply the humiliation involved in the fact that, when the new awards were accorded by Congress after the Spanish war, we were the only seagoing officers in the whole Navy who were excluded. We can never understand why we, as naval officers who are subjected to the same limitations and discouragements and deprivations as all other seagoing officers, should be placed in a position of relative inferiority which not only affects our self-respect, but also curtails our influence and practically casts a reflection upon our labors as representatives of the Christian Church.

Chaplains wanted to be treated on the basis of equality with other staff and line officers. "The matter of pay has been forced to the front," said Hoes, "perhaps logically, but we look upon that as a subordinate matter, and as merely incidental to the one essential point at issue, namely, that, with all deference and respect, we ask for fair play and no discrimination."

The pay scale was discussed in committee. Chaplains asked that they be paid according to the rank held with the same allowance as other officers. The adoption of this suggestion would have greatly increased the pay of the older chaplains. Chaplain Cassard felt that new appointees to the Corps should be commissioned as lieutenants. If this were done, newly appointed chaplains would receive less under the new scale than under the old. He also suggested that in the period of readjustment no chaplain would have his pay reduced but that the new scale would apply only to appointments made after the proposed changes became law.

In the course of the discussion the advisability of chaplains having rank was raised. One of the committee members asked the chaplains whether it might not be well to abolish the chaplaincy altogether and to send civilian clergymen aboard ship to minister to the spiritual needs of the men. "Do you not believe," he asked, "that if you had no rank at all . . . you would command universal respect, and command the love and respect of the naval officers, more than if you attempted to divide with them the mere earthly honors?"

Chaplain Hoes replied: "Under existing circumstances, no sir." And Chaplain Clark added that

> on board ship rank simplifies matters . . . the fact that naval chaplains have rank settles matters that come up necessarily much more than any other method could settle them. For instance, if we did not have rank, there would have to be whole pages of regulations defining our position.

In spite of all the agitation, the session adjourned without approving the bill.

NEW LEGISLATION GIVES INCREASED PAY

On 28 June 1906, Congress passed an Act which contained the following provision:

> That all chaplains now in the Navy above the grade of lieutenant shall receive the pay and allowances of lieutenant-commander in the Navy according to length of service under the provisions of law for that rank, and all chaplains now in the Navy in the grade of lieutenant shall receive their present sea pay when on shore duty: *Provided,* That naval chaplains hereafter appointed shall have the rank, pay, and allowances of lieutenant (junior grade) in the Navy until they shall have completed seven years of service, when they shall have the rank, pay, and allowances of lieutenants in the Navy; and lieutenants shall be promoted, whenever vacancies occur, to the grade of lieutenant-commander, which shall consist of five numbers, and when so promoted shall receive the rank, pay, and allowances of lieutenant commander in the Navy: *Provided* further, That nothing herein contained shall be held or construed to increase the number of chaplains as now authorized by law or to reduce the rank or pay of any now serving.[2]

This law carried two important provisions for chaplains. In the first place, while it did not place the older chaplains on the same pay scale with other naval officers, it did grant a welcomed financial increase. The words "and allowances" meant that the chaplains who qualified would receive not only the base pay of a lieutenant commander, but, also the longevity pay, rental allowance, and other financial benefits that went with the rank. This also became the maximum pay of commanders and captains in the Chaplain Corps.

The new law corrected the unforeseen application of section 13 of the Act of 1899 which credited newly commissioned chaplains, who entered with the rank of lieutenant, with five years service, thus, giving them the pay of a lieutenant commander. Chaplain J. F. McGinty, commissioned on 16 January 1905, was the last to enter with this status.

The second important provision of this law authorized, for the first time in the history of the Navy, the commissioning of chaplains with the rank of lieutenant (junior grade). The first chaplain to enter the Corps with this rank was George E. T. Stevenson, who received his commission 26 April 1907.

The next law which affected the pay of naval officers was the Act of 13 May 1908 which fixed Navy pay at essentially its present (1946) base rates. The annual base pay of ensigns was $1,700; lieutenant (junior grade), $2,000; lieutenant, $2,400; lieutenant commander, $3,000; commander, $3,500; and captain, $4,000. A longevity increase of ten percent of this base pay was granted for each additional five years service up to forty percent. In other words, four increases or "fogies" were authorized. The law also allowed an increase of ten percent to all officers on sea duty or on shore duty beyond the continental limits of the United States. To the general provisions of the law was added the single exception: *"Provided* . . . that the pay and allowances of chaplains in the Navy shall in no case exceed that provided for lieutenant-commanders."[3]

Finally, on 29 August 1916, Congress erased all limitations on the pay of Navy chaplains and granted them full equality with other naval officers. The law passed on that date stated that hereafter "all commissioned officers of the active list of the Navy shall receive the same pay and allowances according to rank and length of service."[4] The law also changed the age of retirement from sixty-two to sixty-four. By the provisions of this Act, all chaplains who held rank

[2] 34 Stat. 554-50. The clause "which consists of five numbers" refers to the Navy custom of printing a signal number opposite every officer's name in the *Navy Register.* This custom began with the 1898 edition. The reference in the law of 1906 to the "five numbers" meant that only five chaplains could hold the rank of lieutenant commander. The law made no reference to the quota of the higher ranks so the limitations set by the Act of 1899 of five captains and seven commanders remained. The signal number is not to be confused with the file number (sometimes erroneously called serial number) given every officer in the Navy. This number appears on the officer's jacket on file in the Navy Department and is stamped on his "dog tag."

[3] 35 *Stat.* 128.

[4] *Ibid.*

above that of lieutenant commander received the full pay of their respective ranks. This reform removed one of the main causes for discontent in the Chaplain Corps. Thus, the way was further prepared for the high morale and splendid efficiency demonstrated by the Chaplain Corps during the First World War.

THE DEMAND FOR MORE CHAPLAINS

It is well to consider at this point, how the expansion of the Navy which took place during Theodore Roosevelt's administration affected the Chaplain Corps. Between 1900 and 1909, the enlisted personnel had increased from 25,000 to 45,000. As there was no corresponding increase in the number of chaplains, it was obvious that the Corps was totally unable to meet the new opportunities and responsibilities. In 1901, even before Roosevelt's expansion program began, all twenty-four chaplains had important assignments. Chaplains Hoes on the *Kearsage* and Wright on the *Massachusetts* were with the North Atlantic Squadron; Dickins on the *Chicago* with the South Atlantic; Brown on the *Iowa* with the Pacific; and Frank Thompson on the *Brooklyn* and Rennolds on the *Newark* with the Asiatic. On ships in the training service were Frazier, Steele, Edmonson, Sykes, Harry Jones, Gill, Reaney, and Helms. Several were on duty at Navy Yards: Tribou at Boston, McAlister at Mare Island, Holway at New York, Morrison at Philadelphia, Isaacs at San Francisco, and Cassard at Newport. Chidwick was attached to the receiving ship *Vermont;* Clark to the Naval Academy, and Royce to the Naval Home (formerly called Naval Asylum) in Philadelphia.

There is evidence that even this distribution of chaplains was inadequate, for in a pamphlet by Strobridge, published around the turn of the century under the title *The Chaplains in the United States Navy, Their Unfair Treatment,* appeared this statement: ". . . at present there are several of our larger ships at sea without chaplains, although one of our leading admirals says this never should be so, and some of our shore stations are also without any."[5]

Another pamphlet, appearing anonymously about the same time, shows the need for more chaplains:

> The Navy Register for Jan'y 1902 shows *twenty-six* ships needing Chaplains, only twelve of which are supplied. There are also *eighteen* shore stations requiring Chaplains.
>
> *At least FORTY-FOUR Chaplains are needed to meet the actual requirements of the service.*
>
> Of all the sea-going officers of the Navy the Chaplains' Corps alone shows no increase in point of numbers for *sixty years.*[6]

In December 1907, when sixteen first-line battleships of the Atlantic Fleet left Hampton Roads for a cruise around the world, only five of the ships had a chaplain aboard. These were: the *Connecticut,* Gleeson; the *Georgia,* Charlton; the *Minnesota,* Evans; the *Rhode Island,* Fleming; and the *Virginia,* Stevenson.[7] A number of other vessels accompanied the battleships, including some units of the Pacific Fleet, bringing the total up to forty-two ships. There simply were not enough chaplains to fill all the billets ashore and afloat.

Throughout the first decade of the twentieth century, Congress refused to approve any of the bills introduced into both Houses which called for an increase in the chaplain quota. *Senate Document* No. 138, printed by order of a 23 January 1903 Senate resolution, recorded correspondence between Secretary of the Navy Long and Chaplain Cassard. The Chaplain proposed an increase in the quota to thirty-four, and Long concurred, stating that he had recommended an increase to Congress. No action resulted, however.

Around 1910, the churches of the United States began to take an active interest in both Army and Navy chaplains. The organization of the Federal Council of the Churches of Christ in America, in 1908, brought a new and a powerful ally to the side of the chaplains.

The House of Bishops of the Protestant Episcopal Church sent a memorial on 30 January 1910 to the Secretary of the Navy requesting a quota increase. On 28 February, the Right Reverend Alfred Harding, Bishop of Washington, followed the memorial with a letter to the Secretary in which he pointed out that, in 1842 the ratio of chaplains to naval personnel had been one to 508, while in 1910 the ratio was one to 2,251. The Bishop, stating that he believed he was speaking "in behalf of the whole religious world of the United States in urging upon Congress the

[5] The Reverend G. E. Strobridge, author, first read the material as a paper before the New York Methodist Preachers Association, whose members authorized its publication and saw that copies were sent to the President of the United States, the Secretary of the Navy, and members of Congress, as well as to individuals. It seems reasonable to believe that Chaplain W. O. Holway, a Methodist, then on duty at Brooklyn Navy Yard, could and may have supplied both facts and arguments to Dr. Strobridge. Copy of pamphlet may be found in Nav. Rec. Coll., Chaplains' File. No date of publication is indicated. Possibly, 1902.

[6] Evidence points to Chaplain Tribou as being the writer, although he could not openly acknowledge authorship.

[7] *Navy Register,* 1908.

necessity of an increase in the number of naval chaplains,"[8] wrote:

It must be evident that the Provision made by Congress for our Naval forces is wholly and grotesquely inadequate to administer to the present Naval Force, including the ten thousand (10,000) men in the Marine Corps, in upwards of three hundred ships of all classes, nine constructive and repairing centers, and thirteen Naval Stations, of which seven are outside the boundaries of the North American Continent.

In July 1912, Admiral Charles H. Stockton (Retired) advocated an enlarged Chaplain Corps through the columns of *The Churchman*. He wrote:

It is to be deplored that in the expansion of late years in the United States Navy, matters of religious instruction and supervision have almost suffered by neglect. . . . The performance of the religious service on board any ships by officers as laymen is exceptional, and the consequence is that Sundays on board a majority of our vessels in commission consists of ceremonial formations in the early part of the day, and idleness or a day for sports on shore in port in the later part of the day . . . so that the old "Articles of War" carry an unfulfilled promise when they state that each commanding officer is required by law to hold divine service every Sunday morning on board his ship whenever the weather and other circumstances will permit.[9]

Admiral Stockton testified that the character and ability of chaplains had improved with the elimination of political influence in appointments and the requirement of ecclesiastical approval.

No chaplain was more indefatigable in his efforts to improve the status of the naval chaplaincy than was Chaplain G. Livingston Bayard. After serving in the Army during the Spanish-American War, Chaplain Bayard accepted a commission in the Navy in 1902. He had the unusual distinction of serving three tours of duty at the Washington Navy Yard: 1907–1908; 1909–April 1914; and October 1914–1917. His residence in Washington and his personal friendship with influential government officials gave him an entree to inner circles which no other chaplain enjoyed. In season and out, Chaplain Bayard pleaded the cause of the Navy chaplain.

The inadequacy of the Corps to meet the needs of the growing Navy is clearly revealed by the following data based on records in the *Navy Registers* of 1913 and 1915:

[8] NRSO: 1572–5.

[9] *Op. cit.*, 20 July 1912, p. 85.

SHIPS	1 Jan. 1913		1 Jan. 1915	
Type:	With Chaplains	Without Chaplains	With Chaplains	Without Chaplains
Battleships, first line	4	10	6	4
Battleships, second line	..	19	7	16
Armored Cruisers	2	8	3	7
Cruisers, first class	1	4	1	3
Gunboat (*Helena*)	..	1	1	..
ASIATIC STATION				
(Ashore)	2		..	
Chaplains at sea or overseas	9		19	
U. S. SHORE STATIONS				
Naval Academy	1		1	
Naval Training Stations:				
Great Lakes	2		1	
Newport	2		1	
San Francisco	1			
Navy Yards:				
Boston	1			
Charleston (then Nav. Sta.)	1			
Mare Island	2		2	
New York	1			
Norfolk	1			
Philadelphia (& Naval Home)				
Portsmouth, N. H.	..			
Puget Sound	..			
Washington, D. C.	1			
Awaiting Orders	1			
Chaplains ashore	14		5	
Total Chaplains	23		24	

WELFARE SECRETARIES ABOARD SHIPS

A complicating factor in the endeavor to increase the allotment of naval chaplains was the introduction of the YMCA or welfare secretaries aboard ships in 1909. The need for chaplains aboard ships and on foreign stations became so acute that the Navy authorized the Army and Navy Department of the YMCA to place a secretary aboard each ship not carrying a chaplain.[10] In 1909, Secretary G. A. Reader of the Y served four months aboard the *Nebraska*. The 1910–1911 *Year Book of the YMCA* reported:

> Following up the special work in the *USS Nebraska*, a shipboard secretary accompanied the *USS Delaware* on its winter cruise to European waters; the commanding officer together with the officers and men were enthusiastic over the results accomplished. During the stay of the Atlantic fleet for target practice at Guantanamo Bay, Cuba, two secretaries conducted a splendid work for the sailors while they were in camp.[11]

YMCA secretaries had served in 1911–1912 with the Atlantic Fleet on the *Delaware* for three months and on the *Kansas* for six weeks. In the Asiatic Fleet, a secretary spent six months on the *Saratoga* and two months on the *Rainbow*.[12] The statistical report for the year indicates that these secretaries conducted fifty-three religious services, all but three aboard ship, with a total attendance of 2,565. They also led fifty-five sessions of Bible classes, supervised libraries, entertainments, recreational activities, and distributed literature. In other words, they were doing what is ordinarily expected of chaplains except administering the sacraments. Even this was possible whenever a secretary was also an ordained man. The reports of the Y secretaries on this part-time duty with the Navy show that their combined services for the previous year about equaled the work of one Navy chaplain.

During 1912–1913, welfare secretaries lived aboard the following vessels for periods varying from two to five weeks each: *Wyoming, North Dakota, Vermont, North Carolina*, and *Prairie*.[13]

Josephus Daniels, who became Secretary of the Navy in 1913 in the cabinet of Woodrow Wilson, looked favorably upon the appointment of civilian welfare secretaries. Daniels, a member of the Methodist Episcopal Church, South, manifested great interest in everything pertaining to the spiritual and moral needs of naval personnel. This, naturally, led to his concern for the improvement of the Chaplain Corps. Never before did Navy Chaplains have such a friend in the Secretary's chair. At the beginning of his official duties as Secretary, Daniels felt that the strength and efficiency of the Chaplain Corps did not measure up to the demands of the Navy. Therefore, he continued for the time being the policy he inherited—that of supplementing their work with that of YMCA secretaries. This aroused suspicion on the part of certain chaplains who were naturally jealous for the prerogatives of the Corps. However, before Daniels left office in 1921, after serving through both of President Wilson's terms, the naval chaplains realized that a great debt of gratitude was due him for the many reforms he had instituted.

Daniels was keenly aware of the Navy's need for religious leaders. His *Annual Report* for 1913, from which the following extracts are taken, brought his recommendations to the attention of Congress:

> It is a reproach to our country that we have only the same number of chaplains in the Navy in 1913 as there were in 1842 . . . . Now the number of officers and enlisted men has grown to 3,600 officers and 61,000 men.
>
> There is need of an immediate increase in the number of men who are charged with the high duty of leading men afloat to a recognition of the truth that man's first and highest obligation is to his Maker. I earnestly recommend an immediate recognition of the need of more leaders in the higher life on board our ships. I recommend additional chaplains and an appropriation to enable the Secretary to employ on every ship that has no chaplain a young religious leader known as a welfare secretary. It has been urged by wise and patriotic leaders that there should be a chaplain for every 1,000 men in the service. The country, will I am sure, warmly approve the suggestion that not less than 60 chaplains or welfare secretaries should be authorized.[14]

Daniels also recommended that a law be passed permitting the appointment of acting chaplains for a three-year probationary period. "The qualities that make a minister successful in a pastorate ashore," wrote the Secretary, "are not always those which equip him for the special sort of service required of a chaplain in the Navy." He felt that, after the acting chaplain demonstrated his ability to minister successfully to sailors during the probationary period, he could then be commissioned.

Regarding the YMCA secretaries, Daniels recommended that such men work under the direction of chaplains. From this report, it is evident that Daniels felt that most chaplains then in service were either

[10] *Annual Survey, YMCA*, 1910, p. 69.
[11] *Op. cit.*, p. 21.
[12] *Year Book of the YMCA*, 1911–12, p. 291.
[13] *Ibid.*, 1912–13, p. 269.
[14] *Op. cit.*, pp. 20–1.

too old or were not trained to take active leadership in welfare activities. Daniels explained:

> The chaplains do a noble work and no commendation could be too high for the deserts of naval chaplains. But, just as in all churches, organizations of societies for young people has been found necessary to draw out all that is best in youths, so on shipboard there is felt the need for a leader, young and enthusiastic, who will enter into the feelings and ambitions of the young men, and direct them in their athletics and entertainments as well as in their religious life. Young Men's Christian Associations have proven the best shore stations for young men in many communities. Other organizations of young men in Catholic and Protestant churches have been helpful influences for giving the right direction to young men looking for exercise and clean living and clean thinking. The Navy should be up with the age. In addition to our chaplains and the good work they accomplish, the best thing that can be done for the young men on shipboard is for Congress to authorize the employment of a welfare secretary on board every ship in the Navy, at naval stations and at other places where young men are trained or enlisted, in addition to the chaplains. There is an indefinable sympathy and interchange between young men that makes it wise to call young men to lead in the forward religious movement on shipboard. The trial of a Y.M.C.A. secretary on a few ships as leader in religious thought, Bible study, athletics, and entertainment, was so successful as to make it certain that enthusiastic young religious leaders are needed on every ship. These consecrated and talented leaders must have the enthusiasm of youth and must themselves love the pleasures that appeal to youth.

Daniels further recommended that such welfare secretaries be given no permanent position and that their tenure of office "be measured by their ability to gain the confidence and cooperation of the young officers and enlisted men."

The attitude of Secretary Daniels regarding welfare secretaries was not shared by chaplains or by members of the Commission on Army and Navy Chaplains of The Federal Council of Churches of Christ in America. A hearing on the Naval Personnel Bill, which included the reforms so ardently desired by the chaplains, was held by the House Naval Affairs Committee on 20 January 1914. The Federal Council, the Protestant Episcopal Church, and the Roman Catholic Church were all represented. All the clergy present were agreed on the following three points: "(1) An increase of chaplains in the Navy so that there shall be one chaplain to every 1,000 of the force; (2) Establishment of the grade of acting chaplain; and (3) Removal of discrimination against chaplains in the matter of salary."[15] According to the Federal Council report: "The Committee did not view favorably the substitution of lay workers for ordained chaplains."[16]

A matter closely related to the presence of welfare secretaries aboard ships was that of having individual civilians or groups from churches go aboard ships for the purpose of conducting Divines Services or other religious activities. During the Spanish-American War, a group of zealous church folk at San Diego, California, who had organized the Floating Society of Christian Endeavor, obtained from Secretary Long permission to hold services aboard war vessels in that harbor.[17] There can be no doubt that many of these individuals and societies conducted their missionary work at sacrifice to themselves and with good results. They carried a religious message and distributed religious literature to naval personnel who otherwise might never have received it.

On the other hand, the insistence of some of these civilian groups to carry on work on vessels where chaplains were assigned brought difficulties. Chaplain E. W. Scott wrote to Chaplain Frazier on 20 December 1919: "I found the other chaplains even more firmly convinced than I that we did not want the Floating Society of Christian Endeavor to come on board to hold services, judging by the experiences some of our ships have had."[18] Civilian groups sometimes included good people who, though sincere, raised problems for the chaplain. It was natural for the chaplain to look upon his ship as his parish and to feel a sense of responsibility about opening the doors to individuals or groups over which he had little control.

THE NEW QUOTA AUTHORIZED

The House Naval Affairs Committee reported in January 1914 in favor of having one chaplain for every 1,250 naval personnel and for the establishment of the grade of acting chaplain, but, made no recommendation regarding equalizing the salaries of the older chaplains with other officers of like rank.

Opponents of the bill in the House, however, before the measure was sent to the Senate, succeeded in striking out all sections dealing with chaplains. The

[15] *Federal Council, Biennial*, 1914, p. 71. Inadequate pay was another point of chaplain dissatisfaction; agitation for this reform was contemporaneous with the demand for more chaplains.

[16] *Ibid.*, p. 72. The Army ratio, fixed in 1918, is one chaplain for 1,200 personnel.

[17] NRSO:7627.

[18] CoC., Scott File.

Senate Committee on Naval Affairs voted to restore the sections and in this the Senate agreed. The bill was sent back to the House twice for conference before an agreement was reached. Finally, on 30 June 1914, the sections concerning chaplains were officially accepted as a part of the Naval Appropriation Act.

Through all the varying fortunes of the bill, Chaplain Bayard and the representatives of the various denominations redoubled their efforts to secure its final passage. Much credit for the enactment of this Act must be given to the Federal Council of Churches of Christ in America. Its *Annual Report* for 1914, which also paid tribute to Chaplain Bayard for his ceaseless efforts in behalf of the bill, summarized the efforts made by this body to create the necessary public opinion which finally moved Congress to act:

> Various sources of influence were sought in securing favorable action, for the opposition, though quiet was widespread — in letters from prominent ministers and laymen to Senators and Representatives; resolutions by general ecclesiastical bodies, such as the Presbyterian and Reformed General Assemblies and the General Conference of the Methodist Episcopal Church, South; personal conferences with Senators and Representatives. The office of the Federal Council sent out several thousand letters to influential men and, thus, secured important endorsement. Also articles in the religious press, furnished from the Washington office, editorially supported, and favorable paragraphs in the daily press helped greatly to carry the legislation through.[19]

No law ever passed by Congress, following the establishment of the naval chaplaincy, has had such far-reaching consequences for the Corps itself and for the spiritual welfare of naval personnel as this Act of 1914. Section 2541–a, which authorized the grade of acting chaplain for all new appointments, will be considered subsequently.

Section 2541–b authorized: "Hereafter the total number of chaplains and acting chaplains in the Navy shall be one to each twelve hundred and fifty of the total personnel of the Navy and Marine Corps as fixed by law, including midshipmen, apprentice seamen, and naval prisoners. . . ." Thus, the limit of twenty-four chaplains in effect for seventy-two years was repealed. The establishment of a ratio instead of a fixed number eliminated the possibility of the Navy expanding beyond the ability of the authorized complement of chaplains to meet the need. Chaplain Bayard, who was responsible for the wording of the sections concerning the chaplains, deliberately included "midshipmen, apprentice seamen, and naval prisoners" as a part of the total naval personnel to be figured when computing the possible number of chaplains to be commissioned.

ENTER THE ACTING CHAPLAINS

The recommendations of Daniels and others regarding appointments to the naval chaplaincy were embodied in the new law of 1914. Under the provisions of section 254–a, chaplains were appointed by the Secretary of the Navy as acting chaplains for a three year probationary period. During this time of trial, an acting chaplain received the rank, pay, and allowances of a lieutenant (junior grade). At the end of the probationary period, which was to be spent at sea, he was examined for "physical, mental, moral and professional fitness" by a board of chaplains and doctors, and if found qualified was commissioned a lieutenant (junior grade).

One of the first results of the establishment of the grade of acting chaplain was that younger men entered the chaplaincy. The average age of chaplains commissioned during the years 1901–1910 inclusive, was thirty-three, whereas the average age of the eleven appointed acting chaplains in 1915 was twenty-nine, and twenty-eight for the seven who entered in 1916.

The last chaplain to come in under the old limit of twenty-four was John Joseph Brady, who was commissioned 11 June 1914 and who was retired 1 November 1934. The first to enter the Navy as an acting chaplain was Irenee Joseph Bouffard, a Roman Catholic, who was appointed 4 February 1915. Eleven others were appointed in 1915, of which number Robert D. Workman (the second acting chaplain and the sixth Chief of Chaplains), Charles V. Ellis, and Herbert Dumstrey were still on the active list in 1946.

Although the Act of 1914 provided for the appointment of acting chaplains and an increase of the total number of chaplains on duty, welfare workers were permitted to work aboard ships at sea for at least two years after the Act was passed. Shipboard work had been continued during 1914–1915 in both the Asiatic and the Atlantic Fleets, with the *Delaware* "officially assigned as headquarters for shipboard work with the Atlantic Fleet."[20] The following secretaries spent a part of the same years aboard the vessels indicated: H. G. Mann, *Texas;* W. E. Herr, *Delaware;* and W. W. Elder, Atlantic Fleet. In addition, one secretary spent six months with the Marines

[19] *Op. cit.*, p. 72.

[20] *Annual Report of the YMCA*, 1914–15, p. 347.

at Vera Cruz, Mexico.[21] The *Annual Report* for 1915–1916 indicated that Herr had spent a full year aboard the *Delaware* and that C. O. Smith had spent three months on the *New Jersey*.

Gradually, however, the number of newly appointed chaplains increased to meet the need. During 1916, seven more acting chaplains were appointed, including Chaplains Lewis and Elder, who were still on the active list in 1946. Chaplain Elder had previously served as a welfare secretary under the YMCA. The accessions of 1915 and 1916 brought the total strength of the Chaplain Corps up to forty on the eve of the First World War.

THE APPOINTMENT AND PROMOTION OF CHAPLAINS

Among other reforms which were effected during these years under review was the establishment of a more careful system of selecting men for the chaplaincy. The old system of giving weight to political considerations had brought a few poorly qualified and unworthy men into the Corps. During the years 1898–1907, four chaplains either were court-martialed or resigned to avoid being tried. Never before or since, in such a short period of time, has so large a proportion of the Chaplain Corps been disciplined. These few focused attention on the necessity of greater care in the appointment of men for the chaplaincy.

No group was more interested in raising the standards of the Corps than were the chaplains themselves. On 20 October 1906, a Board of Chaplains consisting of Clark, Thompson, and Gleeson, appointed by the Secretary of the Navy to make recommendations for the improvement of the Chaplain Corps, made its report. The Board recommended that: there should be a Chief of Chaplains; the Corps be increased to forty; all newly commissioned chaplains be graduates of both college and seminary and that such should receive denominational endorsement; and all candidates appear before a Board of Navy surgeons and a Board of Navy chaplains for their endorsement as to health and other qualifications.[22]

The recommendation of a Chaplain Corps head was one the chaplains themselves repeatedly suggested. Strobridge, when he took up their cause, had enlarged on this point of the chaplains' dissatisfaction:

> A crying demand to-day is that a bureau for the chaplains be organized within the Navy Department. Every other corps is thus provided for. The chaplains have never had an official head from their own number, as is the case with all the other departments. For instance, at the head of the surgeons is a surgeon, at the head of the paymasters is a paymaster, at the head of the naval constructors is a naval constructor. But, over the chaplains is placed not a chaplain, but an admiral of the line — a perfect gentleman, of course, and a competent disciplinarian, but of necessity, out of touch with the chaplains in the tastes and feelings peculiar to their profession. Such a bureau, however, as is here advocated, with the senior chaplain at the head of it, would not only be a righteous, as well as congenial arrangement, but, also would give direction and force to the work of the chaplains.

Chaplain Bayard, too, urged the appointment of a Chief of Chaplains. In response to a request from the Secretary of the Navy for suggestions designed to attain the highest efficiency in the Corps, he wrote, under date of 27 May 1907:

> A wise and good Chaplain holding the confidence and respect of the Secretary of the Navy would understand the laws and limits binding him and adapt himself to existing conditions. Such a man would render valuable service in an advisory capacity both to the Department and to every man in the Corps of Chaplains; especially helpful would he be to the clergymen coming to the Navy direct from civil life with little or no conception of the customs and traditions of the Navy.[23]

A Chief of Chaplains would raise the efficiency of the Corps, not only by supervising the selection and training of chaplains, but, also by making recommendations for a more judicial assignment of chaplains to duty. For years, this important responsibility was handled by a woman clerk in the Bureau of Navigation. Frequently, one of the newest chaplains in the Corps was given duty on a battleship or some other choice billet while others much his senior in rank were given smaller ships or small shore stations.

Chaplain Bayard also recommended that full college and theological education be required of applicants, and felt that, in addition to the examination by a medical man, the candidate should undergo an examination by a "Board of Naval Chaplains, consisting of three members." The ideas in Bayard's paper received a hearty endorsement by the Chief of the Bureau of Navigation, who stated that he was "in favor of any improvement in the present method which will secure desirable and competent men for the corps of naval chaplains."[24]

The combined recommendations from the "Board of Chaplains," from individual chaplains, and from

[21] *Ibid.*, p. 223.
[22] NRSO: 5288.
[23] NRSO: 24620.
[24] *Ibid.*

representative clergymen finally brought about the desired reforms. Article 1606 of *Navy Regulations*, issued 2 January 1909, authorized for the first time a Board of Chaplains to pass on the fitness of all applicants for the chaplaincy. The new article provided:

> His moral character, general fitness, and experiences shall be established to the satisfaction of a board of chaplains, which shall conduct a written examination to determine his mental attainments.

The last of the "politically appointed" chaplains was Sidney Key Evans who was commissioned on 5 December 1907. The first to come before the Board of Chaplains for examination prior to being commissioned, were H. M. T. Pearce, commissioned 5 March 1909, and J. D. MacNair, commissioned 20 May 1909.

In 1905, the Catholic Archbishops of the United States designated the Reverend Alexander P. Doyle, C.S.P., to act as their representative with the Government in the appointment of Catholic chaplains. In those days, before the Military Ordinariate had been established, candidates for the chaplaincy were obliged to procure their faculties from the Bishop of their home diocese. Father Doyle served in his appointed liaison capacity until his death in 1912. He was followed by the Reverend Louis J. O'Hern, C. S. P.[25]

In addition to the steps taken by the Navy to insure a more careful selection of chaplains, individual denominations were also taking steps to examine their candidates. One of the first churches to act was the Methodist Episcopal Church, North. In October 1906, Bishop Earl Cranston of that denomination wrote to President Theodore Roosevelt stating that the Board of Bishops of his church had become concerned over certain unfortunate appointments to the chaplaincy in the Army. The Board appointed a committee, with Bishop Cranston of Washington as the resident member, to work out a method agreeable to the government for granting denominational approval to all Methodist candidates for either the Army or Navy chaplaincy. The Bishop wrote:

> We do not presume to interfere with the Departments, nor to monopolize the right of commendation, but simply offer, in behalf of our own church, *to look carefully into the antecedents, record of service, educational* and other *attainments,* and the *fitness in general* of such men of *our own ministry* as *may be referred* to us by those in authority, as candidates for appointment as chaplains in the army and navy. It is a duty we owe our country and ourselves to do this work if deemed by you *in any instance desirable* or necessary.

Bishop Cranston's letter was forwarded on 24 October to the Secretary of the Navy with a note by the President's Secretary: "The President assumes that the present proportion of Methodist chaplains in the navy will be substantially kept up on the ground that it roughly corresponds to the percentage of Methodists in our population." The President's Secretary added that the President looked with favor upon the suggestions made by Bishop Crandall and recommended that "no Methodist shall be appointed whose name has not been submitted to Bishop Cranston and is not, after proper inquiry, found to be satisfactory by Bishop Cranston's committee."[26]

Evidence of the desire of the Navy Department to keep a certain ratio among the chaplains in the matter of denominational affiliation is also found in a letter written by the Secretary of the Navy on 21 December 1911 to Senator Moses E. Clapp who had requested the appointment of another Roman Catholic.

> I find that the next regular vacancy in the corps of chaplains in the Navy in the Catholic faith will not occur until 1925. It is possible, however, that a vacancy may occur in that faith, either by resignation, or otherwise than by retirement, prior to 1925.[27]

Six of the twenty-four chaplains on duty in 1914 were Roman Catholic.

Even if the Navy had then desired the official endorsement of the other Protestant denominations for men making application for the chaplaincy, there was no ecclesiastical machinery to give such approval. The Protestant churches of the United States were strangely apathetic to the whole subject of the Army and Navy chaplaincy before 1913.

In that year, Chaplain Bayard appeared before a Baltimore meeting of the Federal Council of the Churches of Christ in America and advocated the establishment of a committee for the approval of chaplaincy candidates. The meeting authorized the "Washington Committee on Army and Navy Chaplains,"[28] and appointed Dr. H. K. Carroll its first secretary. Denominational leaders primarily responsible for the action of the Council in setting up this Committee were Dr. W. H. Roberts, stated clerk of

[25] *Handbook of the National Catholic War Council*, 1918, pp. 27–8.

[26] NRSO: 24620.

[27] NRSO: 13179–3.

[28] The Washington Committee was the forerunner of the present General Commission on Army and Navy Chaplains, an independent organization which serves as a clearing house for most Protestant clergymen seeking a commission as chaplain.

the Presbyterian Church, USA; Bishop Earl Cranston of the Methodist Church, North; and Bishop E. Talbot of the Protestant Episcopal Church. One of the most significant results of this Committee is that, at the outbreak of World War I, the Protestant churches of America had at their disposal the necessary machinery for the processing of candidates.

Until Dr. Carroll was able to take over, Chaplain Bayard served unofficially as the secretary, handling all the correspondence of the new organization. He has remained a member of this committee to the time of this writing in 1946. No one person played so long and so influential a role in the history of the General Commission on Army and Navy Chaplains as has Chaplain Bayard.

Dr. Carroll, in a report submitted to the Federal Council of Churches of Christ in America for 1914, stated that both the Army and Navy had welcomed the establishment of an organization to which they could turn for information regarding non-Catholic applicants. Dr. Carroll carried on correspondence with four to five hundred Protestant ministers regarding the Army and Navy chaplaincy. He encouraged the various denominations to set up their own chaplains' commission. Among the first bodies to respond to this suggestion were the Luthern General Council and the Disciples of Christ. "To such commissions," reported Carroll, "all candidates are referred for investigation and recommendation in order that the Secretary of the Navy may be assured that candidates of their respective denominations are worthy of appointment." Carroll felt that this plan tended "to eliminate political influence and to make each denomination in some measure responsible for its candidates."[29] Thus, another reform long advocated by chaplains was accomplished.

Contemporaneous with this improved method of selection was a revision of the Navy's promotion policy for chaplains. When the Act of 1914 fixed the ratio of one chaplain for every 1,250 naval personnel, it also set a new ratio for chaplains holding the higher ranks. Ten percent of the corps were to be captains, twenty percent commanders, and another twenty percent lieutenant commanders. The remainder were to be lieutenants or lieutenants (junior grade). These were the highest percentages in the upper grades ever granted by Congress to either the line or the staff.

The passage of the 1914 Act made some immediate promotions possible. Chaplain Frazier became a captain, and Chaplains Bayard, Stone, Gleeson, and Scott were advanced to commanders, all promotions dating from 30 June, the date of the law. Although they held higher rank, they were still lieutenant commanders as far as pay was concerned. Cassard was promoted to captain in 1915 and three others, Dickins, Charlton, and Patrick, became captains on 29 August 1916. Patrick had been in the service but fourteen years. Chaplain Evans was advanced from a lieutenant (junior grade) to commander within twenty-one months. These quick promotions were necessary to fill the authorized quotas for the various ranks.

Section 2541–c of the Act of 1914 made the following provision for the promotion of acting chaplains:

> Naval chaplains hereafter commissioned from acting chaplains shall have the rank, pay, and allowances of lieutenant, junior grade, in the Navy until they shall have completed four years' service in that grade, when, subject to examination as above prescribed, they shall have the rank, pay, and allowances of lieutenant in the Navy, and chaplains with the rank of lieutenant shall have at least four years' service in that grade before promotion to the grade of lieutenant commander, after which service, chaplains shall be promoted as vacancies occur to the grades of lieutenant commander, commander, and captain: *Provided,* That not more than seven acting chaplains shall be commissioned chaplains in any one year: And *provided further,* That no provision of this section shall operate to reduce the rank, pay, or allowances that would have been received by any person in the Navy except for the passage of this section, and that all laws or parts of laws inconsistent with the provisions of this section be, and the same are hereby, repealed.

The first chaplain to wear the two star insignia of a rear admiral was Thomas A. Gill who attained this rank the day he retired, 8 February 1902. Four other chaplains attained this rank on the retired list during the first decade of the twentieth century. They were Chaplains W. H. Stewart, Donald McLaren, J. J. Kane, and F. B. Rose. These promotions came under provisions of the Act of 3 March 1899 which advanced naval officers who were veterans of the Civil War one step above that held on the active list at the time of retirement.

UNIFORM CHANGES

A part of the story of the chaplains' struggle for full equality with other staff and line officers concerns the uniform. The Addenda to the 1897 *Uniform Regulations* had stripped chaplains of full-dress wear, of the distinctive Navy buttons, and of gold cap and sleeve ornaments.

[29] CoC., Federal Council File.

Secretary Long defended the Navy's policy in an 8 April reply to a chaplain who had evidently written about the unsatisfactory uniform regulations:

> With regard to uniform, the navy chaplain has a single-breasted frock coat of dark navy-blue cloth, and he has a blue service coat exactly like that of other officers also, except that it has black braid on the sleeves instead of gilt braid, and white service coat like that of other officers, except that the chaplain, very properly, does not wear military buttons, as he also does not wear a sword. Both of these would certainly be out of keeping with his profession. His mess jacket is the same. On his collar, he has the silver insignia of his rank and the silver designation of his corps. His cap is the same as other officers, except that it has black braid instead of gilt . . . . I cannot believe that the clerical profession would regard it as anything else than unbefitting if a chaplain were to be arrayed in the full uniform of a naval officer.

Secretary Long had nothing to say about the failure of regulations to prescribe a full-dress uniform for the chaplains.

This letter was ordered by the Senate to be printed on 3 December 1902. It appeared as *Document No. 10,* second session, 57th Congress.

The chaplains, however, in their desire for a "fair and reasonable uniform," took exception to the full-dress regulation. Strobridge revealed their feelings in his pamphlet; he conceded that

> the chaplains were given a satisfactory service uniform, but on dress occasion, they are compelled to wear a single-breasted, long tailed, shad-bellied frock-coat, a sort of monkish gown, sometimes seen on clericals in foreign countries but never worn by American ministers. It is offensively odd, loudly conspicuous, and distasteful to Protestants and Catholics alike.

Strobridge cited instances of the chaplains' being "ignored at all official or dress functions among the officers whether at home or abroad." He recounted Chaplain Clark's humiliation at the opening of the Kiel Canal, because "of this grotesque garb" and expressed indignation over the fact that when

> the great land parade was given in New York in honor of Admiral Dewey, and the nation was thus expressing its joy in the return of the brave warrior, and its gratitude to God on account of the victorious war, not a single chaplain had a part in the parade either as a participant marching on foot, or as a member of the naval service riding in a carriage, and why? Simply, because the uniform was not presentable. The people coming into possession of these facts have a right to ask, "How much longer does the Government propose to continue this policy of abuse and shame?"

He pronounced the regulations affecting the white uniforms of chaplains "a burning shame":

> Their coat and the mess-jacket of the naval cadets are identically the same, and to add yet a keener edge to the humiliation, the white service coat for the chaplains and the jacket of the mess-attendants, i.e., the waiters, are so nearly alike that only an expert can distinguish them.

Shortly after Josephus Daniels assumed the responsibilities of Secretary of the Navy in 1913, Chaplain Bayard, then on duty at the Washington Navy Yard, called on him and laid on his desk three buttons—a a regulation large-size gilt Navy eagle button which all officers were allowed to wear except chaplains, a plain black silk button, and a plain white composition button. "This button," declared Chaplain Bayard picking up the black silk button, "adds distinction to civilian evening dress for the opera or formal dinners or dances but any button not a Navy button on a Navy uniform is discordant, degrading, and in every way objectionable to everyone in the Navy." He continued: "This white button is the same as that worn by bartenders, barbers, waiters, and mess attendants who shine officers' shoes. Only chaplains among navy officers are forced to submit to such indignity and disgrace." Chaplain Bayard found Secretary Daniels receptive to his appeal.

Later in the year, as a result of this incident, the new *Uniform Regulations* were modified so that chaplains could wear the same buttons as other naval officers. They could now wear the "five large-size gilt navy buttons" on their white service coat, but not on their frock coat. Another step toward full equality had been taken by the members of the Chaplain Corps.

The 1913 *Regulations* ruled that chaplains should wear black braid on the sleeve in lieu of the gold braid.

> Sleeve marks on the frock coat shall be as follows: Chaplain — stripes of lustrous black braid of the same size, number, and disposition as for line officers of the same rank.[30]

It was not long before chaplains were referring to the "lustrous black braid" as "mourning braid." With but few exceptions, they wanted the full insignia, without the sword, but with the gold braid of their fellow officers.

[30] *Op. cit.*, p. 34.

OUTSTANDING CHAPLAINS AND THEIR ACTIVITIES

Eighteen new chaplains were commissioned in the first decade of the twentieth century in order to maintain the quota of twenty-four. This represented the heaviest turnover in the chaplaincy in forty years. A seventy-five percent turnover in ten years is evidence of the discontent then existing in the Corps. The introduction of reforms in the second decade, however, was an important stabilizing factor.

Twenty more chaplains entered the services during the years 1911–1916, including eighteen acting chaplains. The goal of one chaplain for every 1,250 men, authorized by the Act of 1914, was not reached for many years. Often the size of the Corps and the number of promotions were limited because Congress appropriated insufficient funds. Another handicap to full strength was section 2541–c of the Act which limited the commissioning of acting chaplains in any one year to seven.

The thirty-eight chaplains who entered the naval service during the years 1901–1916 were divided among the larger denominations as follows: Baptist, 5; Catholic, 12; Congregational, 2; Disciples, 1; Lutheran, 1; Methodist, 6; Episcopal, 6; Presbyterian, 4; and Reformed, 1. Herbert Dumstrey, of the Reformed Church, appointed 15 December 1915, and Paul E. Seidler, a Lutheran, appointed an acting chaplain 15 May 1916, were the first of their respective denominations to become Navy chaplains. Four of these thirty-eight chaplains—Scott, Evans, Duff, and Workman—later became Chiefs of the Chaplains Division.

Of the chaplains who entered the service during these years, the following in order of their seniority, were still on active duty in 1946: Thomas B. Thompson, Robert D. Workman, Charles V. Ellis, Herbert Dumstrey, Roy L. Lewis, and William W. Elder. Chaplain Thompson, commissioned 10 March 1913, became the Senior Chaplain upon the retirement of Chaplain Evan W. Scott in 1940.

Several chaplains who entered the service before 1900 and who continued their naval ministry into the twentieth century, merit special mention because of their long and distinguished record. Chaplain Holway, who was commissioned in 1868, was retired in June 1901. The gentle and benevolent Chaplain Tribou retired in 1910. Of him Chaplain Edel, drawing upon personal recollections, wrote:

> Out of the rich wisdom of his many years of service, he gave advice and counsel to many of the younger members of the Corps in the early days of the First World War, as they came to him for training before being sent to sea. Chaplain Tribou was a scriptural preacher, and his message was uttered in the rich and salty language of the sea, and when stirred it was his habit to burst forth in thunderous approbation or denunciation. The Navy loved him for the splendid catholic humanity of his views and the rare and tender sweetness of his personality.[31]

832159—49——11

Among the most noteworthy chaplains of these years was Henry Howard Clark whose two tours of duty at the Naval Academy, 1890–1894, 1896–1911, totaled nineteen years. He was frequently referred to by his brother chaplains in good-natured jest as "the Angel of the Church that is at Annapolis."[32] The members of the successive classes of midshipmen came to know and to love the genial, warm-hearted man whose unselfish and sacrificing spirit ever prompted him to go out of his way to help one in need.[33] No chaplain of his generation was known by so many naval officers as Chaplain Clark.

Chaplain Clark's ministry at the Naval Academy was broken by a two-year assignment to the *New York*. There, too, the Chaplain endeared himself to the men he served. The mess stewards on 11 October 1896 directed a letter to Chaplain Clark, which said in part:

> We, the undersigned and part of the crew of this great vessel, do most sincerely regret the detachment of your genial and most valuable divine services from our pride of the Navy — the New York . . . .
>
> We profoundly appreciate your great sermon this morning, and it touched the right spot at the right time. It was a well delivered sermon from start to finish, and we are sure that officers and men alike will miss your grand advices.[34]

Another paper that bears witness to Chaplain Clark's effectiveness as a spiritual leader of young men is a petition directed to the commandant at the Academy and signed by thirty-nine midshipmen. It follows:

> United States Naval Academy
> Annapolis, Maryland
> May 31, 1897
>
> Sir:—
>
> We, the undersigned Naval Cadets of the Second Class, have the honor respectfully to request that the sermon "Naval Life: Its Privations and Recompenses" delivered by Chaplain Henry H. Clark to the Naval Cadets on May 30th, 1897, be published and distributed to the Naval Cadets.

[31] Edel, "Navy Chaplains from 1775 to 1917," p. 14.
[32] See Edel article under this title in *Naval Institute proceedings*, Aug. 1942, pp. 1067 ff.
[33] A Communion Rail was dedicated 2 Feb. 1925 in the Academy Chapel in memory of Chaplain Clark.
[34] Clark File. Chaplain's Office, Naval Academy.

A number of books describing midshipman life at Annapolis and boy life in the Navy came from Chaplain Clark's virile pen. These were: *The Admiral's Aid, Boy Life in the United States Navy, Joe Bently, Naval Cadet,* and *Midshipman Stanford.* In his old age, Clark consented to the publication of a number of his sermons which had first appeared in the Sunday editions of the Baltimore Sun. The volume appeared under the title *The Sword of the Nation,* which was also the title of the first sermon in the volume. Such sermon topics and texts as the following are suggestive of both his originality and understanding. "False Bugle Calls"—"For if the trumpet give an uncertain sound, who shall prepare himself to the battle?" (1 Cor. 14:48); "The Word Behind Thee,"—"And thine ears shall hear a word behind thee, saying this is the way, walk ye in it." (Isaiah 30:21); and "Naval Life: Its Trials and Compensations."—"And when we had taken our leave one of another, we took ship." (Acts 21:6). The last mentioned sermon was preached on the last Sunday of an academic year just before the senior class left and before the other cadets departed for their summer cruises.

In December 1899, Chaplain Clark was called to Washington to take part in the ceremonies connected with the re-interment in Arlington Cemetery of the bodies of the victims of the *Maine* disaster.

Although Chaplain Clark was retired on 6 March 1907, he continued on duty at the Academy until 30 June 1911. Through the years he had kept a faithful record of Divine Services held in the Academy Chapel. On the date in which he terminated his active duty, Clark made the following notation in the chapel log:

> Now, after long service at the Academy, I commend those who shall follow me to the One who appoints us to our work and who ever holds our destinies in His hands. We are his workmen.

Three other chaplains commissioned in the seventies continued to serve during the first decade of the twentieth century. They were Chaplains McAlister, Gill and Boorom. After his retirement in 1902, Chaplain McAlister was recalled to active duty in 1914 at Mare Island, serving again in his beloved St. Peter's Chapel. He died at Mare Island on 8 July 1916 and lies buried in the small naval cemetery near the hospital.

Seven chaplains, commissioned in the eighties, saw service in this period. Among these was Chaplain Frank Thompson, son of Bishop Hugh Miller Thompson of the Episcopal Diocese of Mississippi and uncle of Chaplain Hugh Miller Thompson Pearce (commissioned in 1909). Chaplain Thompson served from 1881 to 1920. Included among the seven was Chaplains Hoes, who was on duty at the Navy Yard in Washington from May 1902 to January 1907. This gave him the opportunity to examine Navy Department archives for data about chaplains, and he spent many a spare hour going through hundreds of volumes of bound correspondence which the Department had accumulated during more than a century.

Matthew C. Gleeson, 1903–1927
Wearing uniform of First World War Period.

Among the most able naval chaplains in the history of the Corps was Carroll Quinn Wright who was at the height of his influence during these years. Chaplain Wright had boundless energy and initiative, and was deeply concerned about the welfare of his men. Writing to Chaplain Edel from Washington, D. C., on 21 March 1923, Chaplain Wright gave the following account of the organization of the Silent League and of some of his other activities, while on duty at the Philadelphia Navy Yard:

> In sending you the sketch of my poor life recently, I neglected to mention the fact that, in 1903, at the League Island Navy Yard, I founded or started the Silent League in the U.S. Navy, for the discouragement of Obscenity, Profanity, Gambling and Debt—specially, the first two vices.
>
> There were no meetings, no dues, no reports — each man who, accepted the token (a small metal piece having the initial letters S L on it) was thereby initiated, and

became a league within himself—to enlist others of reliability in the same way—being instructed briefly how to make the approach etc., and admonished that it was a matter left between him and God—that he's to make *his report* on the Judgment Day.

From that time, I think more than ten thousand tokens have been given out, and, possibly, a hundred thousand people reached and influenced by the League.

He also wrote a six-page pamphlet which was published in 1905 aboard the receiving ship *Lancaster* at the Navy Yard, League Island, Philadelphia. This pamphlet, entitled *Instructions for Recruits,* was designed as a brief orientation course for young men entering the Navy. The information and good counsel which Chaplain Wright offered were introduced by the brief statement: "After enlistment in the U. S. Naval Service, every young man should carefully take his bearing, and mark out a course for himself in the new and strange seas in which his life is now cast." Chaplain Wright warned men against certain common temptations and failings of sailors. He emphasized the importance of keeping a clean record and for the information of the recruit, gave the list of penalties usually imposed for various offenses. The pamphlet proved so valuable that the Navy Department issued a second edition for wider distribution.[35]

During the first decade of this century, three chaplains brought out small hymn books and orders of worship. In 1900, Chaplain W. G. Cassard issued his thirty-eight page booklet entitled *Order of Divine Service and Collection of Hymns* for use at the Naval Training Station at Newport, Rhode Island. A revised and enlarged edition appeared in 1905. This was probably the first service and hymn book ever published for the benefit of United States naval personnel. Chaplain C. H. Dickins, in 1904, issued his sixty page book, *A Divine Service for the United States Navy,* which contained prayers, liturgies, selections of Scripture, and hymns. In 1909, Chaplain Wright issued a thirty-four page pamphlet containing an order of service for Divine Worship. This included responsive readings, prayers, the Apostles' Creed, and a selection of seventy-two hymns and a few secular songs. Among the hymns is "Eternal Father, Strong to Save," with only the first and last stanzas.[36]

The first chaplain to use a modern picture machine aboard ship was Chaplain B. R. Patrick, commissioned in March 1902, who took one of the Edison Universal machines aboard the *Yankee* in 1903. The machine cost the chaplain $125. The old flicker movies were such a novelty that they could be shown repeatedly to the same audience. Films then were exceedingly difficult to obtain.

William W. Edel, 1917–1946.
Wearing uniform of First World War Period.

Also among the new appointees of the first decade of the twentieth century was Chaplain Matthew C. Gleeson, a Roman Catholic, who served on the *Missouri* in 1904 at the time of the turret explosion. Chaplain Gleeson distinguished himself in his attention to the wounded and the dying, and was cited for valor by President Theodore Roosevelt in a letter to the Secretary of the Navy dated 21 May 1904.

In the report of Captain Cowles on the late accident aboard the Missouri, particular attention is called to the excellent and gallant conduct of Father Matthew C. Gleeson, the Chaplain of the Missouri. Father Gleeson's behavior throughout this trying ordeal was in every way such as to give legitimate reason for satisfaction and pride to all interested in the Navy, and to all, therefore, who believe that the Chaplains of the Navy should show courage, common sense and efficiency no less than devotion to the spiritual and material well being of those aboard the ships. It is a pleasure to record the gratification I feel in taking notice of what Father Gleeson has done.[37]

[35] Edel Coll., Wright File.

[36] Edel Coll. contains copies of these service books.

[37] Germain, *Catholic Military and Naval Chaplains*, pp. 145–6.

Contemporaries of Chaplain Gleeson tell of his phenomenal memory for names, his quick Irish wit, his ready sympathy, and the hold he had upon every man in the ship's company.

When Congress, in 1914, authorized the Secretary of the Navy to make appointments to the Naval Academy from the ranks of enlisted men, selection to be based on competitive examinations, aspiring candidates turned to the chaplains for help. Two chaplains became particularly active in coaching men. They were MacNair and Brady. MacNair claims that in a year in which only fifteen men were accepted from the entire Navy, six of the eight candidates he coached passed the entrance examinations.[38] In 1919, MacNair was ordered to the Naval Training Station at Newport where he became officer in charge of the newly organized Naval Academy Preparatory Class. Chaplain Brady followed MacNair in 1923, carrying on the work and expanding it. Through the years following the inauguration of this plan, chaplains on board ship and at shore bases have often served as teachers for ambitious young men seeking Academy appointments.

When Chaplain C. H. Dickins was serving aboard the *Florida* (March 1912–June 1915), he replaced the old hand laundry system with modern machinery. The contrasts in degrees of whiteness between the uniforms worn by the sailors of the *Florida* and those worn by men of the other ships in the Fleet soon attracted the attention of the Admiral, who ordered the machines installed on the other vessels.

The chaplains who remained in service during the first decade of the twentieth century, in spite of the discouragements resulting from the inequalities of their official status, were men of consecration and ability. Chaplains of later generations owe much to the faithfulness of these men of the old group of twenty-four who stood by during the years of transition when one by one the long-desired reforms were becoming a reality.

[38] CoC., MacNair File, Letter dated 27 Sept. 1945.

TWO CHAPLAIN HEROES OF WORLD WAR I

Both of whom, having received the Navy Cross for heroism with the Marines in France, were promoted to rear admiral on 16 January 1936 while on the retired list.

John J. Brady, 1914–1934

James D. MacNair, 1909–1930

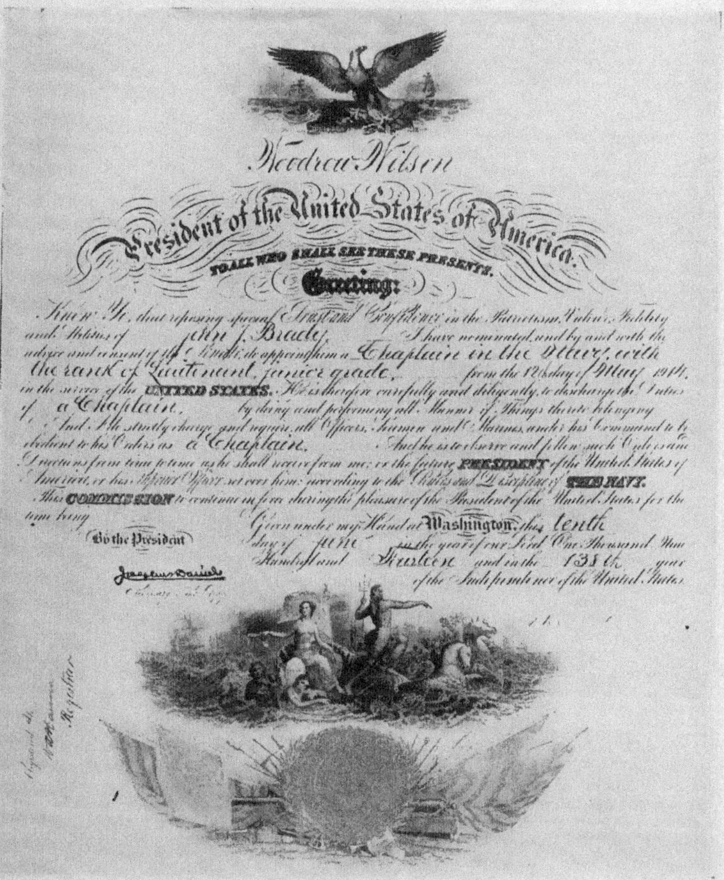
Woodrow Wilson

President of the United States of America.

To all who shall see these presents, Greeting:

Know Ye, that reposing special Trust and Confidence in the Patriotism, Valor, Fidelity and Abilities of John J. Brady, I have nominated, and by and with the advice and consent of the Senate, do appoint him a Chaplain in the Navy, with the rank of Lieutenant, junior grade, from the 12th day of May 1914, in the service of the UNITED STATES. He is therefore carefully and diligently to discharge the Duties of a Chaplain, by doing and performing all Manner of Things thereto belonging. And I do strictly charge and require all Officers, Seamen and Marines, under his Command to be obedient to his Orders as a Chaplain. And he is to observe and follow such Orders and Directions from time to time as he shall receive from me, or the future PRESIDENT of the United States of America, or his Superior Officer set over him, according to the Rules and Discipline of THE NAVY. This COMMISSION to continue in force during the pleasure of the President of the United States for the time being.

Given under my Hand at Washington, this tenth day of June in the year of our Lord One Thousand Nine Hundred and Fourteen and in the 138th year of the Independence of the United States.

By the President

Josephus Daniels

Reproduction of the chaplain's commission granted John J. Brady, 12 May 1914, signed by Woodrow Wilson and Josephus Daniels. This commission is the same as that given all appointees to the Chaplain Corps of the Navy. From original in Chaplain Brady's possession.

THE NAVAL MILITIA AND THE NATIONAL NAVAL VOLUNTEERS

The genealogy of the United States Naval Reserve, established in 1925, is to be traced back to 17 May 1888 when a battalion of Naval Militia was organized in Massachusetts. Other states soon followed the example of Massachusetts, and, in 1894, the Secretary of the Navy was authorized to loan to each state having a Naval Militia certain vessels for training purposes. Membership in the Militia was entirely voluntary and no financial remuneration was available for maintenance of efficiency. The organizations were state controlled. What the National Guard was to the Army, the Naval Militia was to the Navy..

The Naval Militia proved its worth during the Spanish-American War by furnishing about forty percent of the 10,373 additional men taken into the Navy during that struggle. The Federal Government officially recognized the Naval Militia as a vital part of our national defense by an Act of Congress of 16 February 1914.

On 3 March 1915, the organization under federal control of a Fleet Reserve for enlisted personnel who had seen service in the Regular Navy was authorized by law. For a little more than a year, these two naval reserve organizations, one under state control and the other under federal, existed side by side. Then, by Act of Congress of 29 August 1916, the Naval Militia and the Fleet Reserve were merged into the National Naval Volunteers.[39]

The Act of August 1916 also established the Naval Reserve Force with six classes of officers and men. Class 2, into which the Naval Volunteers were later absorbed, included: "Officers and men enrolled for four-year periods and available in case of war or national emergency for seagoing duty on combatant vessels."[40] Members of this class were required to perform sixty days active duty in periods of not less than fifteen days each during their four-year enrollment. Each was to receive two months' base pay per year if the required duty was performed.

[39] Navy Dept. Bulletin, *Information concerning the US Navy*, May 1925, pp. 160 ff.
[40] *Ibid.*, p. 162.

The first chaplain known to have joined the Naval Militia was Edwin B. Niver, whose commission was dated 6 November 1901. The following chaplains were enrolled:

Name	State	Date of Commission	Year of register when name first appeared
Niver, Edwin B	Maryland	6 Nov. 1901	1907
Smith, C. E.	No. Carolina	6 July 1908	1910
Sesnon, R. F.	California	1 Aug. 1909	1910
Ross, Albion H.	Wisconsin	29 June 1909	1911
Olson, Richard	Oregon	1 Aug. 1909	1911
Huske, B. F.	No. Carolina	30 Apr. 1913	1914
McGann, Wm. T.	Florida	Mar. 1916	1917
James, Sidney T.	Washington	21 Sep. 1916	1917[41]

The 1913 *Register* dropped the name of Chaplain A. H. Ross and the 1914 edition omitted that of Chaplain C. E. Smith. Chaplain Niver was credited with nine cruises,[42] which indicates that he took one each year. Only three of the eight Naval Militia chaplains are known to have accepted a commission in the National Naval Volunteers. They were Chaplains Niver, Huske, and James.

Mention should be made of another Naval Militia chaplain who was not listed in the 1917 *Naval Militia Register*. He was James Samuel Day who passed through the National Naval Volunteers and the Naval Reserve Force into the Regular Navy Chaplain Corps. His son, Howard Malcom Day, became a naval chaplain in 1940.

Thus, there was on the eve of the First World War a small band of nine chaplains in the Naval Militia. Perhaps, more important than the number of chaplains available was the fact that the machinery was ready for the enlistment of other reserve chaplains when war was finally declared.

THE CHAPLAINS AT WORK

During these years under review, the correspondence of chaplains, as it had in previous periods, continued to bring into relief not only the lights and shadows of a chaplain's duties, but, also the attitude

[41] *Naval Militia Registers*, 1907, 1909–17. In 1945 these *Registers* were in the custody of Mr. J. R. Durnell, Bureau of Naval Personnel.
[42] *Naval Militia Register*, 1917.

of fellow officers and of the Navy Department. On 16 June 1902, Chaplain Hoes, then on duty at the Washington Navy Yard, wrote to the Department requesting that he be provided with stationery and office supplies. The second endorsement on this request is as follows:

> Respectfully referred to the Bureau of Navigation, for report as to whether a Chaplain is needed at the Navy Yard, Washington; also whether a Chaplain is needed on board any vessel in commission.[43]

Since the number of chaplains was at that time strictly limited to twenty-four, many important shore installations as well as large ships did not have chaplains. Consquently, many naval officers and men had no opportunity to see the contribution a chaplain could offer. After the Act of 1914 fixed the chaplain quota, Secretary of the Navy Josephus Daniels, in his *Annual Report* for 1915, was able to state: "It will soon be possible to assign a chaplain to every battleship."[44]

Chaplains made frequent requests to the Navy Department during these years for stationery, hymnals, books, and office equipment including typewriters. The following from Chaplain Helms, 10 October 1901, is a typical example:

> At present, the only possible way of procuring such supplies, is to beg them or buy them from our individual purses. A reasonable allowance for use in preparing sermons and lectures, for making such official statements and letters as frequently have to be made by Chaplains, and for writing letters home for men and many other purposes too tedious to enumerate, all these supplies are daily needed by Chaplains.[45]

In endorsing the request, the Chief of the Bureau of Navigation recommended that the stationery be supplied all chaplains.

Chaplain Frazier, on 27 January 1902, wrote regarding hymnals, music, and an organ, and expressed "the hope that some provision may be made for supplying such articles." He suggested that an appeal for an organ might be made to the Women's Army and Navy League. Chaplain Frazier's request was endorsed:

> Respectfully returned to the Bureau of Equipment, with authority to furnish hymnals for use on board ship and at naval stations, under the appropriation "Equipment of Vessels."

Thus, this request led to the approval of the purchase of hymnals for all ships and stations needing them. Regarding the request for an organ, the endorsement read ". . . the Department would not like to establish a precedent by making application to a charitable organization." However, it does not appear that the Department approved the idea of appropriating funds for the instrument.[46]

The Board of Chaplains which was appointed to make suggestions for the increase of the effectiveness of the Corps included the following in its report of 20 October 1906:

> In the interest of uniformity, compendious books of worship should be compiled for Catholic and non-Catholic seamen, including prayers familiar to common liturgical usage, containing also prayers specially prepared for the naval service and hymns accepted by all denominations; and in said books there should be a supplement containing concise practical instructions and suggestions as to the standards of conduct and character to be maintained in the naval service, the books to bear the imprimatur of the Navy Department; each man in the naval service who may desire it to be furnished with a copy of either book. In the opinion of the Board, such books would serve a high purpose of religious devotion and moral inspiration and restraint, and we beg leave to suggest that a board of representative Chaplains be appointed for the compilation of such books.

This suggestion was not acted on until 1920 when the Navy joined with the Army in issuing the first edition of the *Army and Navy Hymnal*.

Chaplains called attention to various traditional practices that in any way affected, influenced, or hindered their work. Several criticized the Sunday morning inspection of the crew as imposing an undue burden upon what should be a day of rest and worship. General Order 37, dated 15 September 1909, which corrected the practice to the chaplains' satisfaction, read:

> 1. To the end that Sunday may be generally observed as a day of rest on board ship, it is directed that the duty required of officers and men on that day be reduced to a minimum consistent with the requirements of the service.
>
> 2. The commanding officer's inspection of ship and crew shall not be held on Sunday. The inspection of the ship shall be held on any other day of the week as may be most expedient and the inspection of the crew on Saturday before noon, if circumstances permit; if not, as soon after the dinner hour as practicable.

The custom of offering evening prayer at sundown when the bugle call for hammocks was sounded, gradually died out as a result of the regulation changing compulsory church attendance. "Prayers at evening hammocks" were, however, still being offered

[43] NRSO: 14358.
[44] *Op. cit.*, p. 26.
[45] NRSO: 13179.
[46] NRSO: 13546.

on some ships, and the Board of Chaplains which met in 1906 suggested in their report that

> prayers at evening hammocks should be made explicitly uniform in all ships to which Chaplains are attached. Apart from the religious significance of this service, it has a ceremonial function, which, in our opinion, tends strongly to the aid of discipline and good order. It is a custom inherited from the old Navy, and appears to us an indispensable bequest.

Today, there remains one last vestige of this beautiful tradition. On some ships, following the bugle call to hammocks, the chaplain reads a brief prayer over the public address system.

The directions for the display of the church pennant, which first appeared in the *Navy Signal Code* for 1867, were repeated in substantially the same wording in subsequent editions of the *Signal Book*. The church pennant is non-sectarian. When properly flown, it indicates that Divine Services are actually in progress. Navy customs have a way of moving from sea to shore. Just when the church pennant was first used at a shore station is not known. By the time of the First World War, the custom of flying the church pennant at naval installations in the United States, according to the custom used at sea, was already widely observed.

The importance of the chaplain's work as educational supervisor of boys is found in the testimony of Chaplain Clark before the House Committee on Naval Affairs on 12 January 1903. He stated:

> Mr. Cassard has a thousand boys under his care. He has entire charge of the instruction of these boys in the elementary English branches and has several instructors under him.

Chaplain Cassard was then attached to the training ship *Constellation.*

During this period under review, Navy chaplains were given overseas shore duty for the first time. Chaplain C. M. Charlton was the first of a long line of chaplains to see duty in the Philippines. He served at Cavite from March to December 1903. Chaplain J. B. Frazier initiated the chaplains' work at Tutuila, Samoa, serving there for three years begnning July 1905. The first chaplain to serve at Guam was J. F. Fleming who was stationed there from March to August 1914.

Chaplain B. R. Patrick had an unusual experience in the summer of 1908 when, enroute to his new station at Tutuila, he was shipwrecked with his family on Christmas Island. In those days, the most available means of transportation from San Francisco to Samoa was on a freight steamer. On 6 July, Chaplain Patrick with his wife, two young sons, and a trained nurse, had boarded the S. S. *Aeon.* Chaplain Patrick related the incident:

> On our twelfth day out at 9:00 p.m., hundreds of sea birds screamed about the ship and at 9:30 p.m. the officer of the watch shouted "Breakers ahead", and almost at once the ship hit the reef and was hard and fast aground. Christmas Island was presumably 25 miles west of us, but, we had struck the reef about 3 miles inside the S. E. point of the island (About 90 miles north of the Equator). The engines succeeded in keeping the ship bow on for awhile but by midnight the wind and current had forced the ship broadside to the reef, the engine room was full of water and plate bolts were snapping. The ship seemed to be breaking in two just forward of the bridge. We could make out two lines of breakers, but could discern no land until dawn when the low coral sand of the island became visible. A line was secured from the ship to wreckage ashore and at 6 A.M. our party was landed.
>
> Lumber from the cargo, tarpaulins and food were brought ashore all that day and by night we had three tents, one for our party, one for the ship's officers and one for the Chinese crew. No potable water had ever been found on the island, according to the Sailing Directions, and the ship's chief fresh water tanks had been flooded with salt water. During the first 12 days ashore, we had two showers and saved some water. But for the most part we drank the water from canned asparagus (ten tons of it in the cargo). On the 12th day, the Chief Engineer finished a distiller on the beach and made 25 gallons of fresh water.

In the meantime the Captain of the *Aeon* put to sea in a small motor boat to get aid from Fanning Island, 200 miles distant. Back on Christmas Island, preparations were being hastily made for an event which would not wait even for a shipwreck.

> A small cottage with beds and plumbing from the wreck had been built and was occupied by Mrs. Patrick and Mrs. Riddle [wife of Lieutenant W. K. Riddle]. In this cottage, on 22 September, Jane Clark Patrick was born with Miss Sarah Campbell, nurse attending. Just 24 hours later, the S/S Manuka was sighted coming to our rescue with our Captain Downey and party whom they had picked up at Fanning Island. The Manuka's surgeon came ashore with Captain Downey, but there was nothing for him to do.[47]

FIRST CHAPLAINS ASSIGNED TO THE MARINES

Since the Marine Corps has no Chaplain Corps, its personnel, from the beginning of the naval chaplaincy, have been included within the scope of the Navy chaplains' ministry. The earliest discovered

[47] CoC., Patrick File.

reference to any extensive work by Navy chaplains for Marines appeared in a 4 December 1862 letter which Chaplain C. S. Stewart wrote to a friend. He described his activities with the Marines at the Naval Hospital, Navy Yard, New York:

> I am now established, somewhat permanently, should health and life be spared me, under the same roof with a good portion of my charge and with the Chapel in which I preach to them. The number of patients in the Hospital averages about three hundred. The Soldiers at the Marine Barracks, amounting to some three hundred are also in my *diocese* and I preach to them also every Sabbath in the Mess Hall of the Barracks.[48]

The first Navy chaplain known to have had exclusive duty with the Marines was Chaplain J. F. Fleming, who served on board the *California* in the summer of 1912 when the Marines were sent to Nicaragua. The first detachment of Marines went ashore at Managua on 4 August. Just when Chaplain Fleming joined the landing party, or, when he returned to his ship, is not known. The official "Report of Operation of First Provisional Regiment of U. S. Marines in Nicaragua" contains a reference to the burial of four men on 6 October when Chaplain Fleming officiated. On the 11th of the month he was reported as assisting at the Field Hospital.[49]

The first chaplain known to have been assigned full-time duty with the Marines was Bower R. Patrick who was ordered on 21 April 1914 to the Marine Expeditionary Force of the Atlantic Fleet. This was at the time when friction existed between the United States and Mexico. Chaplain Patrick sailed with the Marines for Vera Cruz from Philadelphia on the *Moro Castle*. He went ashore with the Expedition and, from 14 June to 22 September, maintained club rooms at Vera Cruz for the use and benefit of the enlisted men. Patrick returned to Philadelphia with the Expeditionary Force in the late fall of that year and was detached from this duty on 21 December.

The second chaplain to see duty with the Marines was Leroy N. Taylor who served with the Fourth Marine Regiment from 23 August 1915 to 18 December 1916. This assignment came in response to Chaplain Taylor's request for duty with the Marines at San Diego. "The fact that Colonel Pendleton asked for a Chaplain when the 'Camp' was first established," wrote Taylor to the Department, "is evidence in itself that one is really needed to work among them." Chaplain Taylor accompanied the Marines to Santo Domingo when they were ordered there in 1916. He distinguished himself in an engagement in which the Marines took part and received the following letter of commendation for the Navy Department dated 18 October 1916:

> The Department is pleased to note your excellent work as driver of the motor ambulance and courage in advancing through fire-swept zones when called upon to do so, and informs you that this correspondence will be filed with your official record.[50]

THE WORK OF THE NAVY YMCA

No history of the naval chaplaincy and of religion in the United States Navy would be complete without a consideration of the services rendered by various welfare agencies.

Since its organization during the Spanish-American War, the Army and Navy YMCA has been a faithful partner of Navy chaplains in their work ashore. In those places where its work is limited almost exclusively to naval personnel, the full name is abbreviated in characteristic American style to the Navy Y.

The Young Men's Christian Association has always centered its attention on a three-fold approach to young men—the mental, the physical, and the spiritual, hence, its symbol of the triangle. Travelling libraries were introduced to the Army and Navy by the Y as early as 1902, when some 150 compactly-made wooden boxes, each containing two shelves of books, were circulated.[51] By 1903, over 10,000 volumes were made available to soldiers and sailors through this means. A reading and writing room and a library have been indispensable parts of every Navy Y building. Free stationery with the Y letter-head has carried unnumbered letters from sailor lads to their home folks. Attractive lectures and sightseeing tours have been scheduled. Before movies became available, many YMCA secretaries made extensive use of the stereopticon lantern to illustrate their lectures. For hundreds of thousands of naval personnel such activities have fruitfully absorbed leisure hours and successfully counteracted undesirable influences.

In ministering to the physical needs of the sailor, the Y erected buildings often containing well equipped gymnasiums, swimming pools, bowling alleys, game rooms, cafeterias or restaurants, and dormitories. The first Army and Navy YMCA physical director was employed in 1909 for the staff of the Brooklyn Navy Y. This experiment was so suc-

[48] *Ibid.*, Stewart File.
[49] Nav. Rec. Coll., Register No. 2677, pp. 3, 12.
[50] CoC., Taylor Annual Report File.
[51] Pond, MS., "History of the Army and Navy YMCA.," p. 42.

cessful that appointments were soon made for other Y buildings.[52] All the equipment, as well as the services of the staff, were available to sailors without charge or on a cost basis. A nominal fee was usually collected for dormitory rooms and for food.

Activities sponsored by the Y to meet the spiritual needs of sailors included Bible classes, religious services, distribution of religious literature such as Bibles and Testaments, Sunday morning breakfast clubs at which speakers brought religious messages, and church parties. The Y has made it a practice to work in closest harmony with local churches.

One of the early projects of the Navy Y was the promotion of the Naval Temperance League which had the approval in 1900 of Secretary of the Navy J. D. Long. Members of this League promised to abstain from the use of intoxicating liquors while in the service of the United States Navy. It is reported that nearly 6,000 signed the pledge.

The first Navy Y building, near the Brooklyn Navy Yard, was dedicated 15 May 1902. The second building was located at Vallejo, California, near the Mare Island Navy Yard. This site was donated by Rear Admiral and Mrs. B. H. McCalla who also gave $27,000 toward the cost of the building modelled after the Brooklyn Y. The cornerstone was laid by President Theodore Roosevelt 14 May 1903 and the building was opened 16 July 1904.

On 14 January 1903, the Y initiated its work at Newport in rented quarters at 102 Thomas Street. After eight years, it was moved into a new building on Washington Square which cost over $275,000. This building was a gift of Mrs. Thomas J. Emery of Cincinnati who, at her summer home in Newport, had observed the good work of the Y.

The first Y building erected within a Navy Yard was the "Annex" opened at Norfolk in July 1906. Again, this was possible through the liberality of Miss Helen Gould. The work at Norfolk proved to be so important that it soon outgrew its first quarters. John D. Rockefeller, Sr., was interested and gave $300,000 for the erection of a modern building. On 9 May 1908, Mr. Rockefeller wrote the following for the cornerstone laying ceremonies:

> To the Enlisted Men of the Navy, Gentlemen: It is a privilege to be permitted to give expression in this permanent form to the pride which all Americans feel in their Navy, and to the sense of obligation which we feel to you who have consecrated your lives to our country's defense. The ambitions of our people have always centered in the arts of peace, but, we are not forgetful of the need of strong arms and brave hearts to protect those liberties without which the blessings of peace are worthless. While the purpose of this building is to help you in maintaining that high standard of manhood which is essential in your profession of arms, God grant that you may also find here influences which will help you keep near to Him who is the Prince of Peace. Sincerely yours, John D. Rockefeller.[53]

The building was opened to receive the men of the fleet who returned from their world cruise in February 1909. Formal dedicatory services were held 17 March 1909. Rockefeller supplemented his first gift with another $200,000 for an endowment fund.

The work of the Y in Philadelphia began in October 1906 when a branch was established in rented quarters at 2009 South Broad Street. As these quarters proved inadequate, funds were later secured for the purchase of a larger building at 1301 Shunk Street. It was opened in December 1915. An extension service of the Y was carried on at the Philadelphia Navy Yard in "The Annex," a building which cost about $30,000 and was dedicated 28 April 1911.

In December 1907, just before the Fleet left on its world-encircling cruise, Admiral Dewey wrote to the Army and Navy Department of the YMCA:

> It occurs to me as the departure of the Fleet for the Pacific approaches, that a grand work lies ready for the YMCA to inaugurate for the aid of the enlisted men. Could not some place be secured in San Francisco where our men would find friends to look after their welfare, lodging, etc.? In short, to do the same estimable work that is now being done in the Brooklyn Navy YMCA.[54]

This led to the beginning of the Army and Navy Y in San Francisco. Rented quarters were secured at 82-84 Market Street and were used for eight and a half years.

Navy Y work for the men of the Boston Navy Yard at Charleston was opened in 1909 at the Edward Everett House, 14 Harvard Street. The beginning of the present extensive work at San Diego goes back to temporary activities carried on in a large tent in 1910–1911. On 2 May 1911, the Navy Y occupied rented quarters at 246 Washington Avenue, Bremerton, to serve the men attached to the Puget Sound Navy Yard. In 1913, larger quarters were secured by leasing adjoining buildings.

Thus, during the years 1901–1916, the Y had established branches in the vicinity of the following navy yards and training stations: Brooklyn, Mare Island, Newport, Norfolk, Philadelphia, Boston, San Francisco, San Diego, and Bremerton.

---

[52] *Ibid.*, p. 39.

[53] *Ibid.*, pp. 191 ff.

[54] *Ibid.*, p. 209.

The overseas work of the YMCA began in the Philippines at the time of the Spanish-American War. The first YMCA building in Manila, a two-story structure was secured in 1898 by Lieutenant General Arthur MacArthur. When the United States established a naval base at Cavite, the Y followed. Activities began in a tent and were transferred to a building about 1900. As the work grew, larger quarters were occupied in 1902 and 1903. The Cavite branch was discontinued in 1909. The first building owned by the Y in its overseas activities was erected at Olongapo with the help of the Marines in 1903. This early structure was replaced by a more permanent building in 1912. A secretary remained on duty until about 1914. The Army and Navy Y had adequate quarters in a building at 205 Calle Real in Manila in 1903 and 1906. When the United States Fleet made its famous round-the-world cruise in 1908, the Y rented for temporary use two large buildings on the banks of the Pasig. In 1910, similar temporary service was provided for the men of the visiting Pacific Fleet.

The first continuing work in Shanghai conducted by the Navy Y for American sailors dates back to 1912 when a club house at 7 Nanking Road was opened for their use by the American Women's Club of Shanghai. The Y secretary then attached to the Asiatic Squadron was placed in charge.[55] The *Annual Report* of the YMCA for 1914–1915 stated: "The Asiatic fleet work has so grown that it has become necessary to recognize the necessity for permanent shore work at the headquarters in Shanghai, and the fleet Secretary is now managing this work in a commodious rented building."[56] Quarters were moved to 23 Nanking Road in the fall of 1915.

A Y secretary accompanied the American forces to Peking at the time of the Boxer trouble in 1900. Work with the Legation Guard was begun soon after and continued until 1914. A large clubhouse in the Canal Zone was placed at the disposal of the YMCA in 1911. Other centers for Marines were opened as early as 1909.[57]

Shipboard chapters of the YMCA were organized in 1901 and flourished for several years. Later, the name Triangle Service League was adopted but experience proved that such organizations on board ship were not practicable and this aspect of the project was given up about 1928.

ST. PETER'S, MARE ISLAND

At the north end of the Bay of San Francisco lies what was once an island but is now a peninsula, measuring five miles by two, known as Mare Island. Its nautical name has nothing to do with the Latin word for sea but rather to the mundane subject of General Vallejo's old gray mare which fell off a raft one day while the General was transporting some of his belongings across the Carquinez Straits. He was unable to rescue the mare at the time of the accident and gave her up for dead. Later, she was discovered grazing upon the island—hence the name Mare Island.

By law, all of the islands in San Francisco Bay became government property when California was taken over from Mexico. The island became a naval base in 1854 with Commodore Farragut as the first Commandant. Midway on the small island, in a grove of eucalyptus trees, stands a beautiful wooden building known as St. Peter's Chapel.

St. Peter's, erected in 1901 before the present Chapel at the Naval Academy, is the oldest extant naval chapel in the United States. A bronze plaque in the entrance-way of the chapel at Mare Island bears the following legend:

Erected by the Congress
A.D. 1901,
Through the efforts of
The Honorable George C. Perkins,
United States Senator,
and
The Reverend A. A. McAlister,
Chaplain United States Navy.
Albert Sutton
Architect.

The funds for the Chapel came from the Naval Appropriation Act of 1900 which set aside $5,000 for this purpose. The building, a low rectangular structure with two small rooms, one on either side of the chancel, was built out of California redwood. It has painted shingles now brown with age on the outside walls. A low steeple rises over the squat tower which serves as an entrance-way. The Chapel was dedicated on Sunday, 13 October 1901.

Attractive as the exterior may be, the interior of the chapel is far more so. The skillful modification of the Gothic architecture, the beautiful stained glass windows, the furnishings, and the perfect appointments of the chancel and altar, all contribute to a

[55] *Year Book of the YMCA*, 1911–1912, p. 291.
[56] *Op. cit.*, p. 347.
[57] Information about YMCA buildings at home and abroad obtained from Pond, MS, "History of the Army and Navy Young Men's Christian Associations."

Triptych Window in St. Peter's Chapel, Mare Island, California. This window depicts the calling of Peter and is dedicated to the three navel heroes who played important roles in early California history—Rear Admiral John Drake Sloat, Admiral David Glasgow Farragut, and Commodore Robert Field Stockton.

Interior of St. Peter's Chapel, Mare Island, California.

"religious atmosphere." A mellow-toned pipe organ is included among the furnishings. The sanctuary can seat a maximum of about 200.

Most of the windows are by Tiffany. The first to attract the attention of the visitor is the triptych window which dominates the rear end of the building. It embodies the theme, "The Calling of Peter," and shows the fishermen pulling in their nets from the waters of Galilee. In the center are Jesus and Peter. This triptych is a memorial to the great naval leaders Commodore J. D. Sloat, who raised the American flag over California on 7 July 1846; Commodore R. F. Stockton, who completed the conquest; and Admiral D. G. Farragut, the first Commandant of Mare Island. Over the altar is a rose window, "The Ascension," which is a memorial to the officers and enlisted men of the United States Navy and Marine Corps who have served in the Pacific.

The Chapel also has a number of memorials in the form of carved medallions and bronze plaques. All are dedicated to men, officers and enlisted personnel of the Navy and Marine Corps, with the exception of one window which combines the deep blue of the sea and yellow of the California poppy in a representation of the Madonna and Child. This is dedicated "To the Women of the Navy and Marine Corps."

St. Peter's Chapel was the first chapel built by Government funds on any Army or Navy installation to be used for both Protestant and Catholic worship. However, there was an earlier chapel built by the Government in which both of the great branches of the Christian church held services. This is the chapel built in 1871 at the United States Soldiers Home in Washington, D. C. Records of this church indicate that, as early as 1877, both Catholic and Protestant

Exterior of St. Peter's Chapel, Mare Island, California.

services were being held each Sunday in the building.[58]

When the idea of having one government chapel at Mare Island for the joint use of Catholics and Protestants was first broached, many were hesitant and doubtful. Chaplain McAlister was one who believed that satisfactory arrangements could be worked out so that Catholics and Protestants could worship at different times in the same sanctuary, using the same altar and other furnishings. The experiment proved successful and was further popularized in scores of other government chapels, both Army and Navy, in the Second World War.

Since its erection, St. Peter's has served the procession of naval personnel who have visited or lived at Mare Island. The ledger-size record books, sometimes called the "Chapel Log," tell of weddings performed, baptisms administered, and Divine Services held. Captain Thomas A. Kearney, who once lived on Mare Island and worshipped at St. Peter's, wrote of this Chapel:

> There in the midst of an eucalyptus grove it stands—this navy chapel-church. Apart, and yet so near to the houses around about it; so near to the shops, the ships, the barracks of the men. Its great rose window mirrors back the light of setting sun; its steeple lies half-hidden in the wind-swayed branches of the trees, the steeple cross-crowned, its summoning bell indicates it as a harbor of refuge to those who come in from the sea, an anchorage, good holding ground, a place where weary men may rest before they go back to the sea—St. Peter's Chapel, Navy Yard, Mare Island, California.[59]

NEW CHAPEL AT THE NAVAL ACADEMY

In 1904, the tall spired Gothic church at the Naval Academy, erected in 1869, was torn down and work begun on a magnificent $400,000 structure

[58] CoC., Research File, letter from Major General H. K. Loughry dated 16 Oct. 1945.

[59] Kearney, "St. Peter's Chapel, Navy Yard, Mare Island," *Naval Institute Proceedings*, July 1932, p. 1025.

which remains today as the main part of the present Chapel. The cornerstone was laid on 3 June 1904 by Admiral George Dewey, and the dedication services were held on Sunday, 24 June 1908, with Chaplain H. H. Clark in charge. During the years of construction, Divine Services were held in the old armory adjacent to Worden Field. .

The new Chapel was erected in the form of a Greek cross surmounted by a great dome which rises 150 feet. Seating accommodations were provided for 1,250. The acoustic properties of the new Chapel proved unsatisfactory, for the great dome seemed to swallow up the voice of the speaker and the marble walls of the stately structure added to the difficulty. Since the public address system had not yet been invented, many experiments were tried before the defect was remedied.

Professor Charles A. Zimmerman, bandleader and composer of "Anchors Aweigh," presided at the new organ at the time of the dedication. In January 1913, the casket containing the remains of John Paul Jones, brought from Paris in 1905, was transferred from Memorial Hall to the Crypt of the new Chapel with appropriate ceremonies.[60]

The following description of the stained glass windows is taken from the 1948 edition of the Academy's *Catalogue of Information:*

> Four of its beautiful stained glass windows are memorials to heroic leaders of the past. Significantly, the fifth and newest window shows a young man on Graduation Day. Behind him are years of anxiety and hope, work and play. Before him lies a lifetime of dedicated service. Dressed for the first time in the uniform of an Ensign in the Navy of the United States, he reads the scroll commissioning him as such. High above him the Stars and Stripes flutter in the breeze, but in the background stands the Man of Galilee, pointing to an even higher obligation. It is the graduate's "Invisible Commission," challenging him to consecrate his loyalty to self and to country by supreme loyalty to the God who gave him both.

An undated anecdote relating to compulsory church attendance at the Naval Academy may well belong to this period. According to the story, one sophisticated cadet claimed that he was a Moslem and should, therefore, be excused both from attending Divine Service in the Academy Chapel and from going to any church in the town. The unusual situation was handled with diplomatic firmness in the following manner.

> He was told by his battalion officer that as a devout Moslem he must say his prayers, facing Mecca, every night and morning, and on Sundays he must remain in his room, reading the Koran on which frequent reports would be expected. Conviction involved too great a sacrifice, and the decision was soon reached that it would be better to become conventional once more.[61]

The author of the article from which this anecdote is taken, paid tribute to the several chaplains he had known who had served at the Academy. He wrote:

> The chaplains of the Naval Academy have shown intellectual and spiritual power and they have admirably adapted themselves to the duty. They have made of religion something so natural and manly that strong men could accept it without weakness or compromise. I know of no early Easter morning celebration of the Holy Communion at any of the universities equal to some I have known at the Academy at which fully 600 midshipmen were present and went to the Communion Rail.

IN SUMMARY

A great change took place in the attitude of Navy chaplains toward their work during the years 1901 to 1916, inclusive. The years can well be remembered as an era of reform in the history of the Corps. One by one, a number of annoying discriminations, which existed at the beginning of the century and had adversely affected the *esprit de corps* of the chaplains, were removed by Act of Congress or decree of the Navy Department. Along with the granting of an increase in pay, the establishment of a ratio basis to fix the quota of chaplains, and the modification of certain uniform regulations, came a new policy in their selection and promotion. These combined reforms did much to elevate the dignity and improve the efficiency of the Corps. The YMCA had begun its service to naval personnel on leave and liberty. A few Reserve chaplains were standing by for emergency service.

All these changes took place in the years which were the prelude to World War I. Undoubtedly, the outbreak of hostilities in Europe in 1914 made the United States more aware of the Navy's responsibilities and accelerated some of the reforms so long desired by chaplains. By the end of 1916, there were sufficient signs to warn the most indifferent of the coming conflict. Never before had the United States Navy been so well prepared for such an emergency. Likewise never before had the Chaplain Corps, with its forty men on active duty, been so large and so well regulated.

[60] "The Heart of the Academy," *Shipmate*, June 1942, p. 9.

[61] Alden, "Officers and Gentlemen in the Making," *Naval Institute Proceedings*, Oct. 1935, p. 1497.

CHAPTER ELEVEN

THE FIRST WORLD WAR

1917–1919

The United States entered the war against Germany on 6 April 1917. Although hostilities came to an end on 11 November 1918, the war with Germany and Austria-Hungary was not officially terminated until the approval by Congress of a joint resolution on 2 July 1921. Separate treaties with Germany, Austria, and Hungary were subsequently effected. The year of demobilization, 1919, is included in this chapter dealing with the history of the Chaplain Corps in World War I.

When the United States entered the conflict, the submarine menace was so serious that great emphasis was immediately placed on combating the undersea craft. Construction on battleships and cruisers was stopped during the war and efforts were concentrated on the building and acquisition of light craft adapted for anti-submarine warfare. With the exception of this anti-submarine activity, the units of the United States Navy in World War I had little direct contact with the enemy. A squadron of five battleships, the *New York, Wyoming, Texas, Florida,* and *Delaware* joined the British Grand Fleet at Scapa Flow in December 1917. Before the end of the war three others, the *Nevada, Oklahoma,* and *Utah* were stationed at Bantry Bay in southern Ireland.

Naval forces went ashore in France to man a number of fourteen-inch guns mounted on flat-cars and used on the western front. More than 30,000 Marines were sent, part of whom engaged in the bloody fighting at Chateau Thierry, Belleau Wood, St. Mihiel, and elsewhere.

The main task of the Navy in World War I was convoying to and from France the unprecedented number of troops. In addition to the two large navy transports, the *Henderson* and *Hancock,* the Navy took over some twenty American liners. More than 2,000,000 men were sent abroad, 46 percent in American ships and most of the balance in British vessels.

The first convoy crossed in June 1917, and the peak was reached in the spring and summer of 1918. The United States Navy provided 86 percent of the needed protection. When the war was over, the Navy was able to make the proud boast that not a single American soldier had been lost on the way to France as a result of enemy action.

American naval losses in ships and personnel, exclusive of the Marines, were small. The armored cruiser *San Diego* sank off the entrance of the mouth of New York harbor after striking a mine laid by an enemy submarine. Other losses included two destroyers, two submarines, and a few smaller craft.

NEW LEGISLATION AFFECTING THE CORPS

The declaration of war on Germany made necessary new legislation enlarging both the Army and the Navy. Congress grimly swung into action and on 22 May 1917 passed "An Act to temporarily increase the commissioned and warrant and enlisted strength of the Navy and Marine Corps."[1] The authorized enlisted strength of the active list of the Navy was temporarily increased from 87,000 to 150,000, which included 4,000 additional apprentice seamen. The authorized enlisted strength of the Marine Corps was temporarily increased from 17,400 to 30,000. The Act authorized additional commissioned officers, on a temporary basis, "not above the grades and ranks of lieutenant in the line and staff," to correspond with the temporary increase of enlisted personnel. The authorized strength was later raised as the war demands expanded.

When war was declared, the Navy had 64,680 enlisted men and 4,376 officers. In addition 12,000 Reserves and 10,000 Naval Militia had been enrolled. The Navy took over the Coast Guard with its 590 officers and 3,478 men. This gave a total force of approximately 95,000.[2] The Marine Corps grew from a force of 13,725 officers and men to a total of 75,101

[1] 40 *Stat.* 84–5.

[2] Daniels, *Our Navy at War,* p. 311. Report of Secretary, 1919, p. 173.

including 269 women.[3] By the time of the Armistice, naval personnel, including all categories—Regulars, Reserves, Marine Corps, and Coast Guard—totaled about 533,000. This number included approximately 11,000 Yeomen (F) and 1,713 women nurses. The women yeomen were never officially designated as yeomanettes. The "F" indicated female.[4]

All these developments greatly affected the Chaplain Corps for an enlarged Navy and Marine Corps demanded more chaplains. The Act of 22 May 1917 included the following section:

> Provided further, That temporary chaplains and temporary acting chaplains in the Navy may be appointed for service during the period of the War in the proportion of the personnel of the Navy as now prescribed by existing law.

Upon the passage of this Act, the Regular Navy had four classifications of chaplains in addition to the Reserves. They were: chaplain, acting chaplain, temporary acting chaplain, and temporary chaplain. George Snavely Rentz, who was appointed a temporary acting chaplain on 15 August 1917, was the first and only chaplain to be received in this status in 1917. Forty others were given temporary appointments in 1918. Most chaplains, who entered the Navy during the months of April to October 1918, came under this classification, and many were unaware of any difference between acting chaplain and temporary acting chaplain until they came up for promotion. The temporary acting chaplains then found that they lost precedence to acting chaplains who were promoted first. No temporary acting chaplains were appointed after the signing of the armistice with Germany. Navy Department records disclose no appointments of temporary chaplains.

[3] Daniels, *op. cit.*, p. 216.
[4] *Ibid.*, p. 329.

RESERVE CHAPLAINS IN THE FIRST WORLD WAR

When the United States declared war on Germany on 6 April 1917 there were only forty chaplains on active duty in the Navy. The first immediate and available supply of additional chaplains came from the various reserve organizations. Out of the small nucleus of nine chaplains on the rolls of the Naval Militia at the beginning of 1917, only Chaplains Niver, Huske, James, and Day transferred to the National Naval Volunteers and were thus available for duty. Four other chaplains joined the Naval Volunteers on or before 6 April, bringing the total to eight. Of this number the following five were called to duty the day after war was declared: Gould, Brokenshire, James, Williams and Larned. Only two other names were added to this roll during the remainder of 1917. The following ten Naval Volunteer chaplains were on duty in 1918:[5]

[5] *Navy Register*, 1 Jan. 1918.

	Born	Date of Commission	Date of first duty
Commanders			
Romeo Gould	12 June 1873	6 Apr. 1917	7 Apr. 1917
Ernest LeRoy Paugh	30 Jan. 1890	16 July 1917	23 July 1917
Lieutenant Commanders			
William P. Williams	25 Aug. 1881	12 Mar. 1917	7 Apr. 1917
John I. Brokenshire[6]	1 Aug. 1862	6 Apr. 1917	7 Apr. 1917
Lieutenants			
Edwin Barnes Niver	2 July 1863	1 Nov. 1901	8 June 1917
Bartholomew F. Huske	15 July 1884	30 Apr. 1913	9 May 1917
Sidney Thomas James	9 Dec. 1883	21 Sep. 1915	7 Apr. 1917
James Samuel Day	5 June 1886	30 Nov. 1916	11 Apr. 1917
Albert Cecil Larned	4 Apr. 1883	1 Jan. 1917	7 Apr. 1917
Charles W. Moore	21 Feb. 1886	10 Apr. 1917	13 June 1917

[6] It is interesting to note that Chaplain Brokenshire was fifty-four years old at the time he was commissioned.

852159—49——12

Chaplain Paugh received the rank of commander when he entered the Naval Militia. Chaplains Gould, Williams, and Brokenshire were commissioned as lieutenant commanders. Gould was promoted to commander in June when he transferred to the National Naval Volunteers. There was, evidently, wide disparity in the various units of the Naval Militia in granting rank and promotion.

Class 4 of the Naval Reserve Force, which had been established in August 1916, consisted of "Officers and men enrolled for four-year periods and available in case of war or national emergency for the coast defense and for special technicist duty on shore."[7] There were at least five Reserve Force chaplains who were commissioned and called to active duty in 1917 who belonged to this class and were not, therefore, listed as members of the National Naval Volunteers. These five were:

	Date of Commissions	Date of First Duty
John Nicol Mark..........	3 Apr. 1917	18 Apr. 1917
Thomas Montgomery Mark.	3 Apr. 1917	18 Apr. 1917
Hersey E. Rountree........	31 July 1917	31 July 1917
William B. Ayers	4 Dec. 1917	11 Dec. 1917
Patrick J. Hammersley	10 Dec. 1917	15 Dec. 1917

These men brought the total of Reserve chaplains who were available for duty in 1917 to fifteen.

The National Naval Volunteers and Class 4 of the Naval Reserve Force served a useful purpose in providing a nucleus of Reserve chaplains for World War I. After war was declared, however, the Navy Department laid chief emphasis upon the procurement of Regular rather than Reserve chaplains. New chaplains were encouraged to enter the Navy under the provisions of the Act of 1917. However, this Act did not entirely rule out the further commissioning of Reserve chaplains. Since it gave the status of lieutenant (junior grade) to the acting and temporary acting chaplains, it was to be expected that the younger newly appointed Reserve chaplains should receive the same rank. Such was the case. The first Reserve chaplain to be commissioned lieutenant (junior grade) was Hersey E. Rountree.

On 1 July 1918, Congress transferred the National Naval Volunteers to Class 2 of the Naval Reserve Force. Thus within a period of two years several chaplains were members of three different reserve organizations—Naval Militia, National Naval Volunteers, and Naval Reserve Force. Only one chaplain, Albert R. Parker, joined Class 2 of the Naval Reserve Force in 1918, while sixteen others entered the service by coming under the provisions of Class 4. This brought the total of Reserve Force chaplains who saw service with the Navy in 1917 and 1918 to thirty-two, eleven of whom were in the Volunteers or Class 2 and twenty-one in Class 4.

The most prominent Reserve chaplain was Dr. Henry van Dyke, the famous educator, author, clergyman, and diplomat, who accepted the rank of lieutenant commander on 4 January 1918 and remained in the service until 18 December of that year. Van Dyke served more as a public relations officer than as a chaplain. His brilliance as an orator and his first-hand knowledge of European conditions, gained while serving as United States Minister to the Netherlands and to Luxemburg during the years 1913 to 1917, admirably fitted him to serve in this capacity. He toured the country speaking before civilian audiences and at various naval installations interpreting the issues involved in the war. No other naval chaplain has had such an unique tour of duty.

A number of important lessons were learned through this experience with Reserve officers during the World War I which helped to reshape the new post-war legislation bearing upon this subject. Thus the way was prepared for the much more extensive use of Reserve chaplains in World War II.

FIRST CHIEF OF CHAPLAINS

The appointment of Chaplain John B. Frazier on 5 November 1917 as head of the Chaplain Corps, was made absolutely necessary by the exigencies of war.[8] Ever since 1871 chaplains had urged the Navy Department to appoint one of their number to represent their interests in Washington.

Frazier's appointment was hailed with acclaim not only by chaplains but also by the churches of the United States. Even though five other chaplains then on active duty in the Navy were his seniors, his splendid executive and administrative abilities, added to his keen judgment of human nature, made him a logical choice.

Chaplain Frazier was a Southern Methodist, as was the Secretary of the Navy, Josephus Daniels. The two were fast friends and this often redounded to the benefit of the Chaplain Corps. The Navy De-

[7] Navy Dept. Bulletin, ***Information Concerning the U. S. Navy***, p. 163.

[8] The office of Chief of Naval Chaplains was officially created by law in December 1944 when the rank of rear admiral was attached to the position. However, within the Corps the title "Chief of Chaplains" was universally used before this date.

partment then occupied rooms in the old State, War, and Navy Building, but at first no accommodations were available for the newly appointed Chief of Chaplains. Frazier's first office was in Room 813 of the Woodward Building located at 15th and H Streets, Northwest.

Chaplain Edel, out of his personal memories of Chaplain Frazier, has given the following sympathetic picture of the first Navy Chief of Chaplains and of his pioneer task:

> He was a big, hard-muscled man, with a face that looked as if it might have been chiseled out of stone, and he was as resolute as he looked. But under that flinty exterior there was . . . compassion and tenderness . . . and a most amazing and unpredictable sense of humor. That sense of humor could ridicule and sting, but it was the cut of the surgeon's knife, corrective and healing. And such a man, in 1917, was ready and waiting when the call came to take the helm and steer the Corps through its most difficult days. When he reported for duty at the Navy Department he found that he must develop his own position, continually breaking new ground, solving new problems, laying new plans, continually meeting fresh opportunities and fresh discouragements, finding new friends . . . and through it all he thrust his way, carrying the Corps of Chaplains to new heights of efficiency and prestige. No man less resolute could have done it.[9]

Chaplain Frazier's appointment as the first Chief of Chaplains may be considered the beginning of the Chaplains Division, Bureau of Navigation. He instituted many policies which greatly influenced the work and efficiency of the Corps. He inaugurated the present system of individual jackets for each chaplain which contain correspondence, reports, and other relevant material. Frazier called for monthly reports from the chaplains and exercised a paternal supervision over their work. Backed by the authority given him by Secretary Daniels, he insisted upon the right to make recommendations to the Department for the assignment of chaplains to duty.

As opportunity permitted, he called groups of chaplains together for conference. Chaplain H. R. Davidson in his annual report for 1918 commented on such a meeting:

> In December Captain J. B. Frazier, USN, visited New York on a tour of inspection and at the same time some twenty-five chaplains gathered for a dinner and conference. This was productive of so much excellent advice and wise counsel, I would recommend, if it be not presumptuous that this gathering of chaplains be at least an annual occurrence.[10]

Believing that a higher standard was necessary in the Chaplain Corps than in any other corps in the Navy, Chaplain Frazier wrote:

> A Chaplain is supposed to set an example in moral living, and to do so he must maintain a higher standard than that of the average man. The world expects him to do this, and will not respect him unless he does. Officers and men aboard ship, regardless of religious conviction, demand that he live a life different from their own, or suffer the consequences in loss of influence and usefulness. Not infrequently one will hear the remark:—"I saw the Chaplain do so and so." "You did?" "Yes." "Well, that finishes him so far as I am concerned." The same thing might have been done by other officers without creating the least comment or surprise. This homely illustration but emphasizes the fact that the Chaplains are a class by themselves, not only in what they profess, but in the demands that are made of them. This is as it should be, and is what the clean lived, conscientious Chaplain expects and desires.[11]

Chaplain Frazier carried the heavy duties of his office during the war without assistance other than clerical. Beginning with 25 April 1919, Chaplain J. F. B. Carruthers was assigned as his assistant and served until 3 October of that year. Fortunate indeed was the Corps in having such a strong man as Chaplain Frazier to take over the helm in the days when precedents were being set and principles established. All who followed have been his debtors..

Chaplains who entered the service during the period from November 1917 to November 1921, when Frazier was serving as Chief of Chaplains, have many interesting anecdotes to tell of him. He made it a practice to interview every applicant for the naval chaplaincy and devised unique tests to aid him in determining a man's fitness for the office. Usually, candidates were obliged to remain in Washington three or four days while they were under scrutiny. Protestants were asked to speak at the old City Mission then located at Fourth and Pennsylvania, Northwest. Chaplain Frazier joined the motley congregation gathered in from the streets of Washington and judged the aspiring candidate with critical eyes and ears.

"I want you to speak for twelve minutes," said Chaplain Frazier to one nervous candidate, "and that does not mean thirteen." This unexpected time limit often meant a rapid and impromptu adjustment of an old sermon outline which was a test not every man could meet.

Sometimes after the service Chaplain Frazier

---

[9] Edel, "Navy Chaplains from 1775–1917," pp. 25–6.
[10] CoC., Davidson File.

[11] NRSO: 2650–281.

would push through the circle of hearers who were complimenting the speaker on his sermon, and break the spell by gruffly asking: "Is that the best you can do?" He deflated the ego of one man by bluntly declaring: "I never heard a worse sermon in my life." And of another he asked: "What makes you think you would make a good chaplain in the Navy?" Chaplain Neyman described his manner as being "somewhat frightening and brusque, but really polite and kindly, and certainly wise and practical."[12]

After Edgar W. Davis successfully passed his tests in July 1918 and was assured of an appointment, Chaplain Frazier conducted him into Secretary Daniels' office and in his characteristically gruff manner said: "Well, Mr. Secretary, this is the best I can do for you today."

The Secretary left his chair and, stepping out from behind his desk, put his hand on the shoulder of the young chaplain, saying, "He doesn't seem too bad. What is his church?"

"Methodist, South," proudly replied Frazier. Whereupon Secretary Daniels grasped the hand of the candidate and exclaimed, "There is no need to question further his theology or his ability."

One day an ambitious young wife ushered her husband into Chaplain Frazier's office. While the young man was closeted in the inner office for the interview with the Chief, his wife was nervously anticipating the outcome in the outer room. Finally the separating door opened and the two men emerged. Before either could speak to her, the wife bubbled forth: "Oh, Chaplain Frazier, I do hope you have taken Henry into the Navy. He is so fond of water."

"If that is the case," Chaplain Frazier quickly retorted, "You'd better go home, dig a well, and let Henry jump into it."

When such an incident is told to one of the older chaplains who cherishes personal recollections of Chaplain Frazier, he will invariably hasten to add: "Yes, that's pretty good, but I know a better story about the old man. One day . . ."

Following the death of Chaplain Frazier on 11 November 1939, burial services were held in Arlington Cemetery, led by Chaplain Robert D. Workman, then Chief of Chaplains. In his tribute to the first Chief Workman said:

> Upon the shoulders of our first Chief rested the responsibility of selecting an exceptionally large number of new chaplains, and of establishing our Corps and its responsibilities on a basis such as had never been undertaken prior to that time. The manner in which Chaplain Frazier faced his task and the degree of success which he attained have left us an example and a heritage for which we must ever be thankful.

THE PROCUREMENT AND ASSIGNMENT OF CHAPLAINS

The most pressing problem Chaplain Frazier faced after his appointment in November 1917 was that of chaplain procurement. Between the declaration of war and his appointment only twenty-six acting or temporary acting chaplains were added to the Corps. During December, January, and February. Chaplain Frazier accepted thirty-nine additional candidates. For the two years, 1917 and 1918, a total of 162 chaplains were received into the Corps. Of this number, 107 entered in 1918, almost twice as many as had been commissioned the previous year.

In the procurement of chaplains, the Chaplains Division had the cooperation of Catholic and Protestant churches. On 24 November 1917 the Pope appointed the first Episcopus Castrensis, or Chaplain Bishop, in the United States in the person of the Most Reverend Patrick J. Hayes, then Auxiliary Bishop and later Archbishop of New York.[13] Bishop Hayes organized the Military Ordinariate with headquarters in New York, which endorsed to the Secretary of the Navy all of the thirty-five Catholic chaplains who entered the service during 1917 and 1918. Among this number were Chaplains J. T. Casey, W. A. Maguire, F. L. McFadden, and T. F. Regan, who remained with the Navy to serve in World War II.

Among the Catholic chaplains were seven representatives from the following Orders: Dominican, Franciscan, Benedictine, Jesuit, and Augustinian. The first Order man to enter the service was Q. F. Beckley, O. P., who received his commission as acting chaplain, Regular Navy, on 22 June 1917. Two of the Order men entered the service as Reserve chaplains, the others were commissioned as acting or temporary acting chaplains in the Regular Navy. At the end of the war, the Order men were expected to return to their former duties as soon as possible unless they had special permission which permitted their *exclaustration* during peace time. Chaplain J. F. Underwood, O.S.B., remained in the service until 12 October 1924 and Chaplain J. H. Finn, T.O.R., until 31 December 1928.

---

[12] Neyman, "Manual," p. 39.

[13] Williams, *American Catholics in the War*, p. 238. *United States Catholic Chaplains in the World War*, p. XV.

On 27 March 1917 Dr. Worth M. Tippy began his duties as the successor to Dr. H. K. Carroll, the first secretary of the Washington Committee on Army and Navy Chaplains already referred to. This Washington Committee consisted of two Methodists, two Presbyterians, and two Episcopalians. It was reorganized after the declaration of war to include one member of each of the thirty-two constituent bodies of the Federal Council and was renamed the General Committee on Army and Navy Chaplains. This represented the major portion of the Protestant bodies of the United States.

Arriving in Washington on the eve of the outbreak of war, Dr. Tippy hastened to secure an understanding with both the Secretary of War and the Secretary of the Navy "by which it was agreed that the Federal Council should have authority to investigate and nominate all [Protestant] candidates for chaplains, and that applications already on file in the departments should be sent over to the Washington office of the Federal Council."[14]

According to the report rendered to the Federal Council by the General Committee on 19 October 1918, seventy-eight of its recommended candidates for the chaplaincy had then been accepted by the Navy Department. The Committee was not requested to consider the applications of Reserve chaplains.

On 5 September 1917 Chief Yeoman Alfred James Haines, who had been on active duty as a member of the Reserve Force since 23 April, was appointed an acting chaplain. Haines was ordained by the Methodist Church, North, on 2 September. He was the first of several who, upon meeting certain necessary requirements, transferred from other branches of the Navy to the Chaplain Corps.

Among the Protestant chaplains who entered the service in 1917 and 1918 and remained in the Navy to serve in World War II were Chaplains H. S. Dyer, W. W. Edel, H. G. Gatlin, E. H. Groth, F. H. Lash, M. M. Leonard, R. E. Miller, J. W. Moore, M. H. Petzold, G. S. Rentz, B. D. Stephens, W. P. Williams (all of 1917), E. L. Ackiss, F. L. Albert, E. W. Davis, T. L. Kirkpatrick, C. A. Neyman, H. M. Peterson, R. W. Shrum, A. E. Stone, W. N. Thomas, R. W. Truitt, A. de G. Vogler, and M. M. Witherspoon (all of 1918). Several who resigned their commissions later returned to serve in time of emergency.

Unquestionably, the cooperation of the Military Ordinariate and the General Committee made infinitely easier Chaplain Frazier's task of finding suitable chaplains. In his report for 1917 Dr. Tippy stated that the General Committee had taken aggressive steps throughout the various Protestant churches to stimulate interest in the naval chaplaincy. This positive approach was used "to overcome the use of political influence for the appointment of chaplains."[15] In only one or two cases were appointments to the chaplaincy during World War I made as the result of political pressure. The cumulative insistence of the chaplains themselves and of their respective denominations for high educational and spiritual standards for all new appointees to the Corps was gradually crystallizing into the fixed policy of the Department.

The law stated that no chaplain could be more than thirty-five years old when commissioned. Since the Act of 1914 required a three-year probationary period, during which the acting chaplain served under an appointment, the upper age limit of the candidate was of necessity pushed back to thirty-two. Then, to be doubly sure that there would be sufficient time to clear all technicalities, it became the policy of the Chaplains Division not to appoint any candidate who was more than thirty-one and a half years of age.

The average age for acting and temporary acting chaplains in 1917 was about twenty-seven; in 1918 about twenty-eight. The youngest chaplain to be appointed was John M. Quinn, who was twenty-two years and seven months old when he entered the Navy. Chaplain W. W. Edel was only a little more than twenty-three when appointed. For the most part the Reserve chaplains were older men, their average age being thirty-eight. On 25 April 1917 Congress raised the maximum age limit for Reserve officers on first appointment from thirty-five to fifty. Evidently when a candidate for the chaplaincy was too old to qualify for the Regular Navy, he was encouraged to serve as a Reserve. Chaplain Henry van Dyke entered the Reserve at the age of sixty-five. His appointment was made possible through a special order issued by Secretary Daniels.

Three chaplains, G. S. Rentz, H. M. Peterson, and E. C. Boynton, were thirty-five, thirty-two, and thirty-three respectively when appointed temporary acting chaplains. Chaplain R. J. Davis, a Christian Scientist, was appointed when he was thirty-four. With these exceptions, the general rule of giving no appointment of acting or temporary acting chaplains to men over thirty-one and a half was observed.

[14] *Federal Council, Biennial Report,* 1917, p. 77.

[15] *Ibid.*

In order to facilitate the procurement of chaplains, a circular was prepared to disseminate information about the Navy chaplaincy.[16] This circular reprinted the sections of the Act of June 1914 pertaining to the chaplaincy, outlined the policy of the Navy Department regarding the qualifications of the candidate and the procedure to be followed to secure an appointment, and listed the duties expected of a chaplain as set forth in *Navy Regulations*. It was also stated that chaplains would receive the same pay and allowances as other commissioned officers of their rank.

The circular summed up the nature of a chaplain's work both at sea and on shore as practically identical to "that of a minister in civilian life in charge of a parish on shore." The first of its kind ever issued to encourage qualified clergymen to consider the challenge of the naval chaplaincy, this circular was undoubtedly a great aid in procurement of chaplains during World War I.

When practicable, the newly appointed chaplain was given about a month's shore duty for indoctrination under the tutelage of an older chaplain. He was then sent to sea, in accordance with the Act of 1914 which required that acting chaplains spend their first three years at sea. Temporary acting chaplains came under the same law.

One of Chaplain Frazier's duties as Chief of Chaplains was to assign chaplains to those billets where they would be of greatest service. Under the pressure of immediate war demands, several chaplains still in their probationary period were assigned duty with the Marines in France. Later these chaplains were penalized when they came up for promotion, for such duty, even in the combat area, was not considered a substitute for sea duty. Special provision had to be made by Congress for such cases.

With but few exceptions, only the older chaplains and the Reserve chaplains were available for regular assignments at shore stations. This situation brought difficult problems. The Naval Training Station, Great Lakes, Illinois, for instance, grew to 50,000 during the war. Serving there with Chaplain Frank Thompson, the senior chaplain, were no more than six assistant chaplains, several of whom were there for only a short time. Had it not been for the splendid cooperation received from the local clergy and from the various welfare organizations, the spiritual needs of so large a group could not have been met.

[16] The circular is without date or indication of origin. It was probably prepared by Chaplain Frazier and issued by the Navy Department.

According to Chaplain Frazier, the 199 chaplains on duty at the time of the signing of the armistice were divided as follows:

> Thirteen of these were assigned to duty with Marines in France. One hundred and thirty six served on board battleships, cruisers and transports. Nine were assigned to the various hospital units in France. Four served with the Expeditionary Forces in San Domingo and Haiti, the remainder [37] on shore duty in the United States.[17]

FIRST JEWISH CHAPLAIN

The first Jewish chaplain to be commissioned in the United States Navy and the only one to serve in World War I was Rabbi David Goldberg of Corsicana, Texas, who was appointed acting chaplain on 30 October 1917. The inclusion of a Jewish rabbi in the Chaplain Corps of the Navy came as the result of the intercession of Senator Morris Sheppard of Texas. Having made the appointment possible, Senator Sheppard turned to Dr. Henry Cohen of Galveston, then the acknowledged Dean of the Texas rabbinate, for a nomination. Cohen selected Goldberg.

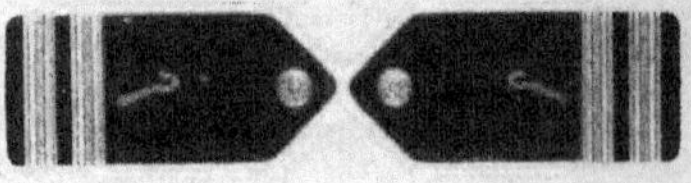

Insignia worn by the Jewish Chaplain, David Goldberg, in the First World War.

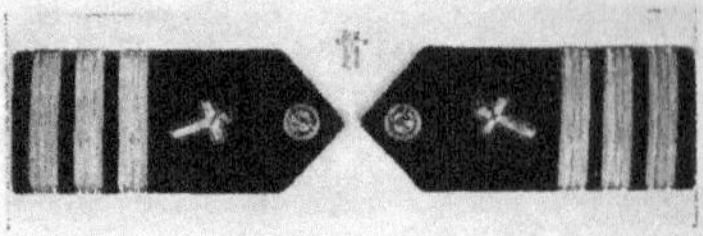

Insignia worn by a Chaplain of the Christian Faith.

The appointment of Chaplain Goldberg as an acting chaplain in the Regular service showed that the Navy had recognized an obligation to meet the spiritual needs of personnel of the Jewish faith. However, the appointment of a chaplain from one of the minority religious groups of the United States introduced new problems of administration for Chaplain Frazier. Did such an appointment establish a precedent for other minority groups? Since the Jewish

[17] Nav. Rec. Coll., Chaplains' File. Frazier did not include the two retired chaplains, called back to duty, in his statistics.

personnel were so widely distributed throughout the Navy, how could Chaplain Goldberg best minister to members of his faith? Where was such a chaplain to be stationed?

The Chief of the Bureau of Navigation wrote to Chaplain Frazier on 14 November suggesting that Chaplain Goldberg be attached to the Washington Navy Yard for a short period of indoctrination and then be given a "roving commission" with instructions to spend two weeks at each of the principal Navy Yards and Naval Training Stations. Chaplain Frazier favored this plan. Secretary Daniels, however, felt that Chaplain Goldberg had not been appointed a chaplain-at-large but had entered the Navy subject to the same rules under which all other acting chaplains served. Under the law governing acting chaplains, a tour of sea duty was obligatory for chaplains in their probationary period and for promotion. Therefore, after spending only a month at the Washington Navy Yard, Chaplain Goldberg was assigned to the *Grant*.

When the novelty of having a Jewish chaplain on board had passed, Chaplain Goldberg began to realize the difficulty of his situation. Writing to Chaplain Frazier on 20 April 1918, he admitted:

> Since my attachment to the U.S.S. President Grant in the capacity of Chaplain, the conviction has grown on me that the position of a Jewish Rabbi called upon to minister to the religious needs of an almost exclusively non-Jewish personnel is anomalous, and indeed intolerable. Of the six-hundred men comprising the crew of our Transport, only five are of the Jewish faith, that is, less than one percentum, and I feel that the services of a Christian Chaplain here would be more acceptable, and indeed more justifiable.
>
> I can provide for recreational facilities as well as could the Christian Chaplain, and I can minister to the educational needs of the men without conscientious entanglements. I can also attract the men to religious services, and my experiences on that score, I can state truthfully, have been very pleasant. But I cannot take the place of a Christian Chaplain, certainly not there where the presence of a Chaplain should be most helpful. The difference between a Christian minister and a Jewish Rabbi, in the deep-rooted belief of the men, is so vast and fundamental, that it leaves me almost beyond the pale, and while in the submarine zone, in the sick-bay, or at the death-bed, I find that the men are left almost without spiritual consolation, as my own spiritual potency there is almost nil.[18]

Chaplain Goldberg felt that he would be more effective were he stationed at one of the large training stations where he would find more men of his own faith. "I could then carry on," he wrote, "under the direction of the Chaplain of such station, social and educational activities for ALL men, while confining the purely religious and ceremonial to those of my faith only."

Chaplain Frazier acted on Goldberg's request by transferring him in June 1918 to the Naval Training Station at Great Lakes. There he remained until he resigned on 11 December 1919.

When Chaplain Goldberg donned the uniform with the Latin cross, the insignia of the Chaplain Corps, he was painfully conscious of the incongruity. However, for the time being there was no alternative. Since he was the first Jewish chaplain in the Navy, he had no precedent to guide him. The newly created Jewish Welfare Board did not then have a committee on the Army and Navy chaplaincy to intercede in his behalf. As a regularly appointed naval chaplain, Goldberg was obligated to wear the prescribed insignia of the Corps. Chaplain Frazier sought to ease the inconsistency by pointing out that he had entered the Corps as a chaplain who was a Jew rather than as a Jewish chaplain. "Anyway," added Chaplain Frazier, "you will probably soon get your orders to go to sea where you will meet only service men. These understand perfectly that you are wearing the insignia of your corps, and not the emblem of your faith."

Upon his first return to the States after making the round-trip to France as chaplain on the *Grant*, Goldberg found a number of letters from several rabbis who were indignant over his wearing Christian insignia. Writing to a friend on 5 April, Chaplain Goldberg patiently pointed out the technicalities of the situation. "I could not remove the insignia arbitrarily without scandal and punishment," explained Goldberg, "No officer could do such a thing arbitrarily."

On 10 May 1918 Chaplain Goldberg requested permission from the Bureau of Navigation to substitute the "Shield of David," (the six-cornered star) the insignia then worn by Army chaplains of the Jewish faith, for the Latin cross. He gave as his reasons:

> (a) It is unnatural for a Jewish Rabbi to be represented by an insignia which is decidedly non-Jewish.
>
> (b) The sight of the insignia renders impotent the influence which I should have with my co-religionists in the service, were the insignia in any way expressive of my vocation.

This request was denied in the reply of the Bureau dated 14 June because the suggested insignia was the

[18] CoC., Goldberg File. Additional information in this section has been taken from this file unless otherwise noted.

same as that used by "officers in the Army doing duty with the General Staff." However, the Bureau invited Chaplain Goldberg to submit other suggestions.

By this time the matter of suitable insignia for the one Jewish chaplain in the Navy had become a matter of concern for the Jewish Welfare Board and was discussed at the Conference of Rabbis held in Chicago 28 June–4 July 1918. The 29 June issue of *The Army and Navy Register* carried the following item under "Navy Uniform Changes":

> Owing to objections on the part of Hebrew Chaplains to wearing the cross, which has distinguished chaplains in both Army and Navy, the suggestion has been made that the Shepherd's Crook be substituted as the insignia of the Navy chaplains, as it has been in the Army, but no action yet has been taken.

The shepherd's crook appears to have been the first definite insignia of the United States Army chaplaincy. It is said to have been in use shortly after the Civil War. Order No. 10 of 1880, War Department, states that the shepherd's crook insignia "may be worn by chaplains." Army chaplains used this insignia until the cross was authorized in 1898.

After noticing the suggestion in *The Army and Navy Register*, Chaplain Goldberg wrote to Chaplain Frazier on 5 July and reported the outcome of his first appeal to the Bureau. Goldberg, emboldened by the suggestion of the *Register* to which he drew Chaplain Frazier's attention, wrote:

> . . . I have not been so presumptuous as to ask for a *general* change of insignia; I only requested it for myself, on the grounds that I have stated. Now that the Bureau had found the insignia of the Shield of David impractical, I will admit that I am not at all disconcerted with the Bureau's reply, for at bottom I would be infinitely more delighted with the device of the Shepherd's crook as a universal insignia than with the Star of David as an insignia of my own, and the Bureau's reply opens the road just to such possibilities.

On the same day Chaplain Goldberg wrote again to the Bureau of Navigation requesting permission to substitute the shepherd's crook for the Latin cross and if that be found impractical then the two tables symbolic of the ten commandments. While Chaplain Goldberg intended his petition to apply only to himself and to other Jewish chaplains who might be appointed, yet the last paragraph of his letter was so worded as to leave the way open to the Bureau for a consideration of an insignia acceptable to both Jews and Christians.

Chaplain Frazier's reaction to Goldberg's suggestion that some insignia of universal design be adopted which could be worn by all "without conscientious entanglements" was immediate and emphatic. ". . . every one of us who professes the Christian faith," wrote Frazier to Goldberg on 8 July, "would resent most bitterly any effort to substitute any other insignia in place of the Cross." He also added: "I feel sure that I speak for the one hundred and sixty Christian chaplains in the Navy, when I say that the Cross is dearer to us than any shepherd's crook . . . could possibly be." Chaplain Frazier, however, had no objection to Jewish chaplains wearing some insignia indicative of their faith.

Chaplain Goldberg hastened to assure Chaplain Frazier that the request for a change of insignia was made only for those of the Jewish faith. On 26 June 1918 Secretary of Navy Daniels put his official approval on Goldberg's request for the shepherd's crook as the Jewish insignia. Shortly thereafter an official directive from the Navy Department confirmed this change.

Since the law governing acting chaplains called for a three-year tour of sea duty, Chaplain Goldberg was ineligible for either a commission as a Regular Navy chaplain or for promotion unless he completed this requirement. His previous experience at sea, however, had demonstrated to him the anomalous position of a rabbi ministering to Christians as regular duty. This situation together with the fact that many Reserve chaplains were returning to civil life in the summer of 1919 made Goldberg feel that he should resign his appointment.

Goldberg in a letter on 27 August laid his problem very frankly before Chaplain Frazier. "If you remain in the service," wrote Frazier in reply, "you will have to go to sea at no distant date and I feel confident that there is no ship to which you could be assigned that carries in her complement more than half a score of your faith."

Goldberg estimated that about two percent of the total naval personnel were Jewish. This was a larger percentage than existed in the pre-war Navy. The fact that he was the only representative of American Jewry in the Corps was an argument in favor of his staying in the Navy if at all possible. During the time Frazier and Goldberg were corresponding regarding this matter, a strong request came from the Jewish Welfare Board for the appointment of additional Jewish chaplains. Frazier disapproved and in this Goldberg agreed. "I am sure, you're quite right in advising against his recommendation," wrote Goldberg on 10 September, "as even the solitary Jewish chaplain already in the service cannot maintain himself without some exception or other." On the other hand Goldberg felt that since it was expected that there would be at least 4,000 men of the Jewish faith in the Navy by July 1920, some provision should be made for their spiritual care. While no Jewish chaplain would have enough work to do if he worked exclusively among men of his own faith on a ship or at a naval training station, Goldberg felt that possibly a billet could be found as the Jewish chaplain of a fleet or in the Chaplains Division of the Bureau of Navigation.

Goldberg submitted his resignation on 21 November 1919 after more than two years' service. These were years of pioneering for he had to set precedents rather than follow them. In spite of the recognized difficulties arising out of the differences of faith, Chaplain Goldberg rendered an unusual ministry to Jews and Christians alike. At the Great Lakes Training Station he served as chaplain of the Detention Barracks and often had as many as 1,200 present for Divine Service, most of whom were Christians. Upon the acceptance of his resignation as an acting chaplain, Goldberg enrolled in Class 2 of the United States Naval Reserve Force. He later transferred to Class 6 and as such was automatically transferred to the United States Naval Reserve when that was formed in 1925. He kept his commission alive until March 1941 when he was honorably retired in the rank of lieutenant commander.

FIRST CHRISTIAN SCIENCE CHAPLAIN

Upon the request of the Christian Science Board of Directors, and the recommendation of such influential people as the Grand Master of the Grand Lodge (Masonic) of Massachusetts, the Governor of that State, and several members of Congress, the first Christian Scientist was appointed an acting chaplain 19 January 1918. He was Richard Joseph Davis of Boston. The appointment was made by Secretary Daniels over the disapproval of Chaplain Frazier.

Even though Davis brought the highest testimonials regarding his character, personality, and ability as a social worker, the fact remained that he was not ordained. He was a layman. With the appointment of a Jewish chaplain in October 1917, the Navy had recognized three main religions in its chaplaincy in the Catholic, the Protestant, and the Jewish faiths. How was the Christian Science chaplain to be classified? Certainly he could not be grouped with either the Catholic or the Jewish faiths. If he were assigned as a chaplain to a ship or station, he would displace a Protestant clergyman and that as a consequence the Protestant personnel would be denied a chaplain capable of administering the sacraments and of serving them in other ways.

Dr. Worth M. Tippy, Executive Secretary of the Committee on Social Action of the Federal Council of Churches of Christ in America, wrote to Chaplain Frazier on 24 January 1918, registering a protest. He declared:

> I called . . . attention to the fact that a Scientist who is also a Layman, as is Mr. Davis, would be unacceptable to both Protestant and Catholic, that he ought not to have an appointment which would make him an official ministrant of religion, on a ship or at a training camp. The only way in which he could serve acceptably would be as an itinerant Chaplain, looking after Christian Scientists in various places.[19]

The Christian Scientists were like the members of the Jewish faith in that while they had several thousand of their number in the Navy, these were so scattered that it would be impossible for any one chaplain to reach effectively even a small percentage of them. The idea of making a representative of one of the religious minorities an itinerant chaplain had been proposed in the case of Chaplain Goldberg and had been firmly rejected by Secretary Daniels. If representatives of the minorities were accepted as chaplains in the Navy, they would then have to accept the usual duty given to other newly appointed chaplains. This meant sea duty.

Chaplain Frazier replied to Dr. Tippy on 25 January saying in part:

> Mr. Davis impresses me as being a fine type of man, and we have talked over carefully the kind of work he will be expected to do. While, of course, he will hold a Christian Scientist Service, he will also hold an undenominational service in which Christian Science will have no part. It is impossible to give him the "roving commission" as you suggest, but in all probability he will be stationed on board a transport used for conveying men back and forth, and on each of these transports in all probability there will be a Chaplain of some other faith. This is the only way that we can possibly provide for a case of this kind.[20]

Another problem arose in regard to this appointment of a Christian Scientist. Davis, while a graduate of the University of Chicago and a qualified reader in the Christian Science Church, did not have the three-year theological education, or its equivalent, required of all other candidates, including those of

[19] CoC., Davis File.
[20] *Ibid.*

the Jewish faith. Would the exception made in the case of Davis set a precedent to be seized upon by other religious minorities? If so, what would happen to the high educational standards which had been set for Navy chaplains after decades of work by those interested in improving the chaplaincy?

After spending an indoctrination period of about three weeks with Chaplain Arthur Stone at the Charleston Navy Yard, Chaplain Davis was assigned to the *Roanoke* of the Mine Force. Here he was most effective in his welfare work and won the praise of his commanding officer. He was faithful, whenever possible, in calling upon chaplains of other faiths, on YMCA workers, and other qualified religious leaders to conduct Divine Service and to administer the sacraments. During the summer of 1918, Davis was given temporary duty ashore in England "to assist the Christian Science Camp Welfare and War Relief Committee in organizing and opening rooms near our larger Naval Bases."[21] Davis served about six months each on the *Supply* and the *Columbia* in 1919. Upon his resignation from the chaplaincy in January 1920, he enrolled in the Naval Reserve Force.

STATISTICS OF THE CHAPLAIN CORPS

If the ratio of one chaplain for every 1,250 naval personnel had been maintained, there would have been approximately 480 chaplains on duty at the time of the peak enrollment immediately before Armistice Day. However, only 203 chaplains were on duty during the years of World War I, which was about forty-two percent of the authorized strength. Not more than 201 were in the Corps at any one time.

The United States Government lacked the techniques and experience of total mobilization for war. Munitions and equipment had to be manufactured; ships had to be built; armies had to be recruited and trained; and all this took time. When war was declared, forty chaplains were already in the Corps. Fifty-five additional chaplains entered during 1917, of which number thirty-nine were acting chaplains, one was a temporary acting chaplain and fifteen were Reserves.

All available Reserve Force chaplains were on duty within four months. Two chaplains, Ayers and Hammersley, entered the Corps through Class 4 of the Reserve Force in December.

[21] *Ibid., Davis* report of 15 January 1919.

During 1918, 107 new chaplains were appointed. Of these, seventeen were Class 4 Reserves, fifty were acting chaplains, and forty were under temporary appointments as acting chaplains.[22] The following chart shows accessions by months and years.

1917:	Jan	Feb	Mar	Apr	May	Jun	Jul	Aug	Sep	Oct	Nov	Dec	Total
Act. Chap.				2	1	8		10	2	3	4	9	39
Tem. Act. Chap.							1						1
Reserves				8	1	2	2					2	15
Total				10	2	10	3	10	2	3	4	11	55

1918:	Jan	Feb	Mar	Apr	May	Jun	Jul	Aug	Sep	Oct	Nov	Dec	Total
Act. Chap.	20	9	3	1		1	1	2	6	2	5		50
Tem. Act. Chap.			1	3	12	4	9	7		4			40
Reserves	2	4	2	1		2	2	1	1	1		1	17
Total	22	13	6	5	12	7	12	10	7	7	5	1	107

Chaplain Truman F. Riddle was appointed an acting chaplain on 19 April 1916 and resigned on 8 January 1918 to enlist as an apprentice seaman. He felt he could be of greater service as a chaplain later, if he could in this way get the point of view of the enlisted man. He served several months in this capacity at the Naval Training Station, San Diego, and on board the *Arkansas,* his shipmates being unaware of his background. Finally, on 5 June 1918 Riddle was appointed a temporary acting chaplain. His experiment is unique in the Navy Chaplain Corps.

Only two retired chaplains were called back to duty. One was the beloved David Tribou, who at the age of sixty-nine resumed active duty first at the Navy Yard, Boston, and then at the Naval Home in Philadelphia. Tribou had been retired in 1910 after thirty-eight years' service. The other retired chaplain called to duty was W. E. Edmonson who served from 3 April 1917 to 31 March 1919.

A summary of the statistics dealing with the strength of the Corps for the years 1917 and 1918 shows:

On duty 1 January 1917	40
Recalled to duty	2
Acting chaplains	89
Temporary acting chaplains	41
Reserves	32
Total	204
Less duplication (Riddle)	1
Net total	203

[22] Statistics do not include Frank A. Smith whose name appears in the 1918 *Register,* but who declined the appointment. There were actually 106 accessions to the Chaplain Corps during 1918; however, 107 commissions were granted, since Riddle was commissioned twice.

The attrition rate during the war was only about one percent. There were no separations in 1917. The Corps lost only three chaplains before 11 November 1918 including the temporary withdrawal of Riddle. Chaplain Brinckerhoff, who reported for active duty on 22 June 1918, was released after ten days' service. The only chaplain to die during the war was Chaplain Simon A. O'Rourke, who was appointed an acting chaplain on 14 June 1918. During his first week's duty at the Navy Yard, Boston, Chaplain O'Rourke fell a victim to influenza. He died 21 September. The first chaplains to be released after the signing of the armistice were Anderson, Schweitzer, and van Dyke, all of whom were discharged in December. The year 1918 closed with 198 chaplains on active duty.

An analysis of the denominational affiliations of the chaplains who served during World War I follows:[23]

	On duty 1 Jan 1917	Joined in 1917	Joined in 1918	Total	Approximate Percentage
Baptist	5	6	14	25	12.5
Catholic	9	12	23	44	22.
Christian		1	..	1	.5
Chr.-Science ...		..	1	1	.5
Congregational	2	2	8	12	6.
Disciples	2	1	5	8	4.
Jewish		1	..	1	.5
Lutheran	1	3	6	10	5.
Methodist ...	9	13	18	40	20.
Moravian		..	1	1	.5
Pro. Epis. ...	6	9	6	21	10.5
Presbyterian .	4	6	22	32	16.
Reformed ...	1	1	1	3	1.5
United Breth.. ..		..	1	1	.5
Universalist ..	1	2	..	3	1.5
	40	57	106	203	

Only one new chaplain, Tipton Lee Wood, formerly a chaplain in the United States Army, joined the Corps in 1919.

FRAZIER'S MANUAL

In the summer of 1918, Chaplain Frazier issued, by authority of the Secretary of the Navy, his forty-four page pamphlet entitled *The Navy Chaplain's Manual.* The booklet carried an introduction by Chaplain van Dyke. Cost of publication was borne by the General Committee of Army and Navy Chaplains of the Federal Council of the Churches of Christ in America and the General War-Time Commission of the Churches. The *Manual* did not set forth a systematic and detailed statement of the chaplain's responsibilities, techniques, or problems, but rather contained the fatherly advice an older and more experienced chaplain might pass on to a newly appointed chaplain assigned to him for indoctrination. Chaplain van Dyke called it "a simple and wise manual of applied Christianity in the service of the Navy." As such it was of great value to the younger members of the Corps.

The kind of personal advice found in the *Manual* is revealed in the following extracts:

> Other officers have their prescribed duties, and when these are completed there is very little more, if anything, that they are expected to do. The Chaplain has few prescribed duties, but if he doesn't succeed in keeping busy, he is branded as a loafer, no matter how well he performs the duties specified in the Regulations. Other Officers have a *definite time* and *definite place,* and hence *definite opportunity,* for their work. In addition, they are clothed with definite authority for the performance of their work, so that no man or officer dares interfere. On the other hand, the chaplain has to *ask* for his time and *ask* for his place and *ask* for his opportunity. In fact, he must be content to gather up the frayed ends and weave them into opportunity. Also he must stand in no one's way, and must conflict with no one's duty. Instead of being able, as are other officers, to tell people what they *"must do,"* he can only persuade, entreat, and exhort, oftentimes in the face of opposition and discouragement.[24]

Under the heading "The Chaplain as a Mess Mate," Chaplain Frazier had the following to say in behalf of tolerance:

> Don't be too sensitive or resentful about what you hear or see in the ward-room or about the ship. Remember that your views about things that may be objectionable to you are not necessarily right; and are most assuredly not always entertained by gentlemen whose belief is just as conscientious as your own. You owe it to them and to yourself to be charitable and reasonable in your attitude towards habits of a lifetime and customs that may not meet with your approval.[25]

The *Manual* contained wise directions as to the relation which a chaplain should have with his superior officers, especially the executive officer. Regarding the importance of paying close attention to personal habits, Chaplain Frazier warned: "It would be difficult to find a place where one's personal habits are subject to closer scrutiny or are more constantly on exhibition than in the cramped quarters of a man-of-war." Chaplains above all officers had to be particular because, wrote Frazier, "Like Caesar's wife, he must be above suspicion."

Regarding Divine Services, Frazier urged that the chaplain should preach on subjects that "are of vital importance to every Christian life." His preaching

[23] Statistics do not include re-appointment of Chaplain Riddle, but do include Chaplains Tribou and Edmonson.

[24] *Op. cit.,* pp. 8–9.
[25] *Ibid,* p. 13.

"should be intensely evangelical, practical, and nonsectarian." Frazier recommended that no sermon be over twenty-five minutes long and that the entire service should not consume more than an hour. "The Chaplain who does not know how to emphasize Divine Worship," added Frazier, "has no business on board ship."

Out of years of experience in visiting men in the sick bay and the brig, Frazier drew his words of sound counsel. He recommended that the chaplain pay daily visits to each place.

Regarding social activities, he stated:

The monotony of ship-board life can never be known except by men who go to sea. The long hours of toil at one task, repeated with each passing day; the trackless waste of waters unbroken by a sail; hearing the same voices and looking into the same faces week after week, with little variety of any kind to break the tedium of the dragging months, make the coming of one who is able to introduce new forms of amusement and entertainment or variety of employment a godsend to any ship. This is distinctly a Chaplain's work, although the regulations do not so state it. He happens to be the one officer aboard who is expected to do things that other officers would not do; and, if he is capable and willing and possesses the initiative, the authorities will be only too glad to have him take charge of this recreational feature of the ship's life.[26]

Chaplains on board ships which carried moving picture machines usually found that they were responsible for all movie programs. Chaplains were expected also to sponsor athletic events, minstrel shows, tournaments of all kinds, church and social parties ashore, sightseeing tours, lectures, educational activities, and in addition be custodian of the library. Sometimes chaplains were expected to issue daily bulletins or edit the ship's paper.

Under the heading "Other People's Business," Chaplain Frazier summed up the chaplain's multitudinous duties:

There is an old saying that the "best way to get along in the world is to attend strictly to your own business." For most men and most businesses this is a wise proverb. The Chaplain, however, who lives up to it — especially when he lets somebody else tell him what his business is — might as well "shut up shop" and go home. The Navy Regulations very wisely do not specify the duties of a Chaplain. The reason for this is that his work is of such a nature that to "build a fence" around it would of necessity so handicap him that the most important of his duties could not be performed. It is my contention that a Chaplain's duties consist of "anything and everything" that he may do in a wise and tactful way for the social, physical, intellectual, moral, and spiritual welfare of the ship's company.[27]

[26] *Ibid.*, p. 30.
[27] *Ibid.*, pp. 37–8.

By way of illustration, Chaplain Frazier cited the case of a certain chaplain at a naval training station who saw the need for an "Official Bureau of Information." He took his idea to the commanding officer who clothed him with the necessary authority. "The result was," wrote Frazier, "that the Chaplain's office became a clearing house for from 5,000 to 10,000 men, many of whose troubles, whether imaginary or otherwise, were explained away by the tactful, wise manner in which they were handled." When this particular chaplain was detached, the commanding officer said: "I can't imagine an officer more essential to the success of this station than the right kind of Chaplain."

Regarding the importance of delivering lectures on venereal disease, the *Manual* stated:

There is one phase of a very difficult but very important subject that can and should be presented by the Chaplain in a series of lectures, delivered once a month. The reference is to social hygiene. The Navy today is made up largely of boys and young men fresh from rural districts and small towns. They have not been exposed to the pitfalls of seaport cities and consequently are unaware of the dangers, physical and moral, that attend association with lewd women. While the Surgeon is the logical person to speak on this theme from a medical standpoint, the Chaplain will find that a wise presentation of its moral, social, and professional aspects will carry just as much weight as does the fear of physical contamination.[28]

Chaplain Frazier suggested that chaplains secure authoritative literature and use moving picture films to illustrate the ravages of venereal disease. He warned new chaplains that occasionally while a ship was in port, especially in a foreign country, some man might "become so taken up by the sights and surroundings of shore that he forgets to come back." If a shipmate carried the word to the chaplain, then it became his duty to go after the "lost sheep" and persuade him to return. Frazier advised the chaplain to protect his own reputation if he went in questionable districts by taking one of the men with him. Frazier mentioned briefly the important task of counselling.

The Chaplain must remember that there are not many people on board ship in whom the one who faces difficulties cares to confide. . . . There are few hearts that do not respond to a manifestation of brotherly interest, and even the hardest, though it may give no outward sign, cannot forget the unselfish interest that another may take in it. Undoubtedly here is a wonderful field and to leave it uncultivated is to fail as a Chaplain. The personal touch between Chaplain and man is the pivotal point on which revolves the entire machinery of his usefulness.[29]

[28] *Ibid*, p. 40.
[29] *Ibid*, p. 43.

The *Manual* closed with the following account of the ideal chaplain:

Not long since, a letter was written by the Commanding Officer of a big ship with reference to a Chaplain who had just been detached, in which he insisted that the Chaplain in question should be returned to his command, because, as he said, "He is my right-hand man—one of the most useful officers on board, and through him I am able to keep in touch with the needs of my crew." On investigation it was found that this Chaplain knew most of his crew by name, was familiar with the peculiarities and disposition of each, was always ready to do anything within his power to minister to their comfort, and as a result of this constant and profound interest on his part the crew would do anything for him that he asked. It is impossible to measure the usefulness of such a man.[30]

Through the medium of such kindly counsel and direction, it is possible to gain an appreciation of what was expected of Navy chaplains in the days of the First World War. Bishop W. F. McDowell of the Federal Council of Churches of Christ called it a "most helpful publication."[31] Chaplain Frazier often commented that the best advice contained in the *Manual* was implied in the single phrase: "The faithful performance of unprescribed duties." For nearly a generation of Navy chaplains this *Manual* remained the only guide for the newly appointed chaplain.

Between the Introduction and the body of the *Manual* appears a poem without a title or signature. However, it was undoubtedly written by Chaplain Henry van Dyke. Through rhythm and rhyme he set forth the ideal conception of the Navy chaplain who in his fight against sin dares to stand alone.

Of better stuff than heroes men are made
Who in this age of deadly, awful strife
Stand all alone, uncrowned, unheralded,
Proclaiming for the noblest things of life.

On battlefield the sound of clashing arms
Makes cowards fight and drives the weakling's blood
Through stagnant veins that ne'er before have known
A near approach to warrior's crimson flood.

But he who fights against relentless foe
When silence reigns and cheer of eager van
Greets not his ear, but steadfast and alone
Drives back the hosts of sin, he is a man.

SUNDAY OBSERVANCE

On 27 October 1917, in reply to a letter sent by the Reverend H. L. Bowlby, General Secretary of the Lord's Day Alliance, Secretary of the Navy Josephus Daniels, wrote:

I am in hearty sympathy with the spirit of the official order issued by President Lincoln in 1862, and believe that always, particularly in a period of war, the American people at home and those in other branches of the service are entitled to the Sabbath as a day of worship.

The policy of the Navy Department is that nothing but necessity shall prevent the men in the navy from having these privileges of the Sabbath. We are appointing additional chaplains so that every man in the navy will have the opportunity to attend divine service on ship board and in our naval training stations.[32]

Upon the suggestion of Mr. Bowlby, President Woodrow Wilson on 21 January 1918 issued the following order relative to Sunday observance in the Army and Navy:

The President, Commander-in-Chief of the Army and Navy, following the reverent example of his predecessors, *desires and enjoins the orderly observance of the Sabbath by the officers and men in the military and naval service* of the United States. *The importance for man and beast of the prescribed weekly rest, the sacred rights of Christian soldiers and sailors, a becoming deference to the best sentiment of a Christian people, and a due regard for the Divine Will demand that Sunday labor in the army and navy be reduced to the measure of strict necessity*. Such an observance of Sunday is dictated by the best traditions of our people and by the convictions of all who look to Divine Providence for guidance and protection, and, in repeating in this order the language of President Lincoln, the President is confident that he is speaking alike to the hearts and to the consciences of those under his authority.[33]

That part in italics includes the full text of the order issued by President Lincoln on 14 November 1862. Strictly speaking, the Sabbath refers to Saturday and Wilson deliberately clarified Lincoln's terminology by referring to Sunday. President Wilson's order appeared as AlNav[34] 12, 1918. Such an order smoothed the way for the regular and faithful performance of Divine Services.

CHAPLAINS AT WORK

"Woeful is the ignorance of the public at large, and, even of most naval officers, of the nature of the work of a chaplain in the United States Navy," wrote Chaplain D. S. Robinson. "How frequently have I had people quizzically inquire: 'Just what are your duties in the Navy?'"[35]

The question may be satisfactorily answered for this period by an examination of the historical

[30] *Ibid*, pp. 43–4.
[31] *Federal Council, Biennial Report*, 1918, p. 171.

[32] *The Lord's Day Leader*, Mar.–Apr. 1918, p. 11.
[33] *Ibid*, Jan.–Feb. 1918, p. 4.
[34] A Navy abbreviation for an "All Navy" directive.
[35] Nav. Rec. Coll., Chaplains' File.

reports submitted by chaplains after the war. Chaplain Frazier had requested these reports of their activities and experiences for historical purposes, and sent them to Naval Records and Library on 6 January 1920.

"A chaplain is primarily appointed," wrote Chaplain Frazier in his summary of the chaplains' historical reports, "to conduct religious services and administer in a religious sense to the officers and crew." In spite of the multitudinous demands of collateral duties, chaplains, with but rare exceptions, placed major emphasis upon their spiritual ministry. The following extracts from a few selected reports are typical.[36]

Chaplain C. M. Charlton's report for 1917 stated:

> I have preached one hundred and sixty different sermons, (repeated a few of them) delivered forty different Bible lectures; delivered twenty-four lectures on secular topics; conducted daily Morning Prayers . . . and frequently administered the Sacrament of Our Lord's Supper.
>
> I have influenced at least two hundred conversions; baptized or assisted at the baptism of twenty; directed and instructed thirty or more for Church Membership.

Chaplain Edel described in an article which appeared in the *Christian Advocate* of 27 June 1918, a Divine Service held on Mother's Day, 12 May 1918, on board the *St. Louis* which was then in the war zone.

> Rigging Church
>
> Above on the quarter deck a bos'un's mate, with a dozen men, was "rigging church," as piano and pulpit, benches and chairs were brought out and put in place. A few turns of line lashed the piano in place, secure from the rolling of the ship, and when Old Glory draped the pulpit, all was ready. The bugle's notes of the "church call" sounded through the ship. With the last note the waiting quartermaster hauled smartly on the halyard and the church pennant shook out its folds to the keen air—the flag of the Master whipping out for all the world to see, the only flag that flies above the Star-Spangled Banner. And though in all the circle of the horizon there were no eyes save those of our own ship to see it, we felt that the eyes of a thousand mothers in a thousand homes two thousand heart-breaking miles away were on us.

Chaplains on board ships, especially on transports carrying troops to France, often held several Divine Services in the morning in order to give as many as possible the opportunity to attend. Such services were held on topside when the weather permitted. In the days before the public address system was invented, chaplains had to depend upon lung power or possibly a favorable wind to help carry the voice. Thus, a chaplain would note the direction of the wind before fixing the place for the pulpit. .

The *St. Louis* made more than thirty Atlantic crossings between October 1917 and July 1919. Convoy duty was wrought with tension; it was dangerous, uncomfortable, wearing, monotonous. One of Chaplain Edel's methods of bolstering morale was to maintain a choir. The *St. Louis* choir was composed of thirty voices. After the armistice many concerts were given before civilian audiences along the East Coast.

Several chaplains reported Sunday afternoon song services on the open deck whenever weather conditions permitted. The sailors loved to sing. Often the ship's band was used in Divine Services. Chaplain B. C. Clausen wrote of an experience aboard the *North Carolina:*

> The bandmaster had always manifested a most enthusiastic willingness to be of help in our services, and I had made use of his organization through the first months of my duty on the ship. But they seemed disposed to drag the hymns always, and they always managed to find a blue chord or two in the most self-respecting old tune. There was a sailor aboard who played the piano magnificently. I was hesitating between two decisions. Should I relieve the band of their duties on Sunday morning, and substitute the more inspiring accompaniments of a well-trained pianist?
>
> In a pile of mail which I was censoring, I came to a pencilled letter. It had been written by a boy just after he had attended church service aboard.
>
> "And O Mother," he wrote, "when they sing, they have a band to play the hymns. It is wonderful. It sounds just like the organ at home in our church. I just couldn't sing. I sat there with tears in my eyes. I could hear that organ at home."
>
> My decision was made for me. If that band of ours could make any boy hear the organ in his own church, nothing could induce me to dispense with it.

In his report for 1918 Chaplain Clausen stated that he had held forty-three services aboard the *North Carolina* with a total attendance of more than 7,000.

Chaplain D. S. Robinson aboard the *Frederick* used "an excellent orchestra" to furnish the accompaniment for singing at his services. He advocated fifteen minute sermons.

> I soon learned that the men liked something snappy and to the point, and that sermons had to be constructed or adapted especially for them. The kind of a sermon one would naturally preach for an ordinary mixed congregation was not at all suited for my congregation of young sailors. The men always joined in the service with a will, singing and reading the responses, and uniting in the prayers when asked to do so.

[36] Unless otherwise noted the quotations used are taken from the individual's jacket in the office of Chief of Chaplains.

Robinson said that his average attendance was about 100 and that "they gave excellent attention, and were as reverent as if they were in a church."

Chaplain Robinson also mentioned one aspect of a chaplain's duty which is taken for granted by members of the Corps but not always appreciated by civilians. "Being a Protestant," he wrote, "I have felt it to be my duty to provide opportunity for Catholic boys to attend mass. Consequently, I arranged for priests to come aboard the USS Frederick. . . ." Catholic chaplains mentioned in their reports of making arrangements for Protestant services aboard their ships. Frequently church parties were sent from one ship to another or ashore so that men of a certain faith might have the ministrations of a clergyman of their persuasion.

The term "General Service" appears to have been first used in the Navy during World War I to describe a Divine Service for all hands. In the absence of a Protestant chaplain, a commanding officer sometimes requested the Catholic chaplain to conduct a General Service especially for the benefit of the Protestants. Some Protestant chaplains, who conducted an early or morning service according to the manner and forms of their own church, in order to make a wider appeal to all, held a General Service as well. Often these were in the nature of evening sings, which were popular on some ships and stations: Chaplin Dickins tells of how 1,500 or more men crowded into one of the temporary wooden recreational centers at Philadelphia on Sunday evenings for such services. Protestant and Catholic chaplains, on alternate Sundays, led these assemblies, at which they also delivered short messages.

The conducting of Divine Services at shore installations brought problems unknown to chaplains at sea. First there was the necessity of finding a suitable place to conduct services. The Government did not build chapels at any naval installation during World War I. Chaplains had to adapt themselves to existing accommodations. When Chaplain H. M. T. Pearce reported for duty at Pensacola, Florida, on 4 December 1917, he found a Naval Air Station in the first stages of development. Facilities for the Chaplain were practically nonexistent. He was told to go ahead and do what he could. Pearce's problems were typical of chaplains at other shore stations, and of these he wrote:

> There was no auditorium where the men could assemble, save the Marine Barracks which would hold but about 400 men, far too small for the 1,500 men then on the station. The Y.M.C.A. Hut was under construction then and was completed in January, 1918. . . . The building could hold not more than 450 men. . . . The most suitable place in these days to hold Church Services was the Mess Hall.

When he arrived, the Air Station at Pensacola was expanding at the rate of two to three hundred men a week and was expected to grow to 10,000. After the Y Hut was completed, Chaplain Pearce moved his Sunday services from the unattractive Mess Hall to the Hut. Later when the Recreation Building was completed, the services were moved to the reading room in that building. Pearce listed the services held on an average Sunday:

> On Sunday morning at 8 o'clock the Roman Catholic men were marched to their church which is just outside the Main Gate. The Y.M.C.A. held their Bible class in their hut at 9 A.M. and the chaplain held his Service at 10 A.M. The Y.M.C.A. also held an evening Service at which the chaplain was sometimes asked to preach.

On board ship, with the possible exception of transports, the number of worshippers was more or less constant. On the other hand chaplains at the larger training stations found that they were working with a constantly changing personnel. At the Great Lakes Station, for instance, as many as 1,000 men arrived a day.

For three weeks the raw recruits were kept under strict discipline in detention. Many were homesick and leisure hours hung heavy on their hands. While there is no indication that church attendance was compulsory in such camps, the chaplains who worked there reported an encouraging response to Divine Services. Chaplain W. H. Wilson wrote of his experiences at Mare Island:

> Detention is a dreary three weeks for the new recruit. . . . It is the Chaplain's opportunity. . . . On Sunday afternoon he assembles a band placed at his disposal and a group of singers who are ready to lend their aid and soon the cheerful music draws the men from their tents and barracks to a sheltered part of the parade ground. The way they sing! and the eagerness which they show in participating in a religious service![37]

Writing on 1 January 1919 from the Naval Training Station at San Diego, Chaplain M. A. Spotts claimed:

> The total attendance at church services for the year was 22,500. The average attendance at church services was 500. All services were voluntary and the boys assembled readily for them.

[37] Nav. Rec. Coll., Chaplains' File.

Chaplain Spotts stated that every man in the Detention Camp was given a testament or a prayer book if he desired one. Several nationally known speakers were invited to the Protestant services and a definite effort was made to influence men to decide for Christ, to receive baptism, and to unite with a church. Spotts wrote:

> During the year, 25,000 decisions for Christ were registered at these special meetings. To follow up and make the work definite, a list of the names with church preference were sent to the local churches, and they in turn invited the boys to unite with the church. More than 1,500 of the men before leaving camp on draft had definitely tied up to a church.

Chaplain B. R. Patrick, writing on 1 January 1918 from the Naval Training Station, St. Helena, Norfolk, Virginia, gave the following statistics of a religious survey taken of 14,789 recruits:

Methodist	3,441
Baptist	3,374
Catholic	2,803
Presbyterian	1,287
"Protestant"	1,179
Christian	669
Lutheran	581
Episcopal	440
Congregational	137
Hebrew	115
Christian Science	19
Mormon	12
Orthodox Greek	1
21 other denominations	157
Unknown or non-members	574
	14,789

Chaplain Frazier, reporting on his activities at St. Helena during the months of 1917 before he went to Washington to be the first Chief of Chaplains, referred to this religious survey and added: "By inquiry from other Training Stations, I find that this proportion is practically the same at all of them."

Two years later chaplains were able to take a religious census of a ship or station by consulting the health records of personnel assigned there. The Bureau of Navigation on 25 October 1919 approved Chaplain Frazier's recommendation that denominational affiliation be indicated on the health records of naval personnel. The old record did not specify with sufficient clarity the religion of the individual; the revised record carried the notation: "Religion (denomination)." In emergency this accurate information often permitted more immediate ministration by chaplains or clergymen of the faith the individual professed. Sometimes religious affiliation was indicated by tabs on the individual's jacket in the personnel office.[38]

Along with conducting the usual Divine Services and taking part in the special programs connected with the religious festivals and certain national holidays, as Thanksgiving, chaplains were busy with those additional duties which come to clergymen everywhere, such as administering the sacraments and officiating at weddings and funerals. Chaplain T. M. Mark writing from Brooklyn on 30 January 1919 referred to the many funerals he conducted during the influenza epidemic: "For almost five weeks there were funeral services every other day." From the Air Station at Pensacola, Chaplain Pearce wrote of the tragic aviation accidents which took so many young lives. "The chaplain always wrote to those dearest to the young men who gave their lives for their country," he stated, "and the appreciative letters in reply brought the certainty that the chaplain is of some use in the naval service."

Catholic chaplains with characteristic faithfulness read Mass and administered the last rites to the sick and the dying. Since there was but one Jewish chaplain in the Navy during the war, chaplains of other faiths ministered to the best of their ability to Jewish personnel or made special arrangements whereby these men could receive the religious ministry of their faith.

The first chaplain, so far as is known, to use an airplane for transportation between posts where Divine Services were held was Chaplain Harry M. Peterson who worked with the Marines on Haiti in the winter of 1918–1919. Of this, Chaplain Peterson wrote:

> Chaplain Frazier informed me in 1921 that I was the first Chaplain to use that mode of transportation in carrying on the Chaplain's work. At that time it took a full hour by plane to fly from Port au Prince to Cape Hatien and only a little less time to fly to Mirabelais, Los Cohobas, and Thomone.[39]

The many material aids for Divine Services which were made available in World War II by the Chaplains Division were not provided to Navy chaplains during World War I. If chaplains felt the need for portable organs, altar pieces, or hymn books, they had to turn to friends, obtain money from the local welfare fund, or pay for such items themselves. Chaplain Pearce at Pensacola wrote of receiving two portable organs, one given by a friend and the other by the War Commission of the Episcopal Church.

[38] CoC., Bureau of Nav. File.
[39] CoC., Peterson File, Letter of 20 Nov. 1944.

Again and again chaplains voiced their appreciation of the splendid cooperation received from local clergy and from the various "War Commissions" of the larger denominations.

In the absence of any official hymn or service book, most Protestant chaplains used *The Service Song Book* published by the YMCA in 1917. Chaplain Scott, writing to Frazier on 19 December 1919, reported that many of the songs included in the YMCA book were strange to the sailors and that too large a percentage were "revival hymns." Chaplain Scott was one of several who urged the compilation of a special hymnal and service book for Navy men.[40]

Chaplain H. R. Davidson summed up his experiences of 1918 in words that would no doubt have brought forth a hearty "Amen" from many of his brother chaplains.

> It has been a rare privilege to me to have had this year's work as a Chaplain. I have thoroly enjoyed the work and in spite of all discouragements sincerely feel that I have been able to make religion real and practical and vital to some of the men . . . . it has been a real privilege to preach to men who do not ask one's denomination nor anything else except that one be able to make a little clearer the instinctive faith they find in their hearts.

COLLATERAL DUTIES

Throughout the years the Navy chaplain has carried on many collateral duties. Many of these duties were made easy by the close proximity in which he lived with his men aboard ship. During World War I, new responsibilities frequently arose which no officer was delegated to handle either by precedent or regulation. In such cases the chaplain was often asked to take over.

From the beginning of the Navy the chaplain had been interested in educational projects. Long before World War I, they had sponsored welfare activities for leisure hours ashore and afloat. Such precedents were accelerated under war conditions until many chaplains found themselves burdened with many extraneous duties, some of which were only remotely connected with their sacred calling. The wise and tactful chaplain knew when to limit these collateral duties and how to turn them to the spiritual benefit of the men.

In his summary of a chaplain's responsibilities, Frazier wrote regarding collateral duties:

> There are however, many duties strictly speaking, not of religious character, but bearing such a close relation to religion in that they administer to the general welfare and contentment of the crew, that the Chaplain's field of usefulness becomes a broad one. Especially was this true during the war, for the Chaplain not only was the Religious teacher, but Welfare officer, Athletic officer, frequently stood lookout watch, was one of Communicating Officers, and Decoding Board. . . . It was up to him to provide entertainment in the way of motion pictures, boxing bouts, minstrel shows or in any other way that would meet the demands of the situation. Vast reading matter was given him for distribution and many of the comforts provided by civilian organizations were passed through his hand before reaching those for whom they were meant. During the Flu epidemic the Chaplain was a friend and counselor to men sick and dying and was a connecting link between them and their homes.

A congested warship is divided into two distinct worlds, for there is a traditional division between the bluejacket and the commissioned officer, between the forecastle and the wardroom. Two officers, because of the nature of their work, break through this wall more easily and frequently than others. They are the doctor and the chaplain.

Navy chaplains in the First World War found that much of their time was spent in listening to the tales of woe poured out by bluejackets in trouble. Most Navy chaplains are quick to affirm that no person can get fouled up as quickly in as many unusual situations as a sailor.

Sometimes these difficulties involved minor points of adjustment to Navy life. The chaplain was one officer in whom the men felt free to confide. It was good just to talk even if the chaplain could do nothing except advise the sailor to keep his chin up and take it like a man. Sometimes the chaplain could help. Often a man was saved for greater usefulness to the Navy by being shifted from one billet to another—the shift being made possible when the chaplain explained the situation to the proper officer.

Sometimes the problem involved money. References to the Navy Relief Society, which was incorporated 23 January 1904 but was not strong or widespread before the First World War, are found in a number of the chaplains' reports for these years. Chaplain Sydney K. Evans writing from the Naval Academy on 1 January 1918 claimed: "Much time has been given to the work of the Navy Relief Society as treasurer of the Naval Academy Auxiliary." Chaplain Wright in his report of activities at the Naval Training Station, San Francisco, dated 4 March 1919 wrote of working with the Navy Relief. Chaplain Bayard, who followed Wright, in his report for 1919 stated. "The Senior Chaplain takes care of all Navy Relief cases."

[40] *Ibid., Scott File*, letter of 12 Dec. 1917.

832159—49——13

Other problems that came to the chaplains' attention involved home difficulties. "I carried on an extensive correspondence with the homes and relatives, wives, and friends of the men," wrote Chaplain Wright in his annual report for 1919. He felt that this service did much to alleviate anxiety and promote contentment both for the men and for the folks back home. Chaplain A. R. Charlton claimed that he kept two stenographers busy simply handling the correspondence that poured over his desk.

Sometimes sordid stories of sexual irregularities which involved sailors were told to the chaplain. Of this Chaplain D. S. Robinson wrote:

> Many cases of sailors wronging girls have come to my notice, and frequently I have been able to straighten the affair out by marrying the couple. Some of these cases are really pathetic, but they are too sentimental and seamy to narrate in detail here. On the other hand, I have been able to save one sailor from being taken in by a loose character who was after the government allotment.

Transport chaplains during 1917 and 1918 devoted the twilight and early hours of the evening to visiting the soldier passengers in the holds. Needless to say, the conversation almost always turned to religion and its meaning. On such visits as many men as possible crowded around to hear the conversation. Questions were plentiful and the answers frank. Often the chaplain was kept until midnight or later.

Many chaplains commented on the importance of personal counseling. Men who never dreamed of turning to a clergyman in their home communities when in trouble, found it easy and natural to turn to their chaplain for advice and help. Chaplain Perry Mitchell was only one of many when he commented:

> The Chaplain's greatest opportunity, however, lies in his personal contact with the men. We have had a monthly average of over fifty visits a day from men who were in need of advice or assistance. Such conferences, as well as visits to men in the Sick Bay, have given the Chaplain a point of contact with practically every man aboard ship.

Closely linked with counseling were the lectures given by chaplains to groups of men, especially to recruits at receiving stations. Chaplain Frazier, who excelled in this particular work, described his experiences at the St. Helena Naval Training Station, Norfolk, in his 1918 annual report.

> By order of the Commanding Officer, the Chaplain was given the duty of delivering to each company as soon as organized a lecture on Naval Customs, etc., also a lecture to each company on moral prophylactics. During the year 286 of these lectures were delivered to companies [assemblies] of men ranging from 85 to 5,000. The Commanding Officer considered these lectures of so much importance that he issued an order that all men of the Station, whether under training or otherwise, should "fall in" immediately after quarters on Saturday morning and hear these lectures. At such times, no building being large enough to accommodate the crowd, the lecture was delivered in the open air.

Chaplains were expected to lecture on the dangers of immorality. The doctor's business was mainly the prevention and control of venereal disease. It was the chaplain's business to go deeper and touch the man's heart and persuade him that there was a better way. Writing of his attempts to save men from the ravages of venereal disease, Chaplain Robinson in his history report stated:

> In this connection I should also mention the attempts I made to save men from the ravages of venereal diseases. I distributed literature on the dangers of these diseases and preached a sermon on "Reaping Wild Oats," in which I pointed out how inevitable are the consequences of immorality, and warned the men against the dangers of licentious living. I know from what men told me that this had a wholesome effect.

Chaplains have always been sensitive to the relation between idleness and wickedness. Keep a sailor busy and happy at worth while activities during his leisure hours and you choke off at the source all kinds of moral irregularities. This is the justification that Navy chaplains give for engagaing in collateral duties. Directly and indirectly by so doing they are helping to build upright character.

A collateral duty which most chaplains had was that of caring for the library and providing clean wholesome reading. On board ship the chaplain was usually given charge of the library. Of this duty Chaplain Riddle on the *New Mexico* wrote that a library

> was furnished, for the most part, by the contributions of civilians. Every effort has been made to make it homelike and cheerful. . . . Through the gifts of the Bureau of Navigation, there is a splendid library of history, travels, and biography; a small collection of educational and school books, and a large assortment of popular, modern fiction. Writing paper and envelopes, stamped with the ship's name, are issued to the crew gratis. At sea the daily Radio news is issued here and a chart gives the daily information of the course of the ship.

Chaplain Robinson purchased $1,200 worth of new books for his ship from funds provided by the Government. These books were issued to the men "with the use of as little red tape as possible." He also reported the organization of classes in mathematics, history, and English to help prepare men for the examinations for Annapolis.

Many chaplains issued daily mimeographed bulletins, especially on board ship where radio news was avidly received. Others who had access to a printing press edited the ship or station paper. Writing of his work in the Canal Zone, Chaplain E. W. Foster included in his history report the following: "I am also the Editor-in-Chief and Business Manager of *Service Life* which was begun through my initiative, and embraces the activities of Service men both in the Army and Navy on Canal Zone."

On the need of wide diversity of activities, Riddle wrote:

> There are no men in the world who go through necessary privations in a more sportsman-like spirit than the American bluejacket. But even he will get stale and stupid, and sometimes morbid, with only drills and watches and ships work. So the problems of entertainment came up to the Officers, and leaders among the men in a serious light. Moving pictures *had* to be put on.

Robinson likewise found moving pictures a necessity in keeping up morale. He was also in charge of musical events, boxing and wrestling contests, and stunt nights with all sorts of fun-making contests. Baseball and football teams were organized to play ashore whenever opportunity afforded.

Chaplain M. H. Petzold outlined the techniques he used to entertain men aboard the *Virginia:*

> I have kept a complete census of the abilities of the men on board, which has enabled me to organize the men of talent and in as far as conditions have permitted have had at least two big entertainments per month and moving pictures nearly every night in port and often while at sea.

Sometimes when there were insufficient funds for athletic and recreational equipment, the chaplain's ingenuity was taxed to obtain the needed gear by devious unofficial ways. A number of transports, on which Navy chaplains served, had been taken over from private lines or were reconditioned German ships. Hence, there were no welfare funds available until such ships had been in commission sufficiently long to build up one of their own. With such funds as were available, however, transport chaplains, with the approval of commanding officers, purchased such items as athletic equipment, books, and magazines, and rented motion pictures. When such funds were exhausted, the chaplain found individuals or groups who were willing to donate the desired items. In a few cases chaplains advanced the money from their own pockets until the ships' finances were in a position to make payment.

Chaplain Riddle described some sightseeing parties he sponsored in France:

> During the last two weeks, while the ship lay at Brest, France, arrangements were made with the French authorities to give our men transportation on the trains to Paris and, as the crew had never been in a foreign port before, every effort was made to make the most of this opportunity. I was sent to Paris in charge of these parties and to provide for the comfort and welfare of the men . . . sight seeing parties averaging seventy men were taken around to such places as the Arch de Triomph, Eiffel Tower, Tomb of Napoleon, the Louvre, and Notre Dame.

Chaplain Leslie Miller on board the *South Carolina* had charge of the "ship's shops" and served as Welfare Officer. Of this he wrote:

> The Chaplain and his Yeoman have charge of the business and of all the ship's shops, such as Laundry, Barber Shop, Cobbler, Tailor, Photographer, and Mechanical Cow. This represents a business of two or three thousand dollars per month, including expenditures and income. We take fifteen percent of the net proceeds of these shops and set it aside for the Ship's Welfare Fund. This money is kept in trust by the Chaplain, and expended for the comfort, pleasure and benefit of the crew. . . .

Several chaplains were closely associated with the Red Cross. Chaplain John M. Mark, writing from Pelham Bay Park on 11 January 1919 stated:

> I was appointed on October 20th, 1917, as Asst. Field Director of the American Red Cross. From January 1st to December 31st, 1918, over 30,000 outfits have been given out from my office, the outfit consisting of sweater, muffler, wristlets, helmet, and comfort bag.

Senior chaplains at shore installations indoctrinated the newly appointed members of the Corps. The coming of such days as Christmas brought extra duties, especially in the hospitals. Chaplain J. F. Fedders writing from the hospital at Great Lakes on 16 February 1919 reported that several "thousands of dollars were expended in bringing Christmas cheer to a population of nearly fifteen hundred boys." Whenever a ship was in port and it was possible to do so, boys and girls from an orphanage or school were brought aboard on Christmas Day in traditional Navy style and given an entertainment that they would never forget. The bluejackets always enjoyed such occasions as much as the children. If a chaplain were aboard a ship where the crew was entertaining a group of children, he would invariably take a major part in the arrangements.

Among the unusual and unexpected duties that devolved upon the chaplain outside the continental limits of the United States was that of censoring mail. Many a chaplain drank hot coffee into the early hours of the morning in order to clear the mail dumped on his desk. Chaplain Truman Riddle converted even this boresome chore into an opportunity through the spirit in which he approached his task. He wrote:

> I suppose every Chaplain in the fleet was put on a censoring board and allow me to express my honest opinion that here he found a great field for usefulness, although it may have interferred, to a great extent, with other activities. As an average I put in from five to eight hours a day on this activity alone while on the U.S.S. Pennsylvania and, many a night previous to getting under-way. . . . But from a moral standpoint, what can be more important than helping the men keep up an uninterrupted communication with the best influences in their lives, their families, wives, and sweethearts?

Many an interesting story grew out of the censorship experiences. According to Chaplain Mitchell, one lad commenting on the activities of the chaplain in regard to providing entertainment, wrote: "We have a Chaplain aboard, in fact, we have all kinds of amusement." Chaplain Edel found the following expression of appreciation for a chaplain in a letter from a seaman to his home: "A chaplain reported on board this ship last week. It makes me feel better just to see him walk by."

Chaplain H. J. Fry summed up the chaplain's attitude to his collateral duties when he wrote:

> He feels that he is saving men and carrying forward the Kingdom of God if by bringing aboard the normal and cultural influences the men had at home, he saves them from such debasing things as foul talk and gambling, and offsets the stress and strain of sea life—a life that breeds the "fling" ashore on the liberty party—a life that naturally neglects man's higher faculties.

And Chaplain Frazier in his brief summary of chaplains' activities sent to Naval Records and Library added:

> Thus the Navy Chaplain serves his government in bringing contentment and happiness to men—in conserving the religious, moral and educational interests of the crew, develops morale.

THE INFLUENZA EPIDEMIC

The influenza epidemic which made its death march around the world in 1918 brought heavy extra duties to many Navy chaplains. Chaplain E. W. Foster describing his experiences at Hampton Roads, wrote:

> Upon my return to duty, the epidemic had spread throughout the camp, more wards were added, and I began working day and night writing letters, telegrams, writing insurance for those who had taken sick from the Detention Camp, and in various ways encouraging them and trying to cheer.

A number of the transports had the misfortune to have an out-break of the Spanish Influenza aboard when they were loaded with troops enroute to Europe. This epidemic apparently varied in its virulence. It was serious afloat in its lightest attack due to crowded conditions from which there was simply no escape once the ship had sailed. Men were stricken down with few advance symptoms and sometimes died in a matter of hours or a day or so. A few ships were reported to have exhausted certain medical supplies and had to receive additional supplies at sea. The supply of coffins on some transports became exhausted and the carpenter's force had to build boxes. Aboard some ships, two and three decks were completely filled with influenza patients lying on mattresses with the break of just a few inches between each mattress for attendants to help the stricken men. The number of patients frequently became so large that Navy officers and enlisted personnel turned to as soon as they came off duty and helped nurse the sick. In some instances, Navy personnel, though running a temperature, refused to turn in the sick-bay because "it will take a bunk from some poor soldier who is sicker than I am." Fortunately, the epidemic burned itself out in a relatively short time and most transports had only about one or two such tragic trips.

Chaplain F. B. Huske, writing from the Naval Hospital at Norfolk claimed that the days when the epidemic was at its height were the busiest of his naval experience. Service men in the First World War sometimes had to be persuaded to take insurance. Many recruits, receiving only thirty dollars a month, objected to the monthly deduction necessary to carry a $10,000 policy. The result was that many bluejackets were without insurance when stricken by the disease. Some dying, were eager to apply for insurance, but had no opportunity to do so. Dependents were thus often left without funds. Chaplain Huske was one of several chaplains who found an opportunity not only to bring spiritual succor to those afflicted but also to aid in the practical matter of helping the sick to take out insurance. He had the following unusual experience to tell:

> It was my especial responsibility to go to those patients who were confined to bed, and many of whom were weak and almost helpless, and to give them the opportunity to take the insurance which their Government desired they should have for the protection of themselves and their families, while they were in their Country's service, and risking their lives for patriotic reasons and motives. Many of these men had a very apparent limit set upon the time in which they might secure this insurance and protection, not only by the requirements of the Insurance Law, but by the sure approach of the Angel of Death. I could often see that if they did not make their application at once, that they could never do so; perhaps the next day would be too late. Thus it was my task to go to the bedside of the ill and dying, and help them to make some provision for their loved ones, even as their Country wished them to do, before they died in their Nation's service. . . . And so it was often the case that a man, weak and low, with the aid of the Chaplain, and the nurse, or a faithful apprentice, would perform his last unselfish and heroic service for his loved ones, and secure for them, by means of this insurance, a regular income for the future years, as the last earthly, conscious act of his life.

In 1918, the policies for which Huske helped sailors apply totaled $4,300,000. "The total amount of money that will be paid to the families of the men out of this number of those who took insurance, who subsequently died," he added, "was a quarter of a million dollars, or more."[41] Many a parent, wife, and child found comfort in this last loving act of one who responded to his country's call and died in "line of duty."

DISTRIBUTION OF RELIGIOUS LITERATURE

A vast amount of religious literature, including Bibles and Scripture portions, was distributed by chaplains and welfare organizations to naval personnel during the First World War. Chaplain Charlton in his report for 1917 said that he had "distributed tons of general reading matter" including "thousands of portions of Scripture, Testaments, and Bibles," and that he had "enrolled nearly a thousand men in the 'Silent League' and hundreds more in the 'Pocket Testament League'."

The American Bible Society distributed nearly 7,000,000 copies of the Scriptures to members of the armed forces during the War. In addition, other societies as well as publishing houses of the larger denominations were busy producing Bibles, Testaments, Scripture portions, and tracts—all for free distribution. Never before in history had such a stream of religious literature flowed from the press.

Chaplain W. H. Wilson paid tribute to the work of the churches in these words:

> The splendid generosity and helpfulness of Church Boards and philanthropic agencies during the period of the war has made possible a helpfulness and Christian service upon the part of the Chaplains which otherwise would have been impossible. Testaments by the thousands, books by the hundreds and periodicals without number have come to the Chaplain's Hut for distribution. In the reception of these gifts the men have displayed the finest spirit of gratitude. In taking a copy of the Scriptures the promise is often made with sincere intent to cherish the Book and read it daily. "That is the first copy of the Bible I have had since I was an apprentice boy eighteen years ago" a CPO said to the Chaplains as he accepted the gift. . . .

Transport chaplains had an unusual opportunity to distribute religious literature to troops on their way to France. They found the men in a receptive mood and with time to read.

CERTAIN COLLATERAL DUTIES DISCOURAGED

Commanding officers, lacking a sufficient number of qualified line officers, sometimes placed chaplains on court-martial boards and assigned them to duty on the decoding watch. When instances of such unusual collateral duties for chaplains came to the attention of Secretary Daniels, he issued orders at once directing the discontinuance of such assignments, except when absolutely necessary.

On 23 January 1918 Daniels wrote to the Commanding Officer of the Naval Home at Philadelphia:

> The Department notes that a chaplain has been ordered as deck court officer for the trial of the above named man. It is considered very undesirable to appoint chaplains to such duty. You are directed that it be not done, except as the exigencies of the service make it absolutely necessary.[42]

Again on 7 March 1918 Daniels wrote of a similar case which occurred at the Washington Navy Yard. Writing to the Commandant, he stated:

> 1. In reviewing the summary court-martial in the case of the above named men, it is noted that an Acting Chaplain, U. S. Navy, has been ordered as recorder.
>
> 2. The Department considers it very undesirable to appoint Chaplains for court-martial duty. You are directed that it be not done except as the exigencies of the service make it absolutely necessary.[43]

[41] Nav. Rec. Coll., Chaplains' File.

[42] NRSO: 27217-2477.

[43] *Ibid.*, 27217-2620.

On 7 February 1919, Special Order No. 65 was issued by the Navy Department which stated: "The Department directs that hereafter Chaplains shall not be detailed to court-martial duty except by direction of the Secretary."

The historical summaries and annual reports of chaplains covering the period of the First World War indicate that many chaplains on board ship were given duty on the decoding watch. Chaplain D. S. Robinson wrote: "I had not been on board the USS *Frederick* long until I was made a member of the decoding watch. . . ." Chaplain H. R. Davidson, reporting on his duties of 1917, stated:

> From the 13th of March I have been on duty as Assistant Communication Officer. This duty involves decoding and coding of messages, custody of secret and confidential publications, and messenger duty ashore.

Chaplain E. L. Ackiss, serving on the *Pocahontas*, was given submarine lookout watches and communication and coding watches. Some chaplains were placed in charge of boats at abandon ship drill.

While such instances do not seem to have come to the attention of the Secretary, chaplains were indirectly involved in the following decision relative to medical officers serving on such boards:

> Upon decision requested as to whether or not medical officers can be legally assigned to duty as coding officers, *Held:* That neither chaplains, doctors, nor members of the Hospital Corps should be assigned to duties of a combatant character. Such members of the staff personnel of a naval vessel are neutralized by the terms of the Geneva Convention, the principles of which the United States accepts for general guidance.[44]

CHAPLAINS WITH THE MARINES

When the United States entered the war against Germany, only one chaplain—E. A. Brodmann—was then serving with the Marines. Brodmann was at the Marine Barracks, Port Royal, South Carolina, where he remained from 19 November 1916 to 5 August 1917. His tour of duty marks the beginning of the unbroken connection of Navy chaplains with the Marine Corps. Chaplain G. S. Rentz relieved Brodmann in August 1917 and remained until October 1918.

In June 1917, Chaplains G. L. Bayard and J. J. Brady reported to the Commanding Officer of the Fifth Marine Regiment then waiting at Philadelphia for overseas duty. This Regiment was composed of companies recalled from Haiti, Santo Domingo, and Cuba. These Marines with their chaplains sailed in the first convoy that crossed to France June 1917.

During the summer of 1917, the Sixth Marine Regiment and the Sixth Machine Gun Battalion were being formed at Quantico, Virginia, from newly organized companies. Chaplain J. D. MacNair was ordered to the Sixth Regiment on 15 September 1917 and Chaplain H. A. Darche reported on 12 October. These chaplains went overseas with their units. In France, the Marine units were joined to form the Marine Brigade which was included in the Second Division of the United States Army. This Division was organized in France in September and October 1917. Thus, the Marines with their chaplains came under Army jurisdiction. Each Marine Regiment had one Catholic and one Protestant chaplain.

The Second Division went into intensive training during the winter of 1917–1918. During these months, the chaplains were active in directing athletics and entertainment, in addition to their usual religious duties. Writing to Chaplain Frazier, Chaplain Brady reported that he had equipped a clubhouse of six rooms "with two pianos, five billiard tables, reading tables, assorted musical instruments for the casual player, numberless magazines, having subscriptions to cover a hundred magazines and papers, and games of all sorts. . . ."[45] Brady received many gifts of money, including a $2,000 contribution from John Cardinal Farley, to cover cost of such equipment.

The chaplains in their reports of activities during these months tell of conducting "happy hours" or "sings" on Sunday afternoon, arranging Christmas parties with French children as the guests, and in general doing everything possible to keep the Marines busy and happy in their leisure hours. American tobacco and cigarettes were then very scarce in France. One of the chaplains won the undying gratitude of his Marine unit when he located a shipment of cigarettes sent to American troops in France by the New York *Sun* which had been stored in a warehouse. While others were arguing how such a shipment should be handled, this enterprising chaplain loaded the tobacco on truck and hastened back to his men where he was received with all of the honors of a conquering hero.

Several changes took place in the chaplain personnel in the early spring of 1918. Bayard was detached from this duty with the Marines and ordered back to the States. In February Chaplain Albert N. Park arrived with the First Marine

[44] *Compilation of Court-Martial Orders*, 1916–37, I:243. Approved by the Secretary of the Navy 9 May 1918.

[45] Nav. Rec. Coll., Chaplains' File.

Replacement Battalion, Chaplain C. M. Charlton with the Second, and Chaplain Q. F. Beckley with the Third. In June, Chaplain MacNair, who had then completed four years of sea or foreign duty, was detached from the Sixth Regiment and was relieved by Chaplain A. C. Larned. In addition to the eight mentioned chaplains, five others saw service with the Marines in France, most of whom were at replacement depots. They were: G. B. Kranz, R. E. Miller, G. F. X. Murphy, G. S. Rentz, and J. C. Short. Before the war ended, these thirteen Navy chaplains served about 31,000 Marines who had been sent to France.

On 17 March 1918, the Fifth Marines moved into the trenches in the Verdun sector and were followed by the Sixth Regiment the next day. Here they remained until 16 May. Although very few men were killed during this period, the trench type of warfare was most trying. The Marines wanted action. They fretted under the restrictions of trench life and grew restless under the difficulties of their living conditions. All this placed a heavy burden upon the chaplains who went with the men into the trenches and shared with them the difficulties of that life. Now and then there was an enemy raid, or an enemy shell dropped into their lines. Once the Germans sent over gas that killed many and incapacitated others.

Describing those days, Chaplain Brady wrote:

> To dispel this depression and gloom required the cheery activity of the Chaplains telling stories, distributing letter paper to write home, holding little impromptu parties with cake and candy, and music from a mandolin or banjo. . . . Every morning Divine services were held in some part of the line. Men went to confession and communion and took a fresh grip on themselves.[46]

The experience in the trenches was but a shadow of what was to come. In the early days of June 1918 the Germans broke through the French line and had a clear road to Paris. The Second Division with its Marine Brigade was hastily thrown across the path of the invader. Then followed the terrible bloody battles of Chateau Thierry and Belleau Wood—the German advance was checked and the enemy thrown back.

During the days of June 6–15 some 451 Marines were killed. Belleau Wood was finally cleared of the Germans on 25 June. "In the terrific fighting in that month," wrote Secretary Daniels, "the Marine Corps lost 1,062 men killed, and 3,615 wounded."[47] This represented casualties of fifty-five percent and was described as "the heaviest losses suffered by any American brigade during a single offensive operation" in World War I.[48] These were days of horror. Men were under fire for a month without being relieved. Some broke under the excessive strain. In the midst of the Marines, suffering with them, sharing the dangers and bearing the strain were Chaplains Brady, Park, MacNair, Darche, and Beckley. Chaplain Charlton was only a short time at the front when he received orders for duty in Paris.

The accounts written by chaplains who were in the thick of the fighting tell of constant ministry to the wounded and refer frequently to burying the dead. Chaplain MacNair's account of his experiences the night of 6 June is typical:

> During the night of the 6th, I remained at Headquarters with a hospital apprentice to take care of any of the wounded that might happen to stray that way and it was a busy night for me. The hospital Corpsmen belonged to the Army and were wholly inexperienced, consequently, the binding up of the wounds was done almost entirely by myself. Seventy-five or eighty of the men came in claiming that they had been gassed. I took care of them as best I could putting them in a quiet place and giving instructions that they wrap themselves up in their blankets and keep warm.

Referring to his experiences in burying the dead, Chaplain MacNair wrote:

> There was no "Burial Corps" in our organization; hence the digging of the graves and the burying was dependent upon volunteer parties of men ordered to play the part of undertaker, grave digger and clergyman. . . .

MacNair gave further descriptions:

> June 10 The day was spent in seeking for the dead and burying them.
>
> The 13th, 14th, and 15th were spent in working around the dressing stations and encouraging the men in the companies as best I could. I buried several men and as Beckley and Darche were with us, those two or three days were not so strenuous as the previous ones. The fighting had not been quite so severe.
>
> On the 18th of June, I met Major Holcomb and asked him how things had been going. He informed me that out of the 1300 original men that he had taken with him in June, he had about 300 left.[49]

MacNair was detached from the Marines on 23 June and ordered back to the States.

Chaplain Park's letter to Chaplain Frazier dated 21 December 1918 carried considerable detail about his experiences with the Fifth Regiment. He wrote:

---

[46] CoC., Brady File.

[47] Daniels, *Our Navy at War*, p. 210.

[48] Metcalf, *A History of the U. S. Marine Corps*, p. 490.

[49] Nav. Rec. Coll., Chaplains' File.

In the Champagne, Ardennes and Meuse fights I had some close calls. The casualties in the first of these were heavy and I soon realized that I had more in my burial detail than some companies had in the line.[50]

Chaplain Brady wrote that during the worst of the fighting the chaplains had scarcely any sleep. Describing the experience of a chaplain, Brady wrote:

He crawled, walked, ran among the fighting men, during these uncertain days finding the wounded and marking the position of a dead Marine by forcing the bayonet of his rifle into the ground so that the butt of the rifle stood upright, cutting away clothes from wounds, even cutting the shredded flesh that held a shattered leg or an arm to the body, sounding out words of encouragement, hearing the whine of the bullet as he dragged a wounded man into a shell hole, dazed and confused by the flashes and explosion everywhere.

And then with the night and a lull in the fighting, the Chaplain with a shovel led off a gang of volunteers to bury those who lay still on the ground. . . .

. . . .

Soon would come a messenger to tell him of some more dead bodies. Dead bodies, dying men, exploding shells, whining bullets, mustard gas. . . . And he of all men had to be cheery, had to perk up and say pleasant words and . . . that in spite of the horribly demoralizing influence of the bloody handling of the bodies of men with whom he had associated and chatted intimately only a few hours before.[51]

Later chaplains were relieved of the task of burying the dead. Brady described the changed policy as follows:

After the battle below Soissons it was realized that the Chaplain and his orderly and a few volunteers hardly made up an adequate regimental burial party and in their place was substituted a line officer and two hundred men. The Chaplain was now called on merely to perform the burial ceremony and he was thus free to go back to his rightful task of sustaining the morale of his men.

Park told of one narrow escape he had when a German shell landed in an old shell hole which he and his Marine "striker" had just vacated. Others in the company, not knowing that the Chaplain had moved and seeing the shell explode, sent a Marine over to investigate. He crawled to the spot, took one good look, and then hastened back with the information: "There was blood in the hole where the Chaplain had been." Thus the word spread through the Regiment that Chaplain Park had been killed. Later much to their relief, it was discovered that he was still alive.

In his letter to Frazier, Park told what really happened. It appears that when a German barrage was laid down near them, Park and his striker sought refuge in a shell hole. "This time," wrote Park, "the shells seemed to light some distance away but they began to creep a little nearer and my striker looked at me with a gleam in his eyes as though he would like to get away." Soon a shell exploded close enough to throw dirt over the two crouching in the hole. The striker's nerve wavered as he suggested: "Let's go, Chaplain." And Park added: "We went." Shortly after, another shell landed in the hole just vacated "My canteen and mess kit were all broken," wrote Park, "my blanket roll was cut by shrapnel, and a can of tomatoes in my kit had been struck and the contents scattered about the hole." It was the contents of this can of tomatoes splattered over the torn remnants of Park's blankets that called forth the remark of the investigating Marine: "There was blood in the hole where the Chaplain had been."

Al Park loved to tell that story in the following years.

Park had a deep religious spirit. Writing to Chaplain Frazier on 3 August 1918, he described his first impressions under shell fire. Park gave the following description of a Divine Service held with the Marines near the front:

I don't suppose I ever felt more like preaching than I did that Sunday morning. At the close of my sermon I called for those who would make a definite stand for Christ and profess His name, and those who would renew their former vows, to raise their hands. It was almost unanimous and reminded me of the service at the station last summer when Lash preached and you called for decisions. In the evening, at the service, I gave the men an opportunity to pray and many responded. Then we had a genuine testimony meeting and Chaplain, I wish you could have heard those men. . . .[52]

Park was the only chaplain to serve in every major engagement in which the Marines participated.

Chaplain Brady wrote of his friend: "Chaplain Park was easily the outstanding Chaplain of the Marines . . . having served with distinction through all major engagements."[53] Although Park was wounded only once and then slightly, the chief result of those terrible days of strain and toil was a nervous state from which he never recovered. Park, the Presbyterian, and Brady, the Roman Catholic, both Navy chaplains serving in the same Regiment under the same conditions, formed a friendship which was dissolved only when Park died in 1944.

[50] CoC., Park File.
[51] *Ibid.*, Brady File.
[52] *Ibid.*, Park File.
[53] *Ibid.*, Brady File. A letter from the Commanding Officer of the Fifth Regiment to the Secretary of the Navy, 26 Nov. 1918, (CoC., Park File) confirms this statement.

Several other Navy chaplains also had narrow escapes. Brady, describing an experience which came to Darche, wrote:

> Chaplain Darche was wounded severely and had to be carried off the field near Verzy. He had become a spectacular figure because of his exploits under fire, paying not the slightest attention to the death being dealt out on all sides of him. Many were the stories of his cool indifference to death. Two Germans had just been taken prisoners and Chaplain Darche was sitting down before them and asking them questions. Suddenly a shell ripped the two Germans into fragments and when two Marines picked up Chaplain Darche, he smiled and said, "There is no use in asking these Germans anything."

On another occasion Darche had a party of German prisoners digging graves when a shell dropped into their midst killing all of the prisoners and knocking the chaplain unconscious.

Chaplain Brady tells of going forward into the danger zone one day with his orderly. Not wishing to expose the orderly unnecessarily, he left him in what was considered a safe place and continud alone. Brady returned to find that a shell had exploded where the orderly was, killing him instantly. The Chaplain, who thought he was taking the greater risk, was untouched. On one occasion Chaplain Brady was knocked senseless by an exploding shell and did not regain consciousness for three days.

The Marines saw action at Soissons in July and later in the Meuse-Argonne offensive. The chaplains went through it all. One day a Marine said to Chaplain Brady: "It certainly made us feel good to see you up there with us in the woods." In addition to Chaplains Park, Darche, and Kranz who were wounded in action, Chaplains Beckley and Larned were gassed.

Four Navy chaplains who were with the Marines during the heaviest fighting received the Navy Cross and these were the first chaplains in the history of the Corps to be so honored. They were Chaplains Brady, Darche, MacNair, and Park. The citation in each case mentioned extraordinary heroism. Other significant comments were:

For Chaplain Brady:

> . . . he exposed himself fearlessly making a complete tour of the front lines twice. . . . He carried out his duties as Chaplain with devotion and was cool under fire.

For Chaplain Darche:

> . . . his efforts in searching for and burying the dead, in giving cheer and spiritual comfort to the fighting troops, in handling working parties and in aiding the surgeons were tireless; and he performed the last rites of the church under fire.

For Chaplain MacNair:

> . . . performing his services in daily risk of death from enemy fire, when his labors in locating and burying the dead and in giving comfort to the wounded were given with fidelity to duty under all conditions.

For Chaplain Park:

> . . . while in charge of the regimental burial detail, voluntarily exposed himself to shell fire to assist two wounded men to a place of safety, and calmly walked around among the men directing them to get under cover, while remaining himself exposed.

Chaplain Brady also received the Distinguished Service Cross, an Army decoration corresponding to the Navy Cross. Since the Marines served in France under Army authority, it was possible for a Navy officer to receive an Army decoration. This citation in part read:

> chaplain, United States Navy, attached to 5th Regiment, United States Marine Corps, 2d Division. For extraordinary heroism in action near Chateau-Thierry, France, June 6-7, 1918. He made two complete tours of the front line under severe fire, carried on his duties as chaplain with untiring service, and ministered to the men of the regiment under unusually trying circumstances. . . .[54]

Three of the chaplains—Brady, Park, and Darche—received the Croix de Guerre. Darche also was decorated with the Chevalier of the Legion of Honor of France. Both MacNair and Darche were awarded the second Division Citation by the Commanding General, Second Division, USMC. These decorations came as the outward signs of the gratitude of France and the United States for unusual heroism and faithfulness to duty. Darche was the most decorated chaplain in World War I.

In addition to the thirteen Navy chaplains who served with the Marines in France, the following eight served with Marines in other places outside of continental United States: H. S. Dyer, W. W. Elder, G. L. Kerns, and M. R. Boynton in Santo Domingo; H. M. Peterson and R. W. Truitt in Haiti; R. L. Lewis in Cuba; and W. McC. Miller on the Virgin Islands.

During the course of the war, some 16,000 Marine officers and men were trained at the large Marine

[54] *The Congressional Medal of Honor, The Distinguished Service Cross, and The Distinguished Service Medal issued by The War Department, since April 6, 1917 up to . . . November 11, 1919*, p. 438.

base developed at Quantico. Here E. B. Niver, a Reserve chaplain, reported for duty on 8 June 1917 and remained for over ten years. He received a letter of commendation from the Secretary of the Navy for his faithful and outstanding services. With him for a time was Chaplain T. B. Thompson. Chaplain J. A. Tomerlin had duty with the Eighth Marine Regiment at Galveston, Texas. More than 6,000 Marines were trained at Parris Island, South Carolina. Serving there were Chaplains E. E. McDonald and R. L. Lewis. This brings the total of Navy chaplains who served with the Marines for varying periods during the First World War up to twenty-six.

Navy chaplains and doctors attached to the Marines found that no regulations existed to direct them in the type of uniform to be worn. Each wore what he thought was best for the occasion. The Navy then had only the blues and the whites, both of which were impractical for combat conditions. Chaplain J. J. Brady in a letter to the author dated 17 September 1945 gave the following description of the uniforms worn by the chaplains assigned to the Marines:

> We wore enlisted men's clothes at first until we located a French tailor who made an officer's uniform for us.
>
> All doctors and chaplains and even corpsmen wore the uniform of the Marines. We had rank insignia on the shoulders, and a cross on one lapel of the blouse and the marine insignia on the other lapel of the blouse. They were not combined into one in any way. There were no rules about the matter. We wore what we thought would best fit the situation. For a while Chaplain Bayard wore an army officer's uniform with the chaplain's insignia. There were no rules and the Marine officers of high rank did not attempt to lay down any rules for the Navy officers. It was just a case of doing what seemed the most polite and considerate thing in the case of outsiders serving with the marines.

A close and friendly relation existed between the Army and Navy chaplains assigned with the Marines in France. Occasionally Army chaplains assisted Navy chaplains in ministering to the Marines. For a time Chaplain Brady had additional duty with the Army which included jurisdiction over about 100 Army chaplains.

SPECIAL WAR SERVICE OF OTHER CHAPLAINS

In addition to the chaplains who served with the Marines in France, several other Navy chaplains saw overseas duty during the First World War. Chaplains W. B. Ayers, E. A. Duff, T. L. Kirkpatrick, W. A. Maguire, F. L. McFadden, and G. E. T. Stevenson, all had shore duty in France during these years at radio stations or naval bases. Chaplains C. R. Blain, R. J. Davis, B. R. Patrick, and A. W. Stone had duty in England or on the Continent. E. W. Foster was the first chaplain to be stationed at the Submarine Base at Coco Solo in the Canal Zone. For a part of the time under review, Chaplain A. W. Stone was at Guam, Chaplain T. B. Thompson at Guantanamo Bay, and Chaplain R. D. Workman in the Philippines.

Chaplains aboard the battleships which composed the Sixth Squadron of the British Grand Fleet based at Scapa Flow saw interesting action. The *Florida*, *New York*, and *Delaware* by skillful maneuvering dodged torpedoes fired by German submarines. Two Navy chaplains were on transports which were torpedoed and sunk by enemy submarines off the coast of France. Chaplain Grover C. Whimsett was aboard the *President Lincoln* when she went down on 31 May 1918 and Chaplain Perry Mitchell was in the *Covington* when she was sunk on 1 July of the same year.

Another to receive the Navy Cross during the First World War was Chaplain W. A. Maguire. His citation tells the story:

> For extraordinary heroism as an officer in a boat sent out from the U.S.S. Christabel to the rescue of men from the Florence H. which vessel, loaded with explosives, was burned in the harbor of Quiberon on the night of the seventeenth of April, 1918. Almost immediately after the outbreak of fire the water in the vicinity of the Florence H. was covered with burning powder boxes, many of which exploded, scattering flames throughout the wreckage. The Officers and the crew of the Christabel's boats drove their boats into the burning mass and succeeded in saving the lives of many men, who but for the help so promptly and heroically given, must have perished in the wreckage.[55]

Two Navy chaplains received decorations from the King of Belgium. They were E. W. Scott, who was made an officer of the Order of Leopold, and P. F. Bloomhardt, who was awarded Chevalier of the Order of the Crown. Chaplains Sidney Key Evans and Charles V. Ellis received Letters of Commendation from the Secretary of the Navy for meritorious services.

Special recognition should be given to the unusual services rendered by Chaplain David Tribou, who in April 1917 at the age of sixty-nine was called from the retired list to serve at the Boston Navy Yard. His previous tour of active duty had extended from the

[55] CoC., Maguire File.

year of his commissioning, 1872, to his retirement in 1910. After spending about nine months at the Boston Yard, Tribou was ordered to the Naval Home in Philadelphia where he soon firmly established himself in the affections of the three hundred occupants. When orders were issued for his detachment in March 1920, many letters of protest from the beneficiaries in the Home were received by Secretary Daniels and Chaplain Frazier. Ninety-eight percent of the occupants of the Home signed a petition to the Secretary of the Navy urging Tribou's retention. One of the beneficiaries wrote to Senator Pat Harrison on 16 March 1920 saying in part:

> Can you believe, dear Mr. Harrison, that for fifty years I had not attended any church! Today I feel differently in matters of religion, so much so that I have not missed once the Service on Sunday from the first week I entered this Home. Why? Because our Chaplain teaches the Gospel. . . .[56]

As a result of these expressions of affection, Chaplain Tribou was permitted to remain. Orders for detachment from the Home were again issued for 30 September 1921 and again a storm of protest was raised. Even though Tribou was then past his seventy-third birthday, he was permitted to remain at the Home until his death on 31 May 1922. He had then completed a total of forty-two years of active service, a record never before or since equaled.

During part of 1919, Chaplain Tribou served as Chairman of a Victory Loan Drive in the Home. On the night of 7–8 October while Tribou was on temporary duty in Washington, the safe in his office in the Home was broken open and $2,667 in cash and bonds belonging to the beneficiaries were stolen. Although held by an investigating board not to have been responsible, Chaplain Tribou repaid every dollar out of his own savings. A bill was introduced into Congress providing restitution to Chaplain Tribou from Government funds. The House of Representatives approved the measure in March 1922, a little more than two months before he died. However, the bill was not passed by the Senate in that session of Congress. The proposal was reintroduced in 1924. The measure was then approved and restitution made to the estate of Chaplain Tribou.

Older chaplains and others in the Navy still cherish the memory they have of one of the most beloved of all Navy chaplains.

[56] *Ibid.*, Tribou File.

THE FIRST FLEET CHAPLAINS

The rapid expansion of the United States Fleet called for a corresponding increase of chaplains assigned to fleet duty. The need was soon felt for an experienced chaplain of mature years to serve somewhat as a bishop to oversee and correlate the activities of all chaplains of the Fleet. The office of a Fleet Chaplain was authorized by Secretary Daniels on 8 June 1918.[57] Daniels in his *Report* for 1919 made the following reference to the appointment of Fleet Chaplains:

> In the assignment of chaplains to duty an effort has been made to so distribute representatives of the various denominations as to give to both Protestants and Catholics the ministrations of clergymen of their own faith. That this plan might be carried out successfully, the Pacific and Atlantic Fleets have each been provided with a fleet chaplain who is on the staff of the commander-in-chief and whose duties it is to direct the younger chaplains in their work, to give the benefit of advice and council, arrange for frequent conferences and for interchange of chaplains so as to provide services for both Protestants and Catholics.[58]

The first to be assigned to duty as a Fleet Chaplain was Matthew C. Gleeson, who was ordered to the staff of the Commander in Chief of the Atlantic Fleet on 16 June 1918. He continued as such until 11 July 1919.

In his annual report dated 14 February 1919, written after assuming his new duties, Gleeson told of some of his achievements. "As I understood my new duties," he wrote, "I was to counsel with, and give such help as was necessary to the various Chaplains of the Fleet, all of whom at this time were comparatively new to the Service." He arranged conferences of chaplains and facilitated arrangements whereby the Protestant and the Catholic chaplains could keep in better touch with the men of their respective faiths. By shifting schedules and by sending chaplains from one ship to another, more men were reached with a spiritual ministry. On several occasions, the Fleet Chaplain was able to intercede on behalf of individual chaplains who felt that too many collateral duties were being given them, thus making it impossible to give adequate time to their religious duties.

Gleeson observed that all the chaplains in the Fleet, with the exception of himself and one other officer, were lieutenants (junior grade). The other officer was a Reserve chaplain with the rank of lieutenant.

[57] *Ibid.*, Sec. of Navy File.
[58] *Op. cit.*, p. 388.

The lack of rank sometimes put chaplains at a disadvantage. For instance, Chaplain Gleeson found that some chaplains were handicapped because they did not have a cabin or an office easily available to the enlisted men. Such accommodations were given other officers who outranked them. Regarding this problem Chaplain Gleeson recommended:

> It was the opinion of many Chaplains of the Fleet, as well as my own, that it would be a wise provision to have a room on each ship near the men's quarters, permanently assigned to the ship's Chaplain, exclusive of what rank he might hold. The Chaplain to have his work efficient and appreciated by the men, must be at their service, not only during office hours, but at all times, and if his cabin is so placed in the Wardroom country that the men will have to pass constantly through the officers' quarters, he will find they will not readily visit him under the circumstances.[59]

Chaplain Gleeson further reported that the plan of having three Protestant chaplains of various denominations and one Catholic attached to each division within the Fleet had been adopted with good results. He felt that every hospital ship should have at least one chaplain and when two of these ships were operating with the same Fleet, one of these chaplains should be a Catholic. "During the Influenza Epidemic," wrote Gleeson, "there being no Catholic Chaplain in the immediate vicinity of the U.S.S. *Mercy*, I was obliged myself to look after the Catholics ill on board, and was so busy during the interval that I had neither time nor opportunity for other work."[60]

On 30 June 1919, the Pacific Fleet was formed and began its movement to the West Coast on 17 July. Chaplain E. W. Scott was appointed Fleet Chaplain on 19 July and held this position until 25 October 1921. He served aboard the Flagship *New Mexico*.

On 21 September 1919, Chaplain Scott sent in a long report of his work. The newly organized Pacific Fleet was given an enthusiastic welcome by coast cities. Scott commented on the fine services rendered by such organizations as the YMCA, the Knights of Columbus, and the Salvation Army in caring for the entertainment of enlisted personnel on shore. He reported that he had held a series of conferences for chaplains at such places as San Diego, Santa Barbara, San Francisco, Seattle, and Bremerton, He wrote:

> I am sure you would be delighted in seeing the zeal and the fine spirit shown by these Chaplains, and in hearing the good words spoken in their behalf by their officers. They are on the job and doing good work and have shown a friendly spirit of co-operation toward each other and a real pride in the Corps.

Chaplain Scott's kindly and efficient supervision resulted in improved working conditions for a number of chaplains in the Fleet. He was instrumental in securing clerical assistance for several and helped others obtain more suitable space for the library and reading rooms on board ship.

Scott made interesting comments regarding the introduction of evening prayers, or hammocks, aboard ships at sea. He reported that Secretary Daniels had observed the revival of this old Navy custom on the *Arkansas* and was greatly impressed. Daniels saw Scott at Tacoma and inquired as to the possibility of making the custom uniform throughout the Fleet. Regarding the practice aboard the *Arkansas* and the attitude of the chaplains, Scott wrote:

> It was suggested by what he saw on the ARKANSAS. It has been instituted there by the Commanding Officer, entirely without suggestion from Chaplain Shrum. We had discussed it in conference and all were agreed that it was not the thing to do. After the suggestion I brought it up in the last conference and we have all agreed "to respectfully adhere to our former opinion." From his experience Chaplain Riddle was most insistent it would antagonize men rather than help them.[61]

The chaplains felt that any plan which involved compulsory attendance at any religious service would do more harm than good. Another factor had also arisen which brought complications—it was the objections of Catholics to attend prayers conducted by Protestant chaplains, and no doubt that of some Protestants to attend those conducted by Catholics. Conditions within the Navy had greatly changed from those existing fifty years or more earlier when practically the whole crew were Protestants.

The institution of the office of Fleet Chaplain (sometimes referred to as Force Chaplain) proved so beneficial to the work of chaplains that it was continued in the Navy thereafter. The work of a Fleet Chaplain set the pattern for the later introduction of District Chaplains. Article 1244 of the 1920 edition of *Navy Regulations* outlined the duties of the Fleet Chaplain as follows:

> (a) To inspect the work of the chaplains at frequent intervals.
>
> (b) To advise, counsel, and offer suggestions to the chaplains of the fleet concerning their work.
>
> (c) To arrange for services on board ships that do not carry chaplains. Also for interchange of chaplains of the different faiths so as to provide services for all.

[59] CoC., Gleeson File, report of 14 Feb. 1918.
[60] *Ibid.*

[61] CoC., Scott File.

(d) To call frequent meetings of chaplains of the fleet for conference, discussion, and adjustment of the matters pertaining to their work.

(e) At the end of each quarter he shall make a written report to the Chief of Naval Personnel concerning the chaplains and their work.

PROMOTION

Not one of the 130 acting and temporary acting chaplains, who were appointed during World War I, was promoted during the years 1917 to 1919 inclusive. On 16 October 1920 the Judge Advocate General wrote to the Chief of the Bureau of Navigation saying: "The Navy Register shows that not one officer of the Chaplain Corps was given a temporary promotion to a higher rank or grade as specifically authorized by act of 22 May 1917 as amended by act of 1 July 1918; and that the Chaplains Corps is the only corps so discriminated against."[62]

The failure of acting and temporary acting chaplains to receive promotion was due in part to the wording of the Act of 1914 and possibly to a misunderstanding on the part of the Bureau of Navigation as to the intent of the Acts of 1917 and 1918 as they applied to chaplains. As previously stated, the Act of 1914 required acting chaplains to serve a three year probationary period at sea with the rank, pay and allowances of a lieutenant (junior grade), and then pass an examination before a Board of chaplains and medical officers before being actually commissioned. The law also stated that only seven acting chaplains could be commissioned in any fiscal year. The statute further provided that an acting chaplain spend seven years which included the probationary period, in the rank or grade of lieutenant (junior grade) and four years as lieutenant before being promoted to the next higher rank.

On 22 May 1917, Congress passed an Act which abolished examinations for promotions in grade in all staff corps. Early in 1918, Frazier raised the question as to whether or not the Act of 1917 abolished examinations for acting chaplains who were eligible for commissions. The Judge Advocate General wrote on 11 February 1918 saying that "the change of status from acting chaplain to chaplain, while a change in grade, is not exactly a promotion in grade." He held, therefore, that acting chaplains should still be required to take the examination as prescribed in the Act of 1914 before receiving a commission.

[62] NRSO: 29226-7. Chaplain W. N. Thomas moved from seventh place in his wardroom mess to fortieth because of the promotion of officers who had been junior to him.

This ruling uncovered an embarrassing situation for the Bureau of Navigation for it was discovered that the provision of the law of 1914 calling for an examination of acting chaplains had not been observed. Chaplain Frazier in a letter to the Bureau of Navigation dated 29 August 1918 declared:

> Notwithstanding the fact that said law became effective in 1914, it was not until March, 1918, that, in conformity to this law, at my request, chaplains due for promotion were required to appear before Examining Boards. Between the date of the passage of the present law governing the promotion of chaplains, and March, 1918, twelve chaplains were promoted, and, so far as I have been able to learn, no Board was ever assembled, and no officially constituted body examined their records or investigated their efficiency as chaplains. The law was simply ignored, the Chaplain's efficiency being considered of so little importance that examination was unnecessary even though required by law. No other Corps could have possibly maintained its efficiency under such practice.[63]

The Judge Advocate General had suggested that the provision of the Act of 1914 regarding the examining Board of chaplains be amended to include one chaplain "of the same denomination as that of the candidate" who was to be examined. The Judge Advocate General claimed that this would eliminate the possibility of controversy arising over chaplains of one of more denominations sitting in judgment on members of other churches. Chaplain Frazier, however, took exception to this point of view in his letter of 29 August. He wrote:

> The Chaplains of the Navy are universal in their approval of the present law. I know of no law relative to the Chaplains' Corps that has been passed within the period of my Naval Service of nearly twenty-four years that has been given such hearty approval. It has been the source of a great deal of disappointment and much comment that the law as passed has not been carried out. We had hoped and believed that the provisions of this law would enable us to procure and retain the high type of Christian gentlemen to which the office is entitled, and which it must have if it is to fulfill its mission.
>
> Chaplains are not examined on their denominational views or church doctrines, but on their efficiency as moral and religious teachers in the broadest sense of these terms, and on their personal conduct and life. The Examining Board of Chaplains does not convene as representatives of any particular denomination, but as advocates of universally accepted standards of morality and Christian religion. There can be no conflict of opinion in such matters between the various religious denominations inasmuch as they are as *one denomination*, all having the same high standards in matters pertaining to the vital principles of morality.

[63] NRSO: 2650-281.

In answer to another communication from the Bureau of Navigation the Judge Advocate General wrote on 24 January 1918 that seven years combined service as acting chaplain and chaplain were still required "before eligibility for permanent promotion to the rank of lieutenant." It was this provision of the Act of 1914 which prevented all appointed during the war who came under that statute from being promoted. The Act of 1 July 1918 authorized temporary promotions without regard to length of previous service. Just why some chaplains were not promoted between the passage of this Act and the signing of the armistice is not clear. According to the Judge Advocate General, the Chaplain Corps was the only corps of the Navy which did not take advantage of the new provisions for promotion.

The following table showing the number of chaplains in the different ranks has been taken from the *Navy Registers* as of 1 January for the years indicated:

	1917	*1918*	*1919*	*1920*
Captain	8	10	14	13
Commander	8	6	2	1
Lieutenant Commander	.	..	..	2
Lieutenant	2	4	4	2
Lieutenant (jg)	4	2	6	11
Total	22	22	26	29
Acting chaplains:				
Permanent	18	56	101	50
Temporary	.	1	40	12
	18	57	141	62
Net total	40	79	167	91

These statistics do not include chaplains of Class 2 and 4 of the Naval Reserve Force. The figures show that promotions were discouragingly few even among the Regular Navy chaplains. Chaplains Gleeson, Stone, Scott, and Evans all received their captaincy on the same day, 1 July 1918. Chaplains Pearce and MacNair tarried but a day each in the rank of lieutenant commander. Pearce was promoted on 8 Februray 1919 and the next day was promoted to commander. On 1 July 1920, he became a captain. MacNair was made a lieutenant commander on 30 June 1919. The next day he was promoted to commander and on 14 July 1920 became a captain.

Promotion among the Reserve chaplains was more liberal, mainly because they were not subject to the laws which governed acting chaplains. Niver and Brokenshire were both promoted to commander; Day and Huske were made lieutenant commanders; and several moved from lieutenant (junior grade) to lieutenant.

The Act of 11 July 1919 which provided appropriations for the naval service for the fiscal year ending 30 June 1920 included the following:

> That officers of the permanent Navy who have served satisfactorily during the war with the German Government in a temporary grade or rank shall be eligible under the provision of existing law for selection for promotion and for promotion to the same permanent grade or rank until July 1, 1920, without regard to statutory requirements other than professional and physical examination. . . .[64]

But since no chaplains, with the exception of a few Reserves, had received such temporary promotions, none qualified.

The fact that no acting or temporary acting chaplains received a promotion during the years 1917–1919 introduced some dissatisfaction among the younger chaplains as the following extract from the annual report of Chaplain H. R. Davidson for 1918 would indicate:

> One other recommendation I would like to make has to do with the present law regarding permanent commissions. Since the present law allows only seven acting chaplains to be fully commissioned in any one year and since I stand 90 in the Corps it will be approximately ten years before I come up for examination for permanent appointment. If one has the professional qualifications required for the Corps he can hardly be satisfied to remain in the service if he has to wait so long for advance in rank and so in pay.[65]

Chaplain Park writing to Chaplain Frazier on 21 December 1918 from France, after a vivid description of his experiences in battle, included the following paragraph:

> In the midst of these experiences, I received a letter from you saying that we would not participate in the temporary promotions. I can't see any justice in that at all. Many a time in those days I wouldn't have given two cents for my chances of living through the day, and now that it is all over, it seems like a wild dream to me. . . . I am proud of what I have done, but it is hard to see everybody else getting promoted and to know that you must stand still.[66]

MORE UNIFORM CHANGES

The whole question of desired uniform changes for chaplains was brought into focus in January 1917, when the *Uniform Regulations* of 1913 were being revised. On 16 January Chaplains Frazier, Gleeson,

[64] 41 *Stat.* 140.
[65] CoC., Davidson Annual Report File.
[66] *Ibid.*, Park File.

and Evans addressed the following communication to Secretary Daniels:

Subject: Change in Uniform.

1. It is recommended that the Corps of Chaplains receive the same consideration in the matter of uniform as do the other staff corps at present, and with that end in view the following changes are suggested:

(1) That the cap, shoulder marks and sleeve braids in the service dress and overcoat be made to conform to those worn by other Staff Corps.

(2) That the frock coat now worn be changed to conform in cut, style and trimmings to that worn by the other Staff Corps.

(3) That the coloring between the sleeve stripes be black and that on the frock coat and evening dress one-half inch above the upper stripe a Latin Cross as now worn be set perpendicularly.[67]

The recommendations were not adopted. The revised *Uniform Regulations* of 1917 still prescribed the lustrous black mohair braid in lieu of the gold braid worn by the other officers. The full-dress uniform was authorized for "all commissioned officers, except chaplains and chief warrant officers."

According to a story handed down by the older chaplains whose memories go back to the days of World War I, Chaplain Henry van Dyke was instrumental in having the gold braid substituted for the black. The story is that Chaplain van Dyke declined an invitation to the Daniels' home for dinner. When pressed for a reason, he told the Secretary that he would not accept such an invitation until "he got out of mourning."

On 7 February 1918 Chaplain Frank Thompson then Senior Chaplain at the Naval Training Station in Chicago wrote to Chaplain Frazier:

I am writing to say that I have just learned a bit of news that is most welcome to me and which I trust is true. It has reference to a proposed change in the Chaplains' uniform, namely, the adoption of gold stripes on the sleeves in place of the black ones worn at present.

. . . .

In my mind it would be a most wise and advisable one as then a Chaplain would be wearing a uniform more in conformity with those of other commissioned officers, and there would be less mistake and confusion as to his official position and rank.[68]

Thompson also expressed his hope that the proposed changes would include the substitution of a gold chin strap on the cap instead of the black mohair and the permission to "obtain some sort of a frock coat more suitable than the single breasted one now in use which, with all due deference to the originator, is, in my humble opinion, a grotesque monstrosity." Undoubtedly Thompson's views accurately reflected the attitude of the Corps.

On 26 June 1918, Josephus Daniels issued "Change in Uniform Regulations, No. 19" which granted chaplains the right to wear the gold braid on the sleeves with "lustrous black cloth" between the gold braid as the distinguishing mark of the Corps. On 27 July, Frazier wrote to his friend Thompson saying:

I have received a copy of the order authorizing the change in uniform and have been instructed to fit the new Chaplains out according to this change.

. . . .

Cap to correspond in every particular with the cap of other staff officers. Shoulder marks to correspond with those of other staff officers except lustrous black cloth between gold stripes. The stripes on the service dress are changed to gold with black between the stripes. I asked for change in the frock coat to correspond to that of the other Corps, but was told by the Secretary that this would not be made until after the war as it would necessitate the use of cloth that might otherwise be used for something else.[69]

On 17 March 1919, the Secretary permitted the chaplains to wear the same frock coat as that worn by other officers. "I believe you were heartily in favor of this," wrote Frazier to Thompson on 23 May 1919. "I feel that we are making progress and that in time there will be no distinction between our Corps and others."

Change No. 25 to the *Uniform Regulations*, dated 16 November 1918, established a new insignia for the Chaplain Corps. The revised applicable articles are quoted below:

Art. 79—The collar shall bear devices indicating rank and corps as follows:

(k) Commissioned staff officers, . . . same as for line officers with whom they rank but the appropriate corps devices, embroidered in gold, surcharged upon the anchors.

Art. 119—Embroidered Corps Devices on Epaulets (cancelled old articles 119-125)

(d) Chaplains: A Latin cross, inclined at an angle of 45 degrees, the top toward the stock of the anchor, embroidered in gold. Chaplains of the Jewish faith may substitute the shepherd's crook for the cross.

Art. 133—Shoulder marks. For all officers, for wear on the white service coat, mess jacket, and overcoat. The corps device will be the foul anchor, in the case of staff officers, surcharged with the appropriate corps device.

[67] Nav. Rec. Coll., Uniform File.
[68] CoC., Thompson File.
[69] *Ibid.*

These revisions meant that chaplains, instead of wearing only the cross behind the rank insignia on the collar, now wore the cross superimposed on the fouled anchor.

Change No. 27 to the *Uniform Regulations* appeared 17 March 1919 and specified the present double-breasted service coat with the roll collar. The insignia of rank and corps was by necessity shifted from the collar of the service coat to the sleeves. The old uniform could be worn until 1 July 1921.

The provision calling for the corps devices surcharged upon the anchors lasted about a year. On 13 November 1919 another change, No. 28, appeared in *Uniform Regulations* which separated the corps device of staff officers from the anchor. The change affecting the chaplains is as follows:

> Chaplains shall wear a Latin cross, embroidered in gold the long arm 1" long, the short arm 9/16" long, and each arm 3/16" wide; to be set inclined toward the rear, the longer arm, making an angle of 60° with the upper stripe of lace.

The new regulations also stated that for the shoulder marks "the rank and corps shall be indicated by stripe of gold lace and corps devices as prescribed for the sleeve of the frock coat."

The first chaplain known to have worn aiguillettes was Chaplain E. W. Scott, the Pacific Fleet Chaplain. In a letter to Chaplain Frazier dated 21 September 1919, Chaplain Scott commented:

> In the next place let me say that I am wearing aiguiel[l]ettes. When I reported I had no intention of wearing them but felt it would be better not to declare myself until I had sized up the situation. After mixing with officers from other ships, strangers, and civilians, I soon came to see the advantage of wearing them, as they established your position at once and relieved you of explaining it. I then referred the question to the Chief of Staff and he said at once that as I was a member of the staff and wore the uniform in all other respects, I ought to wear them, irrespective of what Chaplain Gleeson did. . . . I do not believe it has harmed in in any way, but has helped me in many.[70]

Thus, about twenty years after chaplains were denied the right to wear the full uniform and insignia worn by other officers in the Navy, they regained the privilege of wearing the official button, gold braid, and the frock coat. Only one discriminatory ruling remained and that was the regulation forbidding chaplains to wear the complete full dress uniform. However, the chaplains of 1919 were so pleased to be permitted to wear the same service uniform with insignia and the same dress coat as their brother officers, they said little about the continuing discrimination.

NEW REGULATIONS BEARING ON SOCIAL PROBLEMS

In keeping with the rising prohibition sentiment which was sweeping state after state in the years immediately preceding the First World War, Secretary Josephus Daniels on 1 June 1914 issued his General Order 99.

> The use or introduction for drinking purposes of alcoholic liquors on board any naval vessel, or within any navy yard or station, is strictly prohibited, and commanding officers will be held directly responsible for the enforcement of this order.

The law of 1862 which discontinued the serving of grog and forbade the admittance on board vessels of war of "spirituous liquors", except under the control of the medical officers, was subsequently interpreted by the officers as not applying to wine. Therefore, the officers' wine mess continued on board ships until the issuance of General Order 99.

In *Our Navy at War* Daniels confessed that the order "was not universally popular when it was promulgated," but added that after war broke out "it was recognized that it had contributed to the fitness of naval personnel."[71]

Upon the declaration of war in April 1917, Congress faced the necessity of a draft law. The American public was fearful of the effects military service would have on its young men. Many church leaders and organizations began agitating in favor of the establishment of dry zones in the close proximity of Army and Navy installations which might be established in states which did not have prohibition. As a result, the Selective Draft Act of 18 May 1917 carried provisions requiring the President, as Commander in Chief of the Army, "to make such regulations governing the prohibition of alcoholic liquors in or near military camps and to the officers and enlisted men of the Army as he may from time to time deem necessary or advisable."

This Act was amended by Congress on 6 October 1917 to include the Navy. On 5 March 1918, Navy General Order 373 was issued in accordance with the law. The use of liquor except in private and for medical or religious purposes was forbidden in the dry zones which were established.

70 Co.C., Scott File.

71 *Op. cit.*, p. 344.

This General Order was superseded on 1 April 1918 by General Order 380 which stated in part:

> There is hereby established a zone five miles wide, circumjacent to the boundaries of every place under naval jurisdiction specified below. Alcoholic liquor, including beer, ale, and wine, either alone or with any other article, shall not, directly or indirectly, be sold, bartered, given, served, or knowingly delivered by one person to another within any such zone, or sent, shipped, transmitted, carried, or transported to any place within any such zone: *Provided*, That this regulation shall not apply to the giving or serving in a private home to members of the family or bona fide guests, other than officers or members of the naval forces, any of such liquor as may be on hand in such private home on the date of the taking effect of this regulation, namely, the second day of April, 1918, at four o'clock postmeridian. . . .

In the spring and early summer of 1918 Chaplain van Dyke made a tour of twenty-six naval installations from Massachusetts to California. Secretary Daniels, who placed great trust in van Dyke, requested him to observe a wide range of conditions during his tour. Chaplain van Dyke in his report of 15 July commented on the use of liquor at or near naval installations:

> I saw no cases of men in the service under the influence of intoxicating liquor at any of the stations or in their vicinity. It is beyond question that drunkenness, if not absolutely eradicated, has been reduced to the vanishing point by the regulations recently adopted in the service.[72]

General Order 410 was issued on 2 August 1918 to counteract the growing evil of bootlegging. It read:

> During the present emergency, it is prohibited for any person in the naval service to purchase or accept intoxicating liquor from bootleggers within the proscribed zones, or to have intoxicating liquor in his possession on board any naval vessel, or at any naval station, or at any other place under the exclusive jurisdiction of the Navy Department, except as authorized for medical and religious purposes.

General Order 412 dated 16 August 1918 eliminated that section of Order 373 which stated in part: "*Provided,* That this regulation shall not apply to the giving or serving [of liquor] in a private home to members of the family or bona fide guests." Thus an effort was made to control the use of liquor by so-called "bona fide guests" and even by members of the family in this restricted area.

As far as the Navy was concerned, the various orders issued by the Department made the use of liquor on board ships for drinking purposes illegal and also within the five-mile zone around naval establishments. Chaplains' correspondence for the war years indicates that intemperance among sailors was not nearly the problem it had been in the years previous.

An attempt was made to curb prostitution in the vicinity of naval establishments by General Order 411 issued 3 August 1918.

> Ten miles from any place under naval jurisdiction is hereby designated as the distance determined to be needful to the efficiency and welfare of the Navy, within which it shall be unlawful to engage in prostitution or to aid or abet prostitution or to procure or solicit for purposes of prostitution, or to keep or set up a house of ill fame, brothel, or bawdy house, or to receive any person for purposes of lewdness, assignation, or prostitution into any vehicle, conveyance, place, structure, or building, or to permit any person to remain. . . .

SERVICE ORGANIZATIONS IN THE WAR

After war was declared on Germany, Secretary Daniels was concerned not only with the immediate increase of the Fleet and its personnel, but also with the social, moral, and morale problems which were bound to arise at the training bases and elsewhere. A bill was introduced into the House of Representatives on 28 June 1917 calling for $2,000,000 to be spent on chapels and recreational buildings and grounds at the several Navy Yards. This bill, however, never became law.

In order to care for the growing needs of men, Daniels turned to the only institution which could help, the Army and Navy Branch of the International YMCA. Under date 26 July 1917, he issued General Order 313, the first section of which reads:

> The Young Men's Christian Association, in addition to its large service to enlisted men during times of peace, has greatly increased its facilities and efforts during the present need represented by the increased forces in the Navy and Marine Corps and the calling of the Naval Reserve and Naval Militia. This organization is prepared by experience, approved methods, and assured resources to serve our enlisted men. The results obtained by this voluntary civilian organization are so beneficial and bear such a direct relationship to efficiency, inasmuch as the association provision contributes to the happiness, content, and morale of the personnel, that in order to unify the civilian betterment activities in the Navy and further the work of the organization that has demonstrated its ability to render a service desired by both officers and men, cordial recognition is hereby given the Young Men's Christian Association as a valuable adjunct and asset to the service.

The second part of the Order called upon nava officers "to render the fullest practicable assistance

[72] CoC., van Dyke File.

832159—49——14

and cooperation in the maintenance and extension of the association at the regular navy yards and stations, and at such other stations as may be established on either a temporary or permanent basis." Secretary Daniels then listed the following approved methods of cooperation:

> Authorization by commandants for the erection of buildings at the various navy yards and stations in accordance with instructions already issued, and the provisions of heat and light for said buildings.
>
> Cooperation in facilitating accredited representatives in their access to navy yards and stations and to ships and temporary camps.
>
> The granting of commissary privileges where practicable.
>
> Furnishing where practicable tentage for shelter when in temporary camps.
>
> Transportation on naval craft, when necessary, of secretaries and supplies.

Within a week after the declaration of war on Germany, the Supreme Board of Directors of the Knights of Columbus at a meeting held in Washington, pledged in behalf of its four hundred thousand members "their continued and unconditional support of the President and the Congress" in the war effort. On 24 June, the Directors, meeting in Detroit, authorized the raising of a million dollars to be known as the Knights of Columbus War Camp Fund which was "to be expended for religious and recreational purposes for the benefit of all men in the service."[73]

Other Catholic organizations soon sprang into action to serve Catholic men in uniform. Among these was the Catholic Young Men's Union and the Chaplain's Aid Association. In order to coordinate these activities and to create a responsible central body to serve in a liaison capacity with the Government, the Catholic hierarchy at a meeting held in Washington on 11–12 August 1917 set up the National Catholic War Council. This became the National Catholic Welfare Conference on 24 September 1919.

On 9 April 1917, the Jewish Welfare Board was organized, which constituted the first united effort on the part of the Jewish bodies of the United States to function in behalf of Jewish men in the Army or the Navy. This Board became the official agency for all Jewish welfare work with men in the armed services.

In addition to the three main service organizations already mentioned, there were other agencies which rendered valuable service to Navy personnel during the war years. These included the Salvation Army, American Library Association, YWCA, and service organizations of most of the larger denominations. The Presbyterian Church, USA, claimed that it was the first of the Protestant churches to organize a National Service Commission.[74] While the name varied in the various denominations, the functions of each Commission were much the same. Like the Catholic Chaplain's Aid Association, these denominational agencies furnished religious literature and altar supplies to their respective chaplains. Some churches, as the Presbyterian, gave at least $200 yearly to each of its chaplains in the armed services to be spent as each saw fit. At some naval stations as at Great Lakes where the Lutherans had their own hut, some of the denominations carried on extensive activities.

After each of the larger welfare organizations had conducted its own campaign for funds, it was realized that in the future a united approach to the public would be much more fruitful. Plans were perfected in the summer and fall of 1918 for launching a drive during the week of 11 November under the sponsorship of the "United War Work Campaign." No one then dreamed that the day set for the beginning of the drive should mark the end of the war. It was agreed that the participating organizations should share in the funds collected on the following pro rata basis:

National War Work Council of the Yonng Men's Christian Associations	$100,000,000	58.65%
War Work Council of the National Board of the Young Women's Christian Associations	15,000,000	8.80%
National Catholic War Council (Knights of Columbus)	30,000,000	17.60%
Jewish Welfare Board	3,500,000	2.05%
War Camp Community Service	15,000,000	8.80%
American Library Association	3,500,000	2.05%
Salvation Army	3,500,000	2.05%
	$170,500,000	100.00%[75]

In spite of the temporary disruption of campaign plans due to the frenzy of peace celebrations, the drive "went over the top."

Although the YMCA was the first to receive official recognition from the Government to conduct welfare work on military and naval establishments, such recognition was also accorded the Knights of

[73] Williams, *American Catholics in the War*, pp. 98–9.

[74] *Minutes of the General Assembly*, 1918, p. 71.

[75] Williams, *op. cit.*, p. 206.

Columbus, the Jewish Welfare Board, and other organizations as soon as they were ready to begin work. Secretary Daniels, in his 1917 *Report*, commented on the fine contributions being rendered naval personnel by the various welfare organizations. Referring especially to the evils of drinking and promiscuity, Daniels wrote:

> The Young Men's Christian Association, the Knights of Columbus, the Young Men's Hebrew Association, and like organizations for the uplift of young men have helped to secure to the camps and training stations greater freedom from the evil of drunkenness and immorality than ever before where young men were under training for military service. Much is yet to be done, but with the authority given to the chaplains, the surgeons, and other officers of the Navy, and the cooperation of men and women who see the need of driving out evil by introducing wholesome athletics and diversions, this good work will go on until the day will come when these twin relics of bestiality will be anachronisms.[76]

The years of the First World War were years of material expansion for the YMCA. Several new buildings were rented or erected to serve the needs of naval personnel. A five year lease was secured on a building at 226 Embarcadero, San Francisco, on 1 February 1917. An organization of the Navy Y was effected in Honolulu in March 1917 and a building opened on 25 November of that year. A new building costing more than $225,000 was dedicated in Charlestown, Massachusetts, on 25 May 1918. Permanent work of the Army and Navy YMCA was established in San Diego in the latter part of 1919.

During the summer of 1918 the Y was operating thirty-eight buildings, not including the numerous huts erected on naval and military installations. The Y followed, sometimes even preceded, the Navy to such distant places as the Canal Zone, the Bermudas, and Guam. A Y was opened at Hankow, China, in 1919 for the Navy personnel of the Yangtze Patrol.[77] A tremendous service was rendered by this organization. For instance, between five and six thousand men a day used the Navy Y near the Brooklyn Navy Yard in 1918.

The Y carried its activities to the members of the Marine Corps and the Coast Guard. A Y building was opened at Quantico on 24 October 1917, and soon a second was erected. Two similar buildings were built at Parris Island. Scores of temporary Y huts were erected at the various naval installations. Twenty-three of these huts were established at the Great Lakes Station, one for each regimental area.[78]

One of the most successful projects in Bible study sponsored by the Y among naval personnel was at Great Lakes. There the Y secretaries, in August 1917, working under the supervision of Chaplain Thompson, introduced Sunday morning Bible classes in all regiments. Although some teachers came from the ranks of the sailors themselves, most of them were business and professional men from Chicago, Milwaukee, and other near-by places. These classes continued for more than two years and approximately 600 civilians gave their time and talent to serve as teachers. As many as 200 of these volunteer workers were sometimes on duty on a single Sunday.

The movement started with six classes and an attendance of fifty-three. A year later there were 801 classes with a total attendance of 41,448. A report about these classes and other religious activities of the Y follows:

> At the present time, the total attendance has reached the half-million mark. It is a fact worthy of record that, although about one-half of the Station's population on any Sunday was unavailable, because on liberty or on detail, yet, for an extended period, about twenty per cent of all men at the Great Lakes were regularly attending these Bible classes. At the present, even a larger proportion is reached.
>
> Twice each week—on Sundays and Thursday in the evening—the Y. M. C. A. would hold religious meetings, mostly addressed by visiting ministers and laymen, although occasionally by the secretaries themselves. Considering that attendance at these meetings was purely voluntary, they were remarkably well attended. At the second anniversary of the Y. M. C. A. work, in April, 1919, it was figured out there had been 3,850 meetings held, attended by 502,983 sailors![79]

The Knights of Columbus adopted the motto, "Everybody Welcome," which was prominently displayed in their huts and recreational centers. While their secretaries confined their religious ministry to Catholics, in other activities their huts were open to all. This was a characteristic of the centers conducted by all service organizations. There was a friendly interchange of facilities when need demanded. Many a Catholic priest or chaplain conducted Mass in YMCA huts and often the Jewish Welfare Board found offices for its activities in both the YMCA and the Knights of Columbus huts.

---

[76] *Op. cit.*, p. 64.

[77] Pond, "History of Army and Navy YMCA" contains full information about each of these places.

[78] Harris, *Service for Fighting Men*, Vol. I, Chapter XXIV entitled "The Navy and Marine Corps," gives detailed information on this subject from which the above is taken. See also Goldberg, *Sky-Piloting the Great Lakes*, p. 44.

[79] Goldberg, *op. cit.*, p. 46.

It was the aim of the National Catholic War Council to provide every Catholic soldier and sailor with a New Testament and a prayer book.[80] Catholic chaplains in World War I did not have the facilities to perform three Masses on Sunday as in World War II, but only the customary two. In the absence of chapels on most of the Army and Navy installations, Catholic Divine Services were often held in the K of C huts.

The Knights of Columbus summarize, in part, their activities during the First World War as follows:

> All told, the Knights maintained forty-four buildings in naval and marine training camps, requiring a personnel of approximately 100 secretaries.
>
> Activity inside the naval camps never flagged. Some of these camps were as large as military cantonments. The Naval Training Station at the Great Lakes was one of the largest commands. . . . Here the Knights of Columbus had six buildings in constant operation—the largest number they had in any Naval Camp.
>
> It was at Parris Island that one of the first educational activities ever undertaken in any camp recreational building, was inaugurated. Here in the Summer of 1918, the Knights of Columbus commenced a class in mathematics, in order to encourage young enlisted men to try for commissions. The course was intensive, lasting eight weeks, and the classes numbered from ten to twenty-five.
>
> In Europe the Knights of Columbus won a deserved reputation for the zealous care they gave the men of the navy. In the British Isles the Knights of Columbus clubs were largely patronized by sailors. The club at Aberdeen, Scotland, was headquarters at one time or other, for almost all the men of the Atlantic Squadron.[81]

The K of C also conducted several hospitality houses for naval personnel in Boston, New York, Philadelphia, Detroit, and elsewhere at which many services were given to the men without charge as beds, barber service, laundry, and tailoring. The house in Boston "provided nightly over 700 free beds for men of the Army and Navy." In all these activities the K of C had the indefatigable cooperation of Catholic women in the vicinity of the service centers.

Because of building restrictions, the Jewish Welfare Board was unable to erect all of the buildings needed until May 1918, when restrictions were removed for this organization. This Board conducted activities at forty-four naval stations, at seven of which huts were erected. Special efforts were made to reach all Jewish personnel at the Jewish High Holy Days. Since there was but one Jewish chaplain in the Navy during the war, religious activities for those of the Jewish faith had to be conducted by qualified leaders from among service personnel and civilians. The Jewish Welfare Board was zealous in the distribution of prayer books, Bibles, and other religious aids to members of the Jewish faith.[82]

In addition to the work conducted by the welfare organizations at shore installations, much was done aboard transports. Chaplain H. J. Fry commented on his experience as follows:

> The Government places aboard each transport a "morale unit," composed of an army chaplain, and representative of the Y. M. C. A., the K. of C., the J. W. B. and the Red Cross. These men come aboard ship bringing with them bountiful supplies of all kinds for the soldiers—knitted goods, comfort kits, chocolate and hard candy, "smokes," fruit, games, athletic supplies, writing materials, religious literature, books, magazines and all that can possibly make happier the homeward journey. Through his close relationship with the ship's authorities . . . the chaplain is in a position to co-ordinate and unify the welfare work. . . .[83]

The American Library Association was of outstanding help in its collecting and distributing of books. Transport chaplains could secure from the Association a dozen or more boxes of books for each trip to France. These books were distributed to the personnel aboard and at the end of the voyage were collected, repacked in the boxes, and turned over to the Army.

It is not possible adequately to portray the extent and value of the services rendered to naval personnel during the war by the welfare agencies. Both directly and indirectly they ministered to the religious life and morale of the sailors. Since the number of Navy chaplains for shore stations was strictly limited, the spiritual ministry provided by the civilian groups at such places was indispensable.

ITEMS OF GENERAL INTEREST

A new bugle call for church was gradually introduced into the Navy during the first and second decades of the twentieth century. The first discovered published music of this call is found in the 1902 edition of *Trumpeters Hand Book and Instructor* by William S. Littleton. By the time of the First World War this call was in general use throughout the Navy, yet now and then a bugler would be found who knew only the old church call. (See page 69.)

---

[80] Williams, *op. cit.*, p. 129.

[81] Egan and Kennedy, *The Knights of Columbus*, 1:269–72.

[82] *The Jewish Welfare Board, First Annual Report*, 1919, pp. 86 ff.

[83] Nav. Rec. Coll., Chaplains' File.

According to directions in the *Naval Manual for Buglers,* this call is "sounded as a signal that divine service is about to be held. Aboard ship it is followed by the tolling of the ship's bell. It may also be used to form a funeral escort."[84]

Some chaplains on shore duty in the United States were greatly handicapped in their work by the lack of suitable places to hold Divine Services. Chaplain Curtis Dickins faced this problem at the Philadelphia Navy Yard where the number of recruits increased to 25,000. Dickins appealed to Secretary Daniels for funds to erect a large recreational hall which could be used for many purposes including Divine Services. When Daniels admitted that the Government had allotted no funds for such purposes, Dickins turned to private sources.

He succeeded in raising enough money from friends to purchase two large tents, each measuring 125 by 40 feet. These were used through the summer of 1917 and provided the shelter for Divine Services as well as main recreational center for the Yard. Anticipating the coming of cooler weather, Dickins raised approximately $40,000 from private sources and erected two temporary buildings, each 150 by 50 feet. One hall was made a recreational center with bowling alleys and a canteen. The other was used for assemblies, including Divine Services. In this building a small chapel was arranged at one end called "St. Paul's Chapel" where weddings and other special religious services were conducted. Entertainments were held in this second building each weekday night. At times as many as 1,500 were packed into the hall. (See page 227 for picture of tent-chapel.)

A chapel was secured at the Naval Operating Base at Norfolk when the Navy took over the Jamestown Exhibition Grounds in October 1917. On the property was a Presbyterian church which seated about 100. This became the Protestant chapel and was used as such for the following twenty-five years. The church was located almost exactly on the site of "The Parishe Church Situate at Mr. Sewells-Pointe," which was built in 1636 and which was the first church building erected in the Colonies between Hampton Roads and St. Augustine.[85] Thus, the chapel taken over by the Navy stood on historic ground.

In 1919 a building at the Naval Hospital, Great Lakes, which was formerly a bayonet storehouse, was converted into a chapel and equipped with a frame steeple to give it "atmosphere." It was called "All Souls Chapel." At first this chapel was used only by Roman Catholics. Stained glass windows, pews, and altar furnishings were donated by Catholic churches in Chicago.

Chaplain T. P. Riddle was ordered to the Bureau of Navigation on 3 January 1919 where he remained until 13 May. After a short tour of duty that summer on the Presidential yacht, *Mayflower,* Riddle returned to the Bureau and served until May 1923 as the chaplain representative to the Morale (known also as the Sixth) Division under Captain Claude B. Mayo. Riddle was in charge of the section entitled "Religion, Education, and Libraries," where he had a strategic opportunity to influence naval policies dealing with these important subjects.

During his connection with the Sixth Division, great emphasis was placed on the preparation of suitable educational courses for entlisted men who wished to study for advancement. In 1916 Secretary Daniels through General Order 63 had introduced compulsory education aboard Navy vessels. In time this system became cordially hated by the men, and many officers were out of sympathy with the program. It was felt that a voluntary system would be much more effective in that it would stimulate men to prepare themselves for higher ratings. Promotion was to be placed upon a basis of worth rather than seniority or favoritism. Several welfare organizations as the YMCA, K of C, and Red Cross had turned over a large sum of money to the Navy to be used for morale building. These funds were at the disposal of the Sixth Division.

Some of the first educational courses used in the Navy were bought outright from the International Correspondence School. It was felt by the Morale Division, however, that courses written by Navy men would be far more practical. Experts were found in the various fields, who prepared the courses. These at first were distributed in mimeographed form. The work initiated under Captain Mayo's direction grew from small beginnings to a navy-wide uniform system.

Those who sponsored this voluntary system of study had to overcome the widespread prejudice in the Navy to the former compulsory system. Chaplain Riddle was permitted to continue with this work until "the system caught root and demonstrated its value to the personnel of the Navy." While Riddle did not write any of the courses, he had the good

[84] *Op. cit.,* (1940) p. 37.

[85] CoC., History File, No. 913.

fortune to be located in a strategic place where he helped establish the new educational system.

While Chaplain Riddle was with the Sixth Division, the general policy on libraries was also under consideration. Riddle wrote:

> I worked under Captain Mayo and others to establish a well organized library system supervised by professional librarians. Miss Isabel Du Bois came to the Bureau at that time as an assistant of Mr. Charles Brown, and (prior to obtaining legislation and civil service classification) received her salary first through funds administered by Captain Mayo and later through funds which I obtained from Mr. MacCarl of the Army and Navy Y. M. C. A.[86]

In some respects the chaplains' duty became more important after the armistice than before. Chaplain H. J. Fry described this situation in his historical summary:

> Since the Armistice, the Navy has been facing a condition of special discontentment among the men, many enlisted "for the duration of the war" and no longer feel the urge of patriotic duty, even though the Navy is still on a war footing—further they are constantly taking HOME the men of the Army, and they find many Navy men who were on land duty being discharged—hence the spirit of discontent.[87]

In the various demobilization centers and on board ships Navy chaplains were active in doing what they could to minister to the discharge men. Chaplains at Great Lakes Training Station sent a form letter to the pastors, priests, and rabbis of the returning service men in an endeavor to help link up the returned sailor with his home church. The letter reads:

> Reverend and Dear Sir:
>
> I wish to advise you that .... of your church has been released from active duty in the Navy and has returned to the following address; ....
>
> We earnestly suggest that you and your lay people make every effort possible to tie this man up close to the Church and to enlist in the service of religion and righteousness those qualities of service which he has exhibited here.
>
> You will not find him greatly changed, evangelized, transfigured spiritually. Neither, we think, will you find him debased and deteriorated. He is just the same old fellow, but he is particularly ready now, since everything in his life is in a more or less fluid state, to respond to tactful and wise guidance religiously. We know that he will receive this from you.
>
> Very sincerely yours,
>
> GEO. C. MURDOCH,
> By direction, Captain Frank Thompson,
> U. S. N., Corps of Chaplains.

[86] CoC., Riddle File, letter of 12 Sep. 1945.
[87] Nav. Rec. Coll., Chaplains' File.

The Adjusted Compensation Act which granted a bonus to certain of those who served during the First World War became law in 1924. All who served in the Regular Navy were excluded from the financial benefits of this Act. Thus the temporary acting and the Reserve Chaplains received the bonus while those who were listed as acting chaplains did not, even though the latter had accepted their commissions with the intention of serving only for the duration of the war.

DEMOBILIZATION OF CHAPLAINS

The process of demobilization of the armed forces of the United States began within a few days after the signing of the Armistice on 11 November 1918. Many chaplains were, naturally, eager to return to their former civilian duties as soon as possible and the civilian pastors who had helped at shore stations desired to return to their peace-time routine. Only three chaplains, Anderson, Schweitzer, and van Dyke, however, were released before the end of the year. This brought the total number of chaplains on duty on 31 December 1918 to 198.

Chaplain Frazier writing to Chaplain Frank Thompson, Senior Chaplain at Great Lakes, on 19 April 1919, presented his problem:

> Since that time [Armistice day] twenty-three have been given their discharge and I now have fifty-five applications for discharge, not including the Naval Reserves who will automatically go on inactive duty within six months after the signing of Peace. Those who have applied for discharge are very insistent and some of them have become very much dissatisfied because I have not seen fit to approve their requests. I feel sure that within the next four or five months all these will have to be given their discharge.[88]

A month later Frazier wrote to Thompson announcing that he had "about seventy-five applications for discharge from young Chaplains who have served during the war." Chaplain Frazier realized that the failure of any chaplains, except a few regulars and Reserves, to receive promotion during the war had contributed to their desire to get out of the Navy. "I am sorry," wrote Frazier, "that so many of them see fit to get out as they were a fine bunch of men and would have been a great help to us."

During 1919, eighty-five chaplains returned to civil life and the following five died while in service: John Calvin Ely, Jr., 13 February; William B. Ayers, 8 April; William G. Cassard, 29 June; Isadore C.

[88] CoC., Thompson File.

Woodward, 25 July; and Edward B. Henry, 8 December. The following chart indicates the separations by months.

Jan	Feb	Mar	Apr	May	Jun	Jul	Aug	Sep	Oct	Nov	Dec	Total
3	4	9	8	7	7	7	5	15	14	6	5	90

The *Navy Register* for 1 January 1920 lists twenty-nine commissioned USN chaplains and sixty-two acting chaplains. In this latter group are the names of two men who declined their commissions,[89] thus leaving a total of eighty-nine in the Corps. In addition seventeen Reserve chaplains were still on duty and one retired chaplain, making a total of 107 on the active list when the year 1919 closed.

Twenty-seven of those who were released in 1919 kept up their connection with the Navy by entering the Naval Reserve Force. The Reserve chaplains, many of whom were too old to transfer to the Regular Navy, showed a tendency to remain on active duty as long as possible—only thirteen out of thirty-one who served during the War were released from active duty in 1919.

The eagerness of some of the Reserve chaplains to be released is revealed in an incident involving Chaplain Douglas Horton. During the summer of 1919 each Navy chaplain was requested by Chaplain Frazier to submit three sermons, two of which were to be on topics or texts assigned by the Chaplains Division.[90] A third sermon was to be written on a text chosen by the chaplain. It is reported that Secretary Daniels and Chaplain Frazier were planning to issue a volume of sermons written by Navy chaplains to be used on ships and stations which were without chaplains. Chaplain Horton, who had submitted several requests for release during the spring and summer of 1919, seized upon the idea of the sermon in order to plead his cause further. His third sermon, short and to the point, follows:

"THE IMPORTUNATE WIDOW"

Text: Mt. 7:7 "Ask, and it shall be given you; seek, and ye shall find; knock, and it shall be opened unto you; for everyone that asketh receiveth; and he that seeketh findeth; and to him that knocketh it shall be opened."

I am really beginning to doubt the efficacy of scripture. I have followed out the above precepts to the letter, but without avail. No, I am wrong. I see my error now: I have asked and I have sought, but up to the time of the composition of these sermons I have not knocked. If there are any who think that I have not done sufficient knocking now that these homilies are in type, however, I stand ready and eager to continue, according to gospel.

I wish to call your very earnest attention, my dear congregation, to the well known parable of the Importunate Widow, as it is found in St. Luke, the eighteenth chapter:

"And he spake a parable unto them to the end that they ought always to pray, and not to faint; saying, There was in a city a judge, who feared not God, and regarded not man: and there was a widow in that city; and she came oft to him, saying, Do me justice. . . . And he would not for a while: but afterward he said within himself, Though I fear not God, nor regard man; yet because this widow troubleth me, I will do her justice, lest she wear me out by her continual coming."

This parable repays the closest scrutiny, nay, it is a sermon in itself.

"They ought always to pray, and not to faint." Pass on, O Preacher; this have I written on the tablets of my heart.

"He feared not God, and regarded not man." Possibly this is putting it a bit strong, but we have no reason for believing otherwise. He certainly regarded not some men.

And how apt are the widow's words: "Do me justice." Do me justice. Do me justice. I wonder where she had heard those words. She may have picked them up from those who called on the BuNav in Jerusalem. Possibly her own husband had died in a chaplaincy in the Jewish Navy — of old age, as per regulation.

But here is the gist of the parable: we are permitted to look into the very heart of the judge. He says to himself, "Because this widow troubleth me, I will do her justice." These sweet words cannot be too often repeated. It is an altogether holy and edifying exercise, my dear congregation, to say this sentence over and over again to oneself, especially when one fears that he is doing an injustice. After this manner:

Because - she - troubleth - me - I - will - do - her - justice - because - she - troubleth - me - I - will - do - her justice - because - he- troubleth - me - I - will - do - him justice - I - will - do - him - justice.

Sufficient reiteration of these phrases has often been known to produce an actual desire to do justice.

"Lest she wear me out by her continual coming." This is evidently the only way to do it. I can visualize the letters she must have written:

May:

"I request, etc. . . ."

June:

"I respectfully request, etc. . . ."

July:

"I earnestly request . . ."

August:

"I urgently request . . ."

---

[89] They are A. Frank Smith and John MacWilliams.

[90] Goldberg, *Sky Piloting the Great Lakes*, p. 64, states that the sermons were prepared "By direction of the Bureau of Navigation . . . with a view to publication." Goldberg inserted in his book twelve sermons from Protestant, Catholic and Jewish chaplains, which he described as being "the oddest collection of any that has yet appeared under one binding!"

September:

"I most urgently request . . ."

And so on until the judge saw the light.

O Judge, I most respectfully, earnestly, urgently, and absolutely request that I be dismissed from the United States Navy.

And N. B. — the Lord commended the judge in the parable for granting the request of the Importunate Widow.

With each sermon the chaplain was to submit an appropriate closing prayer. Chaplain Horton's prayer, emphatic and concise, and on a separate page, read:

O Lord, how long? Amen.

Years later when Chaplain Salisbury was helping Chaplain Evans dispose of the piles of sermons in manuscript form which had remained in the Chief's office since the days of Frazier, he found this sermon of Horton's. Attached to it was a telegram, evidently written by Chaplain Frazier, which said in effect: "Were it not for the fact that I would be taking the valuable time of high ranking officers, I would have you court-martialed for insubordination. Resignation accepted." Dr. Horton stated in the fall of 1945 that he never received such a telegram. Nevertheless this unrecommended and unprecedented procedure had its desired effect. Horton's resignation was accepted on 11 October.

In answer to an inquiry sent to the former Secretary of the Navy at his home in Raleigh, North Carolina, regarding this incident, a telegram was received on 16 November 1945 which read in part:

I can not confirm the story about the sermon on the importunate widow but if I had read the sermon I know I would have granted the chaplain's request. I advise you to print the story. It has the ear marks of truth. It is too good to lose.

Josephus Daniels

IN SUMMARY

The Chaplain Corps of the United States Navy reached full maturity during the First World War. The most important gain was the establishment of a Chaplains Division in the Bureau of Navigation with a chaplain serving as Director or Chief. The responsibility of assigning chaplains to duty was given to this Division. A chaplain's manual was issued. Fleet Chaplains appeared. More attention was paid to Reserve chaplains. Most of the discriminations in the matter of uniform were removed. Greater supervision was exercised not only in the procurement of chaplains but also in their activities. Never before were the standards of the Corps so high.

The Chaplain Corps emerged from the war more than twice as strong numerically as when it entered. But more than that, the Corps had won for itself the good-will and respect of the officers and men of the Navy. Never before had so much been expected of the chaplains and never before had so much been accomplished by them.

CHAPTER TWELVE

THE YEARS BETWEEN

1920–1939

The years 1920–1939 were theoretically years of peace but were used by some nations to prepare for another World War so extensive and so devastating that the whole struggle of 1914–1918 appears as an opening skirmish. Even as historians lump together the wars of 1618–1648 as the Thirty Years War, so World War I and World War II may be treated as one struggle with a twenty-year truce between the great conflicts.

The fortunes of the Chaplain Corps are naturally linked directly with the strength of the Navy itself. It is impossible to appreciate the activities and responsibilities of Navy chaplains without some knowledge of the number of ships in commission rating chaplains, the size of the naval establishment ashore at home and abroad, and the number of naval personnel. Therefore a brief review of the national and international developments for these years under review as they affect the United States Navy is in order.

The interlude between the wars began with a hopeful and determined effort to reduce naval armaments by international agreement. People were tired of war. The heavy financial burden of building and maintaining large fleets, added to the debt incurred during the war, was frightening. Even though the United States did not enter the League of Nations, there was a wide-spread demand that this country take the lead in sponsoring international naval disarmament. The time was ripe for action.

On 11 July 1921 the United States issued invitations to Great Britain, France, Italy, and Japan to meet at Washington and join in a Naval Disarmament Conference. Other nations were included in a subsequent invitation. The Conference opened on 12 November at which time Secretary of State Charles Evans Hughes proposed that nearly 2,000,000 tons of naval vessels be removed from the world's chief navies.

The United States agreed to scrap nearly 846,000 tons including fifteen ships in commission and another fifteen then being built or projected. The United States scrapped more tonnage than any other nation.[1] During the years 1919–1926, the United States within treaty limitations added five battleships, two carriers, and ten light cruisers. Several of these vessels were laid down before the conference was held.

Between the years 1927 and 1938, the United States added eighteen heavy cruisers, nine light cruisers and three carriers. Thus in the interlude between the wars the United States commissioned five battleships, five carriers, eighteen heavy cruisers, and nineteen light cruisers. Each of these vessels called for the services of at least one chaplain but the Navy Department would allow the assignment of but one chaplain to each cruiser division. Lest the impression be given that these additions marked a corresponding growth in United States naval tonnage, it should be pointed out that during these years approximately the same amount of obsolete naval vessels was decommissioned or demilitarized.

The strength of the United States Navy in ships and personnel remained fairly constant from 1923 to 1938. In 1923 the tonnage figure stood at a little more than 1,200,000. This fell to about 1,000,000 tons during the years 1932–1938.[2]

A combination of at least two factors kept the naval strength even below treaty specifications. These were the strong peace sentiment which swept the country in the twenties and the early thirties and the financial depression which came in the thirties. There were days when the Navy kept its vessels anchored in harbors because there were no funds available to cover the cost of manoeuvers.

---

[1] Alden and Westcott, *The United States Navy*, p. 371. Two subsequent naval conferences produced no tangible results.

[2] E. J. King, *Our Navy At War*, from chart printed before page 1.

International relations assumed a threatening aspect in the early thirties. In 1931 Japan took over Manchuria. Defying the Treaty of Versailles and the League of Nations, Hitler began to rearm Germany. In 1935 Italy marched into Ethiopia. The refusal of Japan to agree to certain basic principles preparatory to the calling of another naval conference in 1934 was but the prelude to her subsequent announcement that she would drop the 1921 treaty in 1936. The two years' notice required by the Washington Conference had been duly given and the door was left wide open for an unrestricted international race in naval rearmament. The stage was being set for World War II.

The impotence of the League of Nations to solve the international problems was manifest in the continued and ever more brazen acts of aggression on the part of Japan, Italy, and Germany. All alternatives to armed conflict, like the last fleeting patches of sunshine on the landscape before a coming storm, were blacked out by the threatening, billowing clouds of war.

By 1938 it had become apparent in the United States that the old ideal of an effective treaty limitation to naval armaments was but an empty dream. Even though millions of United States citizens still believed that it was possible to remain neutral in the conflict that was just beginning in Europe, there were others who saw the inevitable involvement. At first the European conflict was viewed as a separate and distinct struggle from the one taking place on the other side of the globe in Asia. But the time came when these two became one. The vast geographical distances which separated the two did not divide them ideologically. Back of the material conquests of the aggressor nations lay a basic philosophy, called by different names in the various countries, but fundamentally the same. Dimly, reluctantly, and sometimes in uncomprehending resentment, the peace-loving people of the United States began to realize that the will to peace was not enough. In May 1938 Congress accepted the recommendation of President Franklin D. Roosevelt for a twenty percent increase in the United States Navy exclusive of replacement. The terrible truth was beginning to be realized—another World War was inevitable.

During the summer of 1939, events in Europe rushed to the dreaded climax. Hitler was threatening Poland. England and France warned that an attack on Poland meant war. On 1 September German troops invaded Poland. Two days later England and France followed up their warning by a declaration of war. World War II had started.

On 8 September 1939 President Roosevelt issued a proclamation which recognized the existence of a state of national emergency. This date has been arbitrarily taken as the closing date of this first volume of the History of the Chaplain Corps of the Navy. After 8 September 1939 the United States had a little more than two years to prepare for its role in World War II. The years of peace had ended. The interlude was over.

THE STRENGTH OF THE CORPS, APPOINTMENTS, AND PROMOTIONS

The number of chaplains on active duty remained relatively constant during the years between the wars. Six Reserve chaplains resigned in 1920 and 1921 while eleven transferred to the Regular Navy. The process of demobilization of World War I chaplains was completed by 1 July 1921. A total of 195 different chaplains were on active duty during 1920–1939, of these 101 resigned, died, or were retired, leaving a balance of ninety-four on 8 September 1939. A further breakdown of the figures shows:

		Separations	On duty 8 Sep. 1939
On duty 1 Jan. 1920	107	70	37
Accessions 1920–1939	88	31	57
Total	195	101	94

The following chart shows accessions and separations in the Chaplains Corps by the years indicated:

	1920	1921	1922	1923	1924	1925	1926	1927	1928	1929	1930	1931	1932	1933	1934	1935	1936	1937	1938	1939	Total
Accessions	12	4			7	6	5	2	2	7	7	2	1	4			6	10	6	7	88
Separations	20	13	5	4	5	4	6	5	5	4	5	1	1	7	1	2	2	5	3	3	101
Net Total	99	90	85	81	83	85	84	81	78	81	83	84	84	81	80	78	82	87	90	94	

In addition to the thirty-seven veterans of the Corps of World War I who remained on duty throughout this period was Alfred deG. Vogler, who resigned in 1919 but reentered the Corps in 1921. Thus thirty-eight World War I chaplains were on duty at the time President Roosevelt declared the existence of a state of national emergency. This small group of experienced men formed the nucleus of the vastly expanded Corps of World War II. During these years three World War I chaplains, J. M. Hester, P. L. Mitchell, and J. H. Putnam, entered and left the Corps.

Several new appointees to the Chaplain Corps for the years 1920–1939 were former Army chaplains who saw service during the First World War. They were: Walter L. Thompson, appointed in 1920; Stanton W. Salisbury, 1921; and Joseph E. McNanamy, 1924. Chaplain Salisbury served with the 327 Infantry, 82nd Division, in the St. Mihiel and Meuse-Argonne campaigns. His regiment claims to have been longer under fire, in the Argonne battle, than any other regiment in the United States Army.

The following veterans who became Navy chaplains during these years served in other than a chaplain's capacity in World War I:

Army (7)	*Navy* (9)
H. R. Sanborn, 1921[3]	O. A. Eure, 1924
E. M. Criger, 1924	C. H. Mansfield, 1925
W. H. Rafferty, 1924	E. J. Robbins, 1925
V. H. Morgan, 1928	A. E. Koch, 1926
E. R. LeBaron, 1929	E. O'Neill, 1927
C. M. Moses, 1930	J. Doyle, 1928
F. W. Meehling, 1933	J. F. Hugues, 1930
	D. L. Quinn, 1931
	R. R. Marken, 1933

Marine Corps

G. L. Markle, 1926	W. W. Pierce, 1929

Never during the years 1920–1939 did the actual number of chaplains on duty meet the authorized ratio. In 1920 there should have been 108 chaplains for the 134,956 personnel in the Navy and Marine Corps.[4] Secretary Daniels himself claimed for this year that the "number of chaplains now in the service is not sufficient to meet the demand."[5] The following year Secretary Denby said:

> There seems to be little desire on the part of the clergy to enter this field. The life and work are such as to require men of a particular temperament and equipment. An appeal has been made to all the leading theological seminaries of the country, requesting that tthe attention of their student bodies be directed to the vacancies existing in the Navy. . . .[6]

On 8 September 1939 the total naval personnel, including Marine Corps and Coast Guard was 146,198.[7] Again, if the ratio for chaplains had been maintained, there would have been 117 chaplains instead of ninety-four.

By Act of Congress dated 4 June 1920 officers holding temporary commissions were eligible for transfer to the Regular Navy if found otherwise qualified. The Act stipulated that no chaplain over forty-three could transfer to the Chaplain Corps of the Regular Navy. All temporary appointees were permitted to continue in their respective grades or ranks until 31 December 1921. Temporary acting chaplains when transferring to the Regular Navy were given precedence immediately following acting chaplains. This involved a loss of precedence by chaplains who entered the Corps as temporary acting chaplains to others who entered later but came in as acting chaplains.

The Act of 1920 also provided:

> That, until June 30, 1923, officers of the permanent Navy who have served satisfactorily during the war with the German Government in a temporary grade or rank shall be eligible for promotion or for promotion to the same permanent grade or rank without regard to statutory requirements other than age and professional and physical examination.

As applied to the Chaplain Corps, the law eliminated for those acting and temporary acting chaplains who transferred to the Regular Navy the three years' sea duty that had previously been required before promotion to lieutenant. The provision requiring chaplains to spend four years in the grade of "chaplain," which was equivalent to the rank of lieutenant (junior grade), after completing the probationary period of three years, was also waived.

There were sixty-two acting chaplains, including twelve with temporary appointments, on 1 January

[3] Sanborn served with the Canadian Expeditionary Force 1915–17 and transferred to the AEF in 1917. He rose to the rank of major.

[4] *Report of Secretary of Navy*, 1920, p. 121. Secretary Daniels stated: "The Corps now contains 65 fewer chaplains that we are authorized to appoint." However, neither the Navy nor the Marine Corps was then up to authorized strength.

[5] *Report of Secretary of Navy*, 1920, p. 121.

[6] *Ibid.*, 1921, p. 33.

[7] King, *op. cit.*, p. 13.

1920.[8] By the following year thirty-six of these were lieutenants and the balance who were still in service were commissioned lieutenant (junior grade).[9] With the exception of chaplains who entered the service before the war, and who had completed their probationary period and were thus moved to the grade of lieutenant (junior grade), these promotions were the first that the chaplains who entered the Corps after 1 January 1915 received.

In 1924 and 1925 wholesale promotions came to the Chaplain Corps. The number of captains was raised to fifteen, or ten percent of the authorized strength. Of these fifteen, only three—Scott, Thompson, and Workman—were still on active duty on 8 September 1939. No chaplain, during the period under review, was promoted to this rank after Chaplain E. A. Duff was made captain on 11 September 1925.

Thirty chaplains who had made lieutenant in 1920 were promoted to lieutenant commander in 1924,[10] and twenty-seven more were promoted in 1925.[11] After these wholesale promotions, the relative proportion within the various ranks remained fairly constant with the exception of the increasing number of commanders and the decreasing number of captains.

Another important Act of Congress which greatly affected the fortunes of the Chaplains Corps was the Staff Equalization Bill of 1926. Previously some staff officers had been assigned line officer running mates; now all were assigned running mates. The provision regarding chaplains read:

> All other officers on the active list in the Corps of Chaplains on June 10, 1926, and all officers thereafter appointed thereto shall be advanced in rank, up to and including the rank of lieutenant commander, with the officer of the line with whom or next after whom they take precedence, provided they are found qualified in accordance with law for such advancement.[12]

The assignment of running mates not only guaranteed a proper balance in the various ranks but also insured equality in promotions between staff and line.

Of interest also to the chaplains was section 19 of the Act:

> The provisions of existing law which require acting chaplains to serve for a period of three years on board ship in order to become eligible for commissions as chaplains, and which restrict the number of acting chaplains who may be commissioned as chaplains each year, are hereby repealed, and hereafter all acting chaplains shall be commissioned as chaplains when advanced in accordance with the provisions contained in this Act to the rank of lieutenant.

Thus it was no longer necessary for acting chaplains to spend all of their first three years at sea. This section eliminated for Regular Navy chaplains the grade of lieutenant (junior grade) after 1926.

Special legislation was needed to correct certain legal difficulties which barred the advancement of Chaplain A. E. Stone. Because he was temporarily unable to meet physical requirements for advancement, Chaplain Stone passed the age limit for a permanent appointment to the grade of chaplain. At the request of Secretary of the Navy Curtis D. Wilbur, Congress on 12 June 1926 passed the necessary legislation.[13]

Clarification of the age requirement for acting chaplains at the time of their appointment was made by Congress in 1929, in a law which provided that they should not be less than twenty-one nor more than thirty-five years of age.[14] This meant that the probationary period need not be served before the chaplain became thirty-five. However, it became the policy of the Bureau of Navigation some time about 1934 to set the upper age limit at thirty-four at the time of appointment so as to permit the appointee "to complete thirty years of service prior to attaining the statutory retirement age of sixty-four years."[15]

The law of 16 January 1936 provided that those Navy and Marine Corps officers retired (or thereafter retired) for physical disability who had "been commended for their performance of duty in actual combat with the enemy during the World War" be promoted to the next higher grade with no increase in pay. Under this act Chaplains J. D. MacNair and J. J. Brady, both retired as captains, were promoted to rear admiral and Chaplain A. N. Park was advanced from commander to captain.

Leaving the active service during 1920–1939, because of death, resignation, or retirement, were a number of veteran chaplains who had given many years of faithful service. Among these were Frank Thompson, Carroll Q. Wright, W. G. Isaacs, J. B. Frazier, and Curtis H. Dickens, all of whom were

[8] *Navy Register*, 1920.
[9] *Ibid.*, 1921.
[10] *Ibid.*, 1925.
[11] *Ibid.*, 1926.
[12] 34 U. S. C., 348 r.
[13] 44 *Stat.* 740.
[14] *R.S.* 1396.
[15] CoC., File "Chaplain Corps—Requirements, 1942."

commissioned before 1900. The following, who entered the Corps during the first decade of the nineteenth century, were retired: E. E. McDonald, B. R. Patrick, A. W. Stone, M. C. Gleeson, G. L. Bayard, G. E. T. Stevenson, S. K. Evans, H. M. T. Pearce, and J. D. MacNair. C. M. Charlton resigned. The *Navy Register* for 1 July 1939 carried the record retired list of twenty-one chaplains and three acting chaplains. Among these were two rear admirals, eleven captains, two commanders, two lieutenant commanders, and four lieutenants.

THE RESERVE CHAPLAINS

Reserve chaplains under forty-three years who did not choose to transfer to the Regular Navy, as provided by the Act of 30 June 1921, were permitted to remain on duty until 30 June 1922. The following nine chaplains transferred to the Regular Navy: B. F. Huske, H. E. Rountree, W. P. Williams, J. S. Day, F. L. Janeway, F. E. Moyer, A. E. Parker, P. J. Hammersley, and W. L. Steiner.

Chaplains leaving the service following World War I were invited to join the Reserve Force. Secretary Daniels in his 1920 *Report* stated: "Out of the total number of chaplains (104) resigning from the service since the signing of the armistice, 45 have been enrolled in the United States Naval Reserve Force."[16]

The *Register of the Commissioned and Warrant Officers of the United States Naval Reserve Force*, published under date of 1 January 1921, listed forty-eight chaplains in Class 2, ranking as follows: commanders—3; lieutenant commanders—3; lieutenants—9; lieutenant (junior grade)—33. There was one chaplain in Class 6—John Nicol Mark. All had seen service in World War I.

Even though the experience of World War I had demonstrated the value of a strong Naval Reserve, nevertheless, Congress, in the years immediately after the war, neglected to appropriate the necessary funds for its support. In September 1921 the Navy Department, finding it impossible to carry on the full Naval Reserve Force under existing laws and under the limited appropriations for 1922, abolished all Classes except 1 and 6. Class I included ex-officers of the Regular Navy who enrolled for four-year periods. Class 6 was composed of officers and men who stood by for an emergency and who served in the Reserve without retainer pay. Secretary of the Navy Edwin Denby in September 1921 invited chaplains in Classes 2 and 4 to transfer to Class 6 in order to retain their status in the Reserve. All Reserve chaplains who did not transfer by 30 September of that year were automatically disenrolled. By 21 March 1922 twenty-four chaplains were listed in Class 6 of the Reserve which suggests that only about fifty percent of the chaplains in Class 2 showed sufficient interest to transfer.[17] The commissions of chaplains who entered the Reserve Force on the prescribed four-year term were dated back to the beginning of their Class 2 service.

By 1924, the number of Reserve chaplains had dropped to sixteen. "There has been little opportunity to seek additional chaplains in the reserve force," explained the Secretary, "because of stress on securing men for the Navy, but it is hoped that more attention may be given to this during the coming year."[18]

An Act of Congress on 28 February 1925 changed the designation of the Naval Reserve Force to the Naval Reserve and instituted a complete reorganization. The purpose of this Act was "to create, organize, and train a naval reserve sufficient to provide the supplementary personnel necessary to mobilize the fleet and all its auxiliaries."

According to the Secretary of the Navy's *Report* for 1925, there were only thirteen chaplains in the Naval Reserve Force at the time the old organization was merged into the new.[19] In the list of chaplains were two lieutenant commanders, two lieutenants, and nine lieutenants (junior grade), but the names of these charter members of the Chaplain Corps of the new United States Naval Reserve did not appear. The following list of twelve, lacking one lieutenant commander, has been compiled from service records (* indicates service as a Navy chaplain in 1917, and ** indicates such service in 1918):

[16] *Op. cit.*, p. 560.

[17] CoC., "Reserves" File. Two Reserve chaplains, J. J. Brokenshire and E. B. Niver, were made the subjects of special legislation. Brokenshire was fifty-eight in 1920 and Niver fifty-seven. The latter had been in the Naval Militia and in the later Reserve organizations since 1901. On 1 July 1922 Congress approved the recommendation of the Navy Department that Chaplains Brokenshire and Niver be permitted to transfer to the Regular Navy, but stipulated that neither would be "in the line of promotion and not eligible for retirement." Since both chaplains had been promoted to commander during the summer of 1920, each was permitted to retain that rank after transferring. Brokenshire had gone on inactive duty on 2 July 1921 and did not serve again after his transfer. Niver was on duty until July 1927 when he retired, without pay.

[18] *Report of Secretary of Navy*, 1924, p. 167.

[19] *Op. cit.*, pp. 28, 168.

Name	Denomination	Date of Commission	Rank	Date of Separation
**Dufrane, Leo J.	RC	30 Sep. 1921	Lt. (jg) (G)	9 Oct. 1926
Hartwig, Theodore J.	Luth	31 Oct. 1921	Lt. (jg) (G)	17 May 1939
Claypool, J. V.	Meth	16 Nov. 1923	Lt. (jg) (G)	
Rasmussen-Taxdal, H.	PE	24 Jan. 1924	Lt. (jg) (S)	
Lambdin, C. Howard	Meth	24 Mar. 1924	Lt. (jg) (G)	
Johnson, Frank M.	PE	3 May 1924	Lt. (jg) (G)	5 Aug. 1929
*Larned, Albert C.	PE	12 Jan. 1925	Lt. (G)	28 Jan. 1937
*Putnam, John H. S.	PE	19 Feb. 1925	Lt. Comdr. (G)	20 Oct. 1933
Holt, William T.[20]	PE	27 May 1925	Lt. (G)	
Williamson, Thom J.	PE	1 July 1925	Lt. (jg) (S)	15 Jan. 1930
Wallace, Edward A.[21]	RC	29 July 1925	Lt.(jg) (G)	
*Goldberg, David	Jewish	17 Dec. 1925	Lt.(jg) (G)	

[20] Holt served as a USN Chaplain 1920–4.

[21] Chaplain Wallace (U. S. Army) was badly gassed on 12 October 1918 in the battle of the Argonne Forest. He was erroneously reported to have died. Memorial services were held for him in his regiment and Masses were read for the repose of his soul in his home diocese in Brooklyn. A monument, still standing at 92nd Street and 4th Ave., Brooklyn, erected to the memory of the men of that community who paid the supreme sacrifice, included Chaplain Wallace's name. For years many of his friends believed that he was dead.

As far as available records indicate, Chaplain J. V. Claypool was the oldest senior Reserve chaplain in point of continuous service on 8 September 1939. Chaplain Rasmussen-Taxdal had the longest continuous record with the Reserve of any chaplain on the active list in 1939 since his record went back to 6 June 1918.

The growth of the Chaplain Reserve under the various Chiefs of Chaplains to 8 September 1939 may be observed in the following list. The names are listed chronologically according to the date of commission

Chief of Chaplains, Curtis H. Dickins

Name	Denomination	Date of Commission	Rank	Date of Separation
		1926		
Hughes, E. H.	Meth	17 Jan.	Lt. (jg)	6 Jan. 1931
Barkman, F. T.	Bap	26 Jan.	Lt. (jg) (S)	
Pennington, E. L.	PE	15 Apr.	Lt. (jg) (G)	
Carson, H. R.	PE	28 Apr.	Lt. (jg) (S)	1 Jan. 1934
Montgomery, C. L.	UBreth	9 Nov.	Lt. (jg) (G)	17 May 1931
Humphreys, F. L.	PE	10 Dec.	Lt. Comdr. (S)	22 Oct. 1931
		1927		
Seagle, N. A.	PE	19 Jan.	Lt. Comdr. (S)	1 Jan. 1933
Rifenbark, M.	PE	27 Jan.	Lt. (S)	
Rosa, M. J.	RC	28 Mar.	Lt. (S)	1 Sep. 1930
Hodgkins, H. B.	PE	14 June	Lt. (jg) (G)	
Keith, G. A.	RC	28 July	Lt. (jg) (G)	
Gerhart, L. F.	Luth	14 Nov.	Lt. (jg) (G)	
*Darlington, G. S. B.	PE	27 Dec.	Lt. (G)	2 Mar. 1932
		1928		
Morgan, J. H.	PE	10 Feb.	Lt. (jg) (G)	31 Mar. 1931
Lever, C. M.	PE	20 Apr.	Lt. (jg) (G)	29 June 1932
**Burke, T. J.	RC	23 May	Lt. Comdr. (G)	
**Callender, S. J.	Meth	1 July	Lt. (G)	

Chief of Chaplains, Sydney K. Evans

Name	Denomination	Date of Commission	Rank	Date of Separation
		1930		
*Huske, B. F.	PE	8 Sep.	Lt. (S)	
Pearson, H. J.	PE	12 May	Lt. (jg) (S)	
Miller, H. J.	PE	26 June	Lt. Comdr. (S)	23 Oct. 1935

Name	Denomination	Date of Commission	Rank	Date of Separation
		1931		
Bentley, C. E.	PE	2 Feb.	Lt. (S)	
Parkman, H. C.	PE	6 Mar.	Lt. Comdr. (S)	1 Sep. 1936
Kroll, L.	PE	5 June	Lt. (S)	13 Jan. 1937
Lockhart, M. W.	PE	28 Oct.	Lt. (S)	
Straus, H. C.	Jewish	15 Dec.	Lt. (G)	
		1932		
Shultz, P. T.	PE	17 Mar.	Lt. (jg) (G)	10 Mar. 1933
Shearer, L. A.	PE	22 Mar.	Lt. (jg) (G)	
Doty, W. P.	PE	25 Apr.	Lt. (jg) (G)	16 Mar. 1938
Kopf, C. H.	Cong	3 Aug.	Lt. (jg) (S)	
Branch, J. M	Bap	3 Sep.	Lt. (S)	16 Mar. 1934
		1933		
deForest, W. J.	PE	13 Mar.	Lt. (S)	
Zimmerman, J. D.	PE	25 Apr.	Lt. (jg) (G)	
Martin, A. O.	Meth	21 Sep.	Lt. (jg) (G)	
Drury, C. M.	Presby	11 Dec.	Lt. (S)	
		1934		
Carson, K. B.	Presby	19 Jan.	Lt. (jg) (S)	
Jackson, R. W.	PE	21 Jan.	Lt. (jg) (S)	
White, R. J.	RC	16 Mar.	Lt. Comdr. (S)	
Hults, C. L.	PE	17 Apr.	Lt. (jg) (G)	
Ockenga, H. J.	Presby	15 June	Lt. (jg) (G)	
Parker, C. L.	PE	13 Aug.	Lt. (jg) (G)	
Kennedy, W. L.	Presby	22 Nov.	Lt. (jg) (G)	
Hanson, H. F.	Luth	12 Dec.	(Lt. (S)	
		1935		
Philips, Z. B. T.	PE	12 Jan.	Lt. Comdr. (S)	1 June 1939
Jones, E. R. L.	PE	23 Jan.	Lt. (jg) (S)	
Tucker, A. C.	PE	11 Feb.	Lt. (S)	
Zinn, W. R.	Disc	8 June	Lt. (S)	
McCartney, A. J.	Presby	27 June	Lt. Comrd. (S)	
		Chief of Chaplains, Edward A. Duff		
Peterson, A.	Unit	10 July	Lt. (jg) (S)	
Haylor, W. L.	PE	5 Aug.	Lt. (jg) (S)	
Vollmer, M. A.	PE	30 Aug.	Lt. (jg) (G)	
Armstrong, J. G.	PE	20 Dec.	Lt. (S)	
		1936		
Lever, C. M. (1928)	PE	12 Mar.	Lt. (jg) (S)	
Parker, J. T.	Bap	27 Apr.	Lt. (jg) (G)	
Gill, E.	Bap	19 Oct.	Lt. (S)	
Wagener, J. P.	RC	26 Oct.	Lt. (jg) (S)	
Bloomquist, R. T.	PE	14 Dec.	Lt. (jg) (G)	
Tennyson, M. G.	PE	28 Dec.	Lt. (S)	

1937

Name	Denomination	Date of Commission	Rank	Date of Separation
Olton, R. M.	PE	12 Jan.	Lt. (jg) (G.)	
Carney, T. A.	RC	5 Apr.	Lt. Comdr. (S)	
Craft, L. B.	Meth	27 Apr.	Lt. (jg) (G.)	

Chief of Chaplains, Robert D. Workman

Name	Denomination	Date of Commission	Rank	Date of Separation
Buchholz, G. W.	PE	16 July	Lt. (jg) (G)	
Nelson, C. W.	PE	13 Sep.	Lt. (S)	
Clyne, F. B.	RC	27 Sep.	Lt. Comdr. (S)	
Collins, W. E.	RC	1 Dec.	Lt. (S)	

1938

Name	Denomination	Date of Commission	Rank	Date of Separation
Sheehy, M. S.	RC	26 Mar.	Lt. (S)	
Williams, F. R.	Presby	1 July	Lt. (jg) (G)	
Nee, P. A.	RC	1 July	Lt. (jg) (G)	
Thomas, L. O.	PE	9 Aug.	Lt. (jg) (S)	
Camerman, J. E.	RC	13 Sep.	Lt. Comdr. (S)	
Witwer, A. M.	Meth	14 Dec.	Lt. (S)	
Cook, O. B.	RC	27 Dec.	Lt. (jg) (S)	

1939

Name	Denomination	Date of Commission	Rank	Date of Separation
McColgan, G. P.	RC	12 Apr.	Lt. (S)	
Winter, W. W.	Presby	5 Aug.	Lt. (jg) (G)	

An analysis of the accessions and separations by years follows:

	'25	'26	'27	'28	'29	'30	'31	'32	'33	'34	'35	'36	'37	'38	'39	Total
Accessions	12	6	7	5		3	5	5	4	8	9	6	7	7	2	86
Separations		1			1	2	4	2	4	2	1	1	2	1	2	23
Net Total	12	17	24	29	28	29	30	33	33	39	47	52	57	63	63	63

Thus sixty-three Reserve chaplains were available for duty when the state of national emergency was declared on 8 September 1939. The following chart shows the religious affiliations of the Reserve chaplains:

Denominations	Separated before 8 Sep. 1939	Available on 8 Sep. 1939	Total
Baptist	1	9	10
Catholic	2	7	9
Congregational		1	1
Disciples		1	1
Jewish		2	2
Lutheran	1	2	3
Methodist	1	6	7
Prot. Episcopal	17	27	44
Presbyterian		7	7
Unitarian		1	1
United Brethren	1		1
	23	63	86

It is to be noted that a disproportionately high percentage of Reserve chaplains came from the Episcopal church. Reserve chaplains were fairly well distributed through the various naval districts. The following charts shows this distribution and also the classification by rank on 8 September 1939:

	Lt. Comdr.	Lt.	Lt. (jg)	Total
Wash. D. C.	3	1	1	5
1st N. D.	2	3	3	8
3rd N. D.	1	5	3	9
4th N. D.		2	4	6
5th N. D.		2		2
6th N. D.	1		1	2
7th N. D.		2		2
8th N. D.	1	2	2	5
9th N. D.	1	4	5	10
11th N. D.		3	1	4
12th N. D.	3	2	2	7
13th N. D.	1			1
14th N. D.		1		1
15th N. D.			1	1
	13	27	23	63

A number of men who had served in either the Army or the Navy during the First World War, in some capacity other than that of a Navy chaplain, entered the Reserve Chaplain Corps during these years. Among them were the following five who served as Army chaplains: E. A. Wallace, H. R. Carson, C. M. Lever, J. E. Camerman and F. B. Clyne.

Others who had seen service included:

Army (8)	*Navy* (10)
J. G. Armstrong	H. F. Hanson
W. J. deForest	H. B. Hodgkins
C. M. Drury	C. H. Lambdin
L. F. Gerhart	C. V. Montgomery
M. W. Lockhart	E. L. Pennington
A. O. Martin	H. Rasmussen-Taxdal
C. W. Nelson	M. G. Tennyson
A. C. Tucker	R. J. White
	T. J. Williamson
	W. R. Zinn

Marine Corps
J. V. Claypool

Including the eight ex-Navy and the five ex-Army chaplains, a total of thirty-two veterans of the First World War entered the Reserve. The fact that more than one-third of the Reserve Chaplains were ex-servicemen is significant. It may be presumed that their interest in the chaplaincy can be traced to an appreciation of the spiritual needs of servicemen gained through personal experience.

In his annual report as Chief of Chaplains, S. K. Evans on 10 July 1934, made the following comment on the work of some Reserve chaplains:

> The Peace-time value of the Chaplains of the Naval Reserve is witnessed in Haiti where during interregnums between Chaplains (due to shortage in numbers) twice had the local Reserve Chaplains carried on the work temporarily with great credit. In the Canal Zone, for some months, there being no Chaplain available for detail at Coco Solo, the Reverend Robert W. Jackson, Lieutenant (jg) ChC-V(S), USNR has not only conducted Divine Services regularly on Sundays but has been active daily in the interest of the Navy personnel.

Reserve chaplains were encouraged to take correspondence courses in *Navy Regulations,* naval history, and kindred subjects. A few were able to take two to four weeks training duty a year sometimes without pay. Reserve chaplains were sometimes invited to go on summer cruises, and when appropriations were not available, paid their own mess and travel expenses to and from the port of embarkation. No allowances were made for uniform. In some of the larger naval districts, meetings of Reserve officers were held once or twice a month at which lectures on aspects of naval affairs were usually given. Reserve chaplains in the vicinity were expected to attend such meetings. In some naval districts no Reserve chaplain was recommended for promotion who had not completed the prescribed reading and/or who had not shown interest in the Naval Reserve by attending the meetings in his district.

832159—49——15

SUGGESTED ESTABLISHMENT OF A CHAPLAINS' BUREAU

In his 1920 *Report* Secretary Daniels strongly urged the establishment of a Chaplains' Bureau in the Navy Department. The Chaplains' Division, of which Chaplain Frazier was the first Director, was a part of the old Bureau of Navigation and later became a part of the Bureau of Naval Personnel. Secretary Daniels pointed out that the valued services of chaplains in the World War demanded fuller recognition. "We accord appropriate rank to our military and naval leaders," wrote Daniels, "and we can not consistently deny this to spiritual and moral leadership."

After more than seven years as Secretary of the Navy and out of his intimate knowledge of the work of the chaplains, Josephus Daniels submitted the following recommendation:

> For this reason I urge the creation of a chaplains' bureau in the Navy, to be headed by a chief chaplain, who, while exercising the duties of that office, shall have the rank of a rear admiral of the lower half. This would give the chaplain corps the standing that it deserves and would enable us to put this entire department on a permanent basis, which would greatly increase its effectiveness, evidencing our appreciation of the invaluable work being done by the faithful chaplains who for many years have devoted their lives to the Navy and are a part of it, and attracting to the service more ministers of higher type, who would find here even broader opportunities for usefulness than they now have in civil life.[22]

The General Committee on Army and Navy Chaplains of the Federal Council of Churches of Christ in America was much interested in the proposed Chaplains' Bureau. Writing to Secretary Daniels on 11 October 1920, Bishop W. F. McDowell stated:

> I write now to give my unreserved endorsement of the two bills and the statement of reasons. In addition, will you let me say, as representing the Churches, that it is my firm conviction that it is but a fair representation of the Churches that this office should have as their representative the dignity of the added rank of a Rear-Admiral of the Navy of the lower half. The Churches, in view of their experience during the war, have acquired an interest in the welfare of the Navy as of the Army that did not exist five years ago. There has come to the Churches a new consciousness of the Navy and the Army and we are exceedingly anxious to meet the responsibility of which we have become aware. In order to be able to do that, it is our judgment that such legislation as is here proposed is really necessary.[23]

The Senate Naval Affairs Committee, acting upon

[22] *Op. cit.*, p. 120.
[23] CoC., Research File.

Secretary Daniels' recommendation, drew up a bill to create a Chaplains' Bureau and sent it to the Secretary for approval. On 22 January 1921 Daniels returned the bill to the Chairman of the Committee with a few changes. The suggested law then read as follows:

> Be it enacted by the Senate and House of Representatives of the United States of America in Congress assembled, That hereafter there shall be in the Department of the Navy a separate bureau, to be known as the Chaplains' Bureau, which under the authority of the Secretary of the Navy shall direct and supervise the work of the Chaplains' Corps, and shall perform such business of the Department of the Navy as the Secretary of the Navy shall judge to be expedient and proper: Provided, That the chief of the Chaplains' Bureau in the Department of the Navy shall be appointed by the President by and with the advice and consent of the Senate from the list of officers of the Chaplains' Corps of the Navy not below the rank of commander on the active list.[24]

Separate bills calling for the establishment of the Chaplains' Bureau were introduced in Congress, House Bill 2494 and Senate Bill 152. Each bore the title: "To establish in the Department of the Navy a bureau to be known as the Chaplains' Bureau, and for other purposes." The bills also called for a redistribution of the ranks of chaplains as follows: captains, 6; commanders, 12; and the balance in the lower ranks.

After the bills had been introduced into both Houses of Congress, the General Committee on Army and Navy Chaplains issued a folder over the name of Dr. E. O. Watson, its secretary, which listed ten reasons why the proposed legislation should be supported. It was pointed out that since the Army had created such a department, the Navy should do likewise. The Committee felt that the creation of such a Bureau would promote the efficiency of the Corps by giving greater recognition to the work of the chaplains. "The proposed legislation," wrote Dr. Watson, "will give the Corps the recognition already given other staff corps and afford it all the advantages realized by these from such establishment."[25]

When the administration changed in 1921, Edwin Denby relieved Josephus Daniels as Secretary of the Navy. Without the friendly interest and powerful influence of Secretary Daniels, the proposed legislation was doomed to failure. On 31 May 1921 Admiral Thomas Washington, Chief of the Bureau of Navigation, writing to the Judge Advocate General recommended the deletion from the proposed legislation of all references to creation of a Chaplains' Bureau. He wrote:

> A bill creating a Bureau to be known as the Chaplains' Bureau is inadvisable for the following reasons: First, it will be the cause of considerable additional expense to the Government, made necessary by more officers performing shore duty and the employment of additional clerical force; all of which will be without adequate return. Second, it will create and establish another bureau dealing with personnel matters, which can only result in duplication of work.[26]

It was also pointed out that since the Chief Chaplain of the Army held only the rank of colonel, it would not be fitting to give the Chief Chaplain of the Navy the rank of rear admiral.

Members of the General Committee on Army and Navy Chaplains waited on Secretary Denby in behalf of the proposed legislation. On 10 June 1921 Denby explained his position in a letter to the Committee:

> My dear Dr. Watson:
>
> When visited the other day by Bishop Harding, Bishop McDowell, and Dr. Radcliffe and yourself, I promised to write you my opinion regarding the proposed establishment of a Chaplains' Bureau in the Navy Department. I then explained to you my views on this subject and I am happy to place those views in writing for transmittal to the organizations you represent.
>
> . . . .
>
> The chaplain service is now administered by a division of the Bureau of Navigation and a chaplain is specially detailed as the head of this division. This organization has in the past proved satisfactory. The present head of this organization is the Reverend J. B. Frazier who has served nearly four years in that position, so that he is now due for assignment to other duty with the Fleet or elsewhere and he will shortly be so detailed and another chaplain assigned as head of the chaplain division. There is no reason to believe that the administration of the service will in the future be in any way less satisfactory than it has been in the past.[27]

The proposed legislation failed to pass. However, many chaplains and many of their friends in responsible positions in denominational circles believed that the efficiency of the Chaplain Corps would be increased if a separate Chaplains' Bureau were established such as was done for Navy doctors when the Bureau of Medicine and Surgery was created.

THE SUCCESSION OF CHIEFS

Six chaplains, all of whom entered the service before 1917, served as Director or Chief of the Chaplains Division during the years reviewed in this chapter. They were:

---

[24] NRSO: 15721.
[25] CoC., Research File.

[26] NRSO: 15721.
[27] HRSO: 26255-671:1.

J. B. Frazier,	Meth.,	5 Nov. 1917–29 Nov. 1921
E. W. Scott,	Cong.,	21 Nov. 1921–14 July 1926
C. H. Dickins,	Episc.,	1 June 1926–24 July 1929
S. K. Evans,	Episc.,	15 July 1929–22 July 1935
E. A. Duff,	Cath.,	19 July 1935– 1 June 1937
R. D. Workman,	Presby.,	18 June 1937–

The Navy Department had no fast rule regarding the length of service of a Chief of Chaplains. With the appointment of a new Secretary of the Navy in March 1921, it was to be expected that there would follow a change in the Chief of Navy Chaplains. Denby, in a letter of 10 June 1921, wrote that such a change "is now due."[28] In November of that year Chaplain Frazier was relieved by Evan Walter Scott.

No other Chief during this period under review left as deep an impress upon the Corps as did John Brown Frazier. As the first Chief, he set precedents and established traditions. He cast the administrative functions of that office into a mold which guided his successors. His greatest bequest to the Corps was the Corps itself. Largely through his efforts, the Corps was lifted to its unprecedented high standard of educational training, of spiritual power, and of native ability. When Frazier turned over the responsibilities of his office to Chaplain Scott, approximately sixty percent of the total Corps were men of his choosing.

Chaplain Frazier returned to Norfolk where he served at the Naval Training Station until 1 June

[28] *Ibid.*

SIX CHIEFS OF NAVY CHAPLAINS

Dates indicate length of duty as chief

John Brown Frazier, 1917–1921

Evan Walter Scott, 1921–1926

Curtis Hoyt Dickins, 1926–1929
(By Harris and Ewing)

Sydney Key Evans, 1929–1935

Edward Aloysius Duff, 1935–1937

Robert DuBois Workman, 1937–1945.

1925. He retired in September of that year after thirty years' service and settled on a farm, which he called the Anchorage, near Keswick, Virginia. There he spent the intervening years until his death on 11 November 1939.

The second Chief of Chaplains was E. W. Scott who had just completed a two-year tour of duty as Fleet Chaplain of the Pacific Fleet. Scott's term of almost five years was during a period of readjustment for the Navy. The Staff Equalization Bill, which meant so much to the Chaplain Corps, became a law during the closing months of his responsibilities as Chief.

Scott was followed by Curtis H. Dickins in the summer of 1926. Chaplain Dickins was the only one of the six Chiefs who was at the same time the senior chaplain on active duty. Dickins attracted the attention of Secretary of the Navy Curtis Wilbur by his fine record as senior chaplain of the Philadelphia Navy Yard during two tours of duty. Dickins first served at the Yard for about five and a half years beginning in February 1916. After serving as Fleet Chaplain aboard the *New Mexico* and *California* in 1921–1923, he returned to Philadelphia. Chaplain J. D. MacNair carried on the fine work at the canteen and recreational center during the two years Dickins was absent.

The financial success of the canteen, which was conducted under the supervision of Chaplains Dickins, MacNair, and Dickins in turn, was so phenomonal that a new modern brick recreational building was erected with the accumulated profits of the years. This new building with equipment cost about $228,000 and was donated to the Government in March 1926. The commercial aspect of this project, begun by Chaplain Dickins during World War I as a service to enlisted personnel, had grown to such proportions that the Chaplain was relieved of the responsibility of its management after the opening of the new building. However, the Recreational Center at Philadelphia Navy Yard remains as a splendid tribute to the initiative and business acumen of two chaplains, Dickins and MacNair.

When the new building was dedicated on 24 May, Secretary Wilbur was present. During the festivities, Wilbur called Chaplain Dickins aside and informed him that he was to be transferred to new duty. Dickins expostulated: "But, Mr. Secretary, this building is just opened. I want to stay here." Then the Secretary explained that the new duty was to be Chief of Navy Chaplains. Dickins carried to his new office the same capacity for achieving results that had characterized his earlier tours of duty. After serving three years as Chief, he retired on 24 July 1929.

The fourth Chief was Sydney Key Evans who had served twelve years at the Naval Academy, including two tours of duty, the first in 1915–1920 and the second in 1921–1929. In an article for the April 1931 issue of the *Army Chaplain,* Chaplain Evans commented on the vitality of the religious life of the midshipmen at Annapolis:

> After more than twelve years of duty as Chaplain of the Naval Academy, he believes that there is no other place in America where Christianity—as distinguished from a sectarian Christianity—is more in evidence in the things which count. In the closing years of his duty there, the writer regularly welcomed more than seven hundred Mid-shipmen at the voluntary Communion service held in the Chapel at 7 a.m. on Easter Sunday. In addition, hundreds of others went to Mass and other Communion services in the city of Annapolis the same morning.[20]

The statement reflects credit on the effectiveness of Chaplain Evans' work at the academy. After serving six years as Chief, he was retired 1 October 1935. His successor was Chaplain Edward A. Duff who took over the responsibilities as Chief in July of that year. Chaplain Duff was on board the *Olympia* which brought the body of the Unknown Soldier from France to the United States. The body was buried with fitting ceremonies in Arlington Cemetery on 11 November 1921, the third anniversary of the signing of the armistice. Chaplain J. B. Frazier, then Chief of Chaplains, had a part in the services. Chaplain Duff wrote a lecture on the "Unknown Soldier" which he delivered in many parts of the country. It is estimated that approximately 300,000 people heard this lecture. After serving as Chief for about two years, Duff requested sea duty and was assigned to the *California* as Fleet Chaplain of the Pacific Fleet.

The sixth Chief was Robert D. Workman who took over the office on 18 June 1937. He had served as an enlisted man in the Marine Corps for four years beginning in February 1905. After his discharge as a sergeant, he began his education for the Christian ministry. Having been commissioned a chaplain in 1915, Workman had already served twenty-two years in Corps when he was called to its highest office. His two years and more service before the declaration of a state of national emergency were years of training for World War II, a duty which proved to be the most exacting ever laid upon any Navy Chaplain.

[20] *Op. cit.,* p. 6.

The duties of the Chief of Chaplains have never been officially defined. In 1938 Chaplain C. A. Neyman drew up a "Provisional Outline of a Manual for Chaplains of the United States Navy" in which he submitted the following:

Par. 113

The duties of the Chief of Chaplains have not been outlined or specified by law. But orders issued from time to time, combined with established procedure and custom, have made the following a part of the routine of his office:

(1). To interview and correspond with prospective candidates for appointment as acting chaplains; and to recommend to the Secretary of the Navy the examination of those candidates considered qualified when vacancies are to be filled.

(2). To serve as President of the Examining Board which is appointed to conduct the mental and professional examination of candidates.

(3). To recommend to the Chief of the Bureau of Navigation the assignment of chaplains to particular ships and stations.

(4). To receive and examine the monthly and annual reports made by chaplains of the duties performed by them; and to report to the Secretary of the Navy, annually or as required, the work done by members of the Chaplain Corps.

(5). To maintain contact, by personal interview or correspondence, with the members of the Chaplain Corps; and to advise, counsel, and guide them in the performance of their duties.

(6). To assist in the examination of chaplains selected for promotion; and to participate in that selection.

(7). To assist in the selection, enrollment, instruction, and supervision of Naval Reserve chaplains.

(8). To advise the Chief of the Bureau of Navigation and the Secretary of the Navy, as required, in all matters pertaining to the policies and details of the employment of chaplains in the United States Navy.[30]

During these years under review, the office of the Chief of Chaplains was in the Navy Building on Constitution Avenue. While the office force usually consisted of only one clerk, nevertheless, the Chief, in carrying out his duties, had the assistance of the chaplain on duty at the Washington Navy Yard and, for a period, the chaplain on duty in the Sixth Division of the Bureau of Navigation. For example, Chaplain R. W. Shrum, who was assigned to the Yard and the *Mayflower,* the President's yacht, spent a good portion of his time with the Chief of Chaplains Dickins. On 2 August 1939, when the international situation became more clouded, Chaplain W. H. Rafferty received orders not only to the Navy Yard but also additional duty orders as assistant to the Chief of Chaplains.

There were responsibilities which belonged to the Chief's office other than those outlined by Chaplain Neyman. These included speaking engagements and official appearances which were expected of the Chief of Chaplains by the Navy Department and the public. He served as chaplain to the entire Navy Department and often conducted naval funerals in Arlington National Cemetery in Washington. These burials gradually increased during the years under review. In his annual report for 30 June 1938, Chaplain R. D. Workman stated that a chaplain's ministry had been provided through his office for 207 Navy funerals. The chaplain at the Navy Yard conducted many of these services. The Chief usually officiated at the funerals of the higher ranking deceased officers.

On 8 September 1939 the Chaplains Division, on the eve of the greatest expansion in its history, had the benefit of the accumulated experience of six Chiefs over a period of twenty-two years.

DEVELOPMENTS WITHIN THE CORPS

Several developments which added to the efficiency of both the individual chaplain and the Corps as a whole took place during these years. Since the number of chaplains on duty after the war was approximately twice that before the war, conferences of chaplains where ideas could be exchanged and constructive criticism offered became advisable. Secretary Daniels reported in 1920: "A conference to which all the chaplains on duty on the east coast were ordered was held in Washington May 12, 13, and 14, 1920."[31] This was the first regional conference called by the Chief of Chaplains. Writing to Chaplain Frank Thompson on 24 March 1923, Chaplain E. W. Scott said:

I am sure you will be interested to know that Chaplain Dickins arranged for a conference of all the Chaplains in the combined fleets, for two days, March 9th and 10th. Twenty-two Chaplains were present, and three personal letters have come in, speaking very enthusiastically of the conference. It is a wonderful thing to have so many Chaplains get together. . . .[32]

Regional conferences of chaplains of the Fleet and chaplains on shore who were stationed in the vicinity of the Fleet anchorage were often held. Such a conference took place at Rancho Sante Fe, near Long

[30] Neyman, "Manual."

[31] *Report of Secretary of Navy,* 1920, p. 561.

[32] CoC., Thompson File.

Beach California, 25 and 26 November 1935*. The series of papers which were presented at this meeting were edited and published by Force Chaplain T. P. Riddle in a book entitled *Religion and Welfare in the United States Navy.* A similar conference was held at Long Beach on 9–11 February 1937. Such gatherings proved of great value in helping chaplains become acquainted with one another. Even so, the Corps had grown to such an extent and the chaplains were so scattered that it became impossible for each chaplain on active duty to know all the others. Chaplains Kirkpatrick and Albert, who had both entered the service during the First World War and had both attained the rank of commander before this period under review closed, met but once and then incidentally during these years.

The normal tour of duty for chaplains during these years was as follows:

(a) Acting chaplains and chaplains with the ranks of lieutenant and lieutenant commander: Three years at sea, alternating with two years on shore.

(b) Chaplains with the ranks of commander and captain: two years at sea, alternating with three years on shore.

There were exceptions to this general policy. Sometimes captains remained on shore for periods longer than three years. Chaplains, regardless of rank, served eighteen months in Samoa and two years on the Asiatic station.[33]

The title of District Chaplain appears for the first time during these years. Although the district organization of the Navy dates back to 1903, it took the First World War to clarify and enlarge the administrative functions within the prescribed areas. The original vague boundaries of these twelve districts were expanded to include the whole of the United States. The Second Naval District was abolished in 1919, its territory being incorporated into the First and Third Districts. The senior chaplain at district headquarters became recognized as District Chaplain, a term that occurs as early as 1923 and by 1930 was in general use.[34]

The District Chaplain became to chaplains on shore what the Fleet Chaplain was to chaplains at sea; in general the duties of the latter foreshadowed those of the District Chaplain. During this period the duties of the Fleet Chaplain appeared for the first time in *Navy Regulations,* but this counterpart on shore was not mentioned. His duties, however, were outlined in some district regulations.

For several years during the twenties the Navy Department gave a few chaplains the opportunity to take postgraduate work on a duty status at institutions of higher learning. The chaplains were permitted within certain limitations to choose their own schools. Each was given only one such opportunity and this was usually sandwiched in between tours of duty. Some took a short refresher course of three months while others remained for a full academic year. Of this plan Secretary Denby observed in 1924:

> During the year the policy of assigning chaplains to postgraduate work at universities has been continued and four have been so assigned and have completed their work, one receiving degree of master of arts. It is contemplated continuing this policy, both for benefit to the corps and the chaplains themselves, and for its effect through chaplains bringing students and churchmen into more intimate knowledge of the chaplains' work in the Navy.[35]

The 1920 edition of *Navy Regulations,* the first to appear after 1913, carried the following new paragraphs under article 1245, "Duties of Chaplains":

> (11) Before the ship reaches port, he shall ascertain the names of the men on board whose families reside in the immediate neighborhood of said port, and, after conference with the men, if they so desire, he shall pay an official call on their homes, with the object of establishing a feeling of good will between the ship and the home.
>
> (12) He shall report to the commanding officer the names of the families visited.

This ruling placed a heavy responsibility upon chaplains, especially when their ships called at such ports as San Diego where many enlisted men had their homes. The *Regulations* were revised again in April 1939. Paragraph eleven was revised to read:

> (11) He shall, in case of sickness, death, or other emergency, call on the homes of men whose families reside in the vicinity of the ship. In addition to emergency calls, he shall, when occasion offers, make such calls on families as he may deem desirable for the development of a sense of interest by the ship in the welfare of the men and of their families.

The twelfth paragraph in the 1920 edition was omitted in the revision of 1939 and the following substituted:

> 12) Report to the Bureau of Naval Personnel all marriages, funerals, and baptisms at which he officiates, giving names, dates, and places.

This record of marriages, funerals, and baptisms,

[33] Neyman, "Manual," Paragraph 118.
[34] CoC., L. N. Taylor File, letter of 1 Feb. 1923.
*See picture on page 239.
[35] *Report of Secretary of Navy,* 1924, p. 167.

kept in the Chaplains Division, has grown in importance and value with the passing years.[36]

PACIFISTS ATTACK THE CHAPLAINCY

War was a new experience for the vast majority of the citizens of the United States when the course of events swept this country into the European struggle. The short episode of the Spanish-American War was just a skirmish; only a few old veterans remembered the bloody fighting of the Civil War. Hence the people were unsophisticated about war and were subject to waves of emotion. Frequently prejudices and hatreds of everything German, including the music and language, were fanned into mob violence. On the other hand many youths of the country enlisted in the armed forces with high idealism. This was a war to end war. This was a war to make the world forever safe for democracy.

Then came the aftermath. Several propaganda stories used to inflame public opinion were debunked. There was a reaction against certain provisions of the Versailles Treaty. Many began to feel that the war guilt did not rest exclusively upon Germany, but that most nations involved had some share in it. The high hopes of the League of Nations seemed blighted when the United States refused to join. Young men who left for the front with bands playing drifted back shaken by the dreadful realities of modern war. A spirit of disillusionment revealed itself in the press, over the radio, and was found even in the pulpit. Old slogans which meant so much during the war were ridiculed and derided. A great conviction seemed to grip the country: "Never again."

The reaction to this common conviction was not always the same. Some individuals worked more earnestly for international cooperation, while others advocated complete isolationism. In this period of reaction there developed in many Protestant church circles, in the name of a more devout acceptance of the teachings of Christ, the philosophy of pacificism. There were many shades of this doctrine. The extremists maintained that it was never right to use force even to defend one's own home against a criminally-minded intruder. Pacificism was frequently preached with evangelistic fervor especially to young people in the colleges. Many in deep sincerity made solemn pledges that under no condition would they engage in war.

The fundamental ideas of the pacifists naturally clashed with the basic ideas of the Army and Navy chaplaincy. The influential *Christian Century,* one of the most widely read Protestant religious publications, became a leading advocate of the pacifist position. A number of incidents occurred beginning in 1924 which focused criticism on the chaplaincy in general and on Navy chaplains in particular.

The first of these, typical of others, involved Chaplain A. N. Park, Jr. When Chaplain Park attended the University of Chicago as a graduate student on a duty status in 1923–1924, a Quadrennial Student Volunteer Convention was scheduled to meet in Indianapolis during the Christmas vacation. Park had been a Student Volunteer in his undergraduate days and had taught in a mission school in Egypt. Upon his request, the Navy Department granted him permission to attend the Indianapolis Convention at his own expense. The Registrar of the Convention was advised of this action and expressed his pleasure. "I feel," he wrote, "that there are a great many points of common interest and common service between chaplains of the navy and missionaries of the foreign field."[37]

The Convention attracted about six thousand delegates. Under the sponsorship of the Fellowship of Reconciliation, a pacifist organization, several side-meetings were held on the subject of war. Park attended whenever he could. He entered into the discussions, frankly taking issue with the pacifists. At one of these meetings he heard a "wild-eyed Hindus" student refer to the American flag as "a piece of rag." This statement aroused Park, as well as other delegates, and resulted in "many men leaving the room."

The presence of a Navy chaplain was somewhat disconcerting to the pacifists as was reflected in a news report of the Convention which appeared in the 10 January 1924 issue of *The Christian Century.* In the following week's issue an editorial appeared under the heading: "Why This Continued Espionage." The editor stated: "Reports of the recent Student Volunteer Convention emphasize the presence of police and naval officer" in the meetings where pacificism was discussed. The editor questioned: "Who sent these men to these meetings?" On 26 January Chaplain E. W. Scott, then Chief of Chaplains, wrote to the editor of *The Christian Century* giving the facts as presented above. His letter appeared in the issue of 6 February.

A second incident involved an ex-Navy chaplain. The 6 November issue of the same publication carried an article entitled "The Making of a Pacifist"

[36] Letters are frequently received in the Chaplains Division from interested parties requesting information from this file.

[37] CoC., Research File.

by J. W. Decker, then a missionary under the American Baptist Foreign Mission Society at Ningpo, China. Decker had served as an acting chaplain in the Navy during World War I. He made twelve trips to France on a transport. Writing in retrospect of his experience, Decker declared: "My whole heart and soul espoused the cause" The "loyalty that I then gave to the cause of the allies was complete," he said, and there was never "a flicker of doubt till long after I had taken off my naval uniform." In all earnestness he had considered the World War as "a holy cause."

During the following five years, a transformation had taken place. Of this he wrote: "The holy grail, which seemed almost in our hands, vanished. During these years of disillusionment, I have slowly come to an avowed pacifist position." He traced the reasons for the change of attitude. War was " a depraved and cruel deceiver." "It is the destroyer of everything man deems worthwhile." He would have nothing more to do with it. "I have therefore registered a solemn vow," he declared, "that I will never take part in, or support in any way, the prosecution of another war." He added: "When the next epidemic of flag waving, patriotic lying and enemy hating begins I trust I shall have gained sufficient immunity to protect me against the contagion."[38]

Two weeks after Decker's article appeared, *The Christian Century* carried an editorial entitled "Get the Churches out of the Chaplaincy Business." "It is high time," said the editor, "for the church to quit standing sponsor for Army chaplains." It may be assumed that he referred to Navy chaplains as well. "Ever since the organization of the Federal Council of Churches of Christ in America," he wrote, "this business of feeding Christian ministers to the war system has been one of its chief and proudest functions." He added: "We believe this entire activity of the Federal Council is pagan business and not Christian." The editor claimed that the endorsement of chaplains for the armed services was tantamount to tying "the church to the chariot of Mars." He did not mean that men in the armed services should be denied a spiritual ministry, but insisted that chaplains go "as Christian ministers not as army officers." The editor did not know about, or did not mention in his editorial, the Navy's unsuccessful experiment with such civilian ministers on board ship during the years 1913–1917.

A third incident involving a Navy chaplain in a clash with pacifists occurred in May 1926 when Chaplain J. W. Moore was sent to present the opportunities of the naval chaplaincy to the students at Yale Divinity School. A letter signed by Oliver M. Zendt appeared in the 27 May issue of *The Christian Century* asserting that "a navy chaplain from the war department spoke in the divinity school chapel for the purpose of enlisting men as navy chaplains." The fact that Zendt confused the War Department with the Navy Department is immaterial. He claimed that sixty-five students in the Divinity School had signed a resolution which included the following: "We . . . believe that religion should not be subject to political dictation." And "we believe that they [chaplains] should not wear a military uniform, be paid by the federal government, be under military control, or in any way be a part of the fighting machine."

The strength of pacificism in many of the larger seminaries made it increasingly difficult to recruit strong men for the Navy chaplaincy. While the thinking of the pacifists was not always consistent nor uniform throughout the country, yet their influence became so strong in the decade 1925–1935 that Army and Navy chaplains were often on the defensive. Variously worded resolutions were introduced in denominational and interdenominational assemblies which were designed to call chaplains out of uniform and to divorce the church completely and absolutely from any connection with the chaplaincy in the armed services.

---

[38] There is a sequel to this story. Dr. Decker, later Foreign Secretary of the American Baptist Foreign Mission Society (Northern Baptist) and now (1946) one of the Secretaries of the International Missionary Council, in the light of international developments and experience and of more mature consideration, abandoned his pacifist position. Writing to the author on 25 Oct. 1945, Dr. Decker explained that his former position was based "on the fundamental thesis that war represented the ultimate in evils, and therefore must be unqualifiedly rejected by the Christian." He then added: "I began to doubt my basic thesis in connection with my experience in China from 1931 on. During the months and years following the Moukden incident of September 18, 1931, I saw China the victim of an unprovoked and utterly ruthless and cruel aggression. China appealed to existing treaties and to the League of Nations; that is, to the legal machinery which had been set up to make such international crimes impossible. Her appeals were in vain, and it was only as she resisted the invaders, in defense of her homes and firesides, that hope appeared both in her own national situation and in the regard and attention won from other nations. With the Chinese young people I had to consider the question of what a Christian ought to do in the face of such an attack, and I was compelled to conclude that a Christian would be right in resisting." (CoC., Decker File.)

Many of the pacifists who were in China and witnessed the rape of Nanking in 1937 were among the first to change their views.

When World War II came, Dr. Decker sanctioned military service for two of his sons who became officers in the Navy.

In the 16 January 1935 issue of *The Christian Century* the editor returned to his theme, "The Chaplaincy Question." He admitted that he had been discussing the chaplaincy in his publication "for more than a decade." "We look with shame," he wrote, "upon the blind servility with which the Christian church gave itself to the government of the United States in 1917 and 1918." And he declared: "The churches are steadily making up their minds not to act that way again." The fact that within ten years the churches of the United States sent into the armed services the unprecedented number of more than 12,000 of its finest clergymen would appear to contradict the editorial claim.

In this editorial, the position was taken that all discussion as to whether or not a chaplain should wear a uniform or receive his pay from the government was "sophomoric." The real issue, in the editor's opinion, was "whether the Christian church expresses the mind of Christ when it recruits its ministers for the military status of the chaplaincy." The editor advocated the discontinuance of ecclesiastical endorsement by the General Commission on Army and Navy Chaplains or any of its affiliated denominations. Thus he would have thrown overboard that for which Navy chaplains had prayed and labored for decades.

Another example of the pacifist attitude is found in an incident involving Chaplain Charles Ellis, who was called upon to offer the prayer at the launching of the carrier *Yorktown*. The New York *Times* of 5 April 1936 carried a news story of the event which contained the following sentence: "Christened and baptized by Mrs. Franklin D. Roosevelt and blessed by Navy Chaplain Charles V. Ellis . . ."

An article entitled "Blessed by Chaplain Ellis" appeared in the 6 May issue of *The Christian Century* over the name of Alfred Schmalz. In mock seriousness the author asked: "Do all warships get 'blessed'? Is that part of the navy Chaplain's regular religious duties?" The author continued: "The national guard or the R. O. T. C. may even ask me in some time to bless a gun or two, and I would not know in my ignorance where to turn for the proper ritual." In simulated earnestness, Schmalz requested: "I hope you will help me out, Mr. Ellis . . . I am awfully stupid about this sort of event." The author concluded with the following exhortation:

> But never mind, Mr. Ellis. We mustn't let our sentiments turn us from the stern call of duty. Bless us another airplane carrier, will you?
>
> Yours in Christ.

The article was naturally disturbing to Chaplain Ellis who sent it with a copy of the prayer used on the occasion of the launching to Chaplain E. A. Duff, then Chief of Chaplains. When the reporter of the *Times* in confused terminology referred to the ship as being "blessed," he was referring to the prayer offered at the launching. The prayer is free from any blatant militarism implied by pacifist critics. Indeed it can be given as a model suitable for such occasions. The complete prayer follows:

> Almighty God, our Heavenly Father, the Giver of every good and perfect gift, we thank Thee for the material and spiritual blessings of life, and for the wondrous way Thou hast led us as a nation during the past—through shadow and sunshine Thou hast been our refuge and strength.
>
> We remember with a deep sense of gratitude the sturdy characters who, through toil and sacrifice, founded this great Republic—"the land of the free, and the home of the brave"—and we pray for Thy continued favor and guidance in all the affairs of our beloved country.
>
> In the exercises of the hour, which will send forth this vessel from the ways, destined to join our First Line of Defense, we would humbly seek Thy approval, lest our efforts come to naught. God grant that she may be a champion of justice and righteousness—a living exponent of the principles near and dear to our hearts—and if the occasion should arise, may she strike terror in the hearts of all who dare to molest or disturb our God-given freedom.
>
> Let Thy blessings rest upon the President of these United States, the members of his cabinet, the members of the Legislative and Judicial departments, and all those charged with the responsibility of leadership. May they lead us in the paths of righteousness and peace! Through Jesus Christ, our Lord and Saviour we ask these favors, Amen.

The occasion came, much sooner than many, who witnessed the launching of the *Yorktown*, expected, when she stood in that thin "First Line of Defense" and struck terror in the ranks of the enemy. In the Battle of the Coral Sea the men and planes of the *Yorktown* helped turn the tide of battle. In that epic fight, the gallant ship was damaged by a bomb which left a gaping hole in her deck and forty-four dead. Even though partly disabled, the *Yorktown* left one battle and speeded 5,000 miles in less than a month to play her supreme role in the decisive Battle of Midway. There she was sunk on 6 June 1942. According to Admiral Nimitz the Battle of Midway was the turning point of the war.

Religious publications other than *The Christian Century* also joined in the criticism of the chaplaincy. The Catholic and Episcopal churches were for the most part unaffected by the pacifist doctrines. The

CHART · SHOWING · DISTRIBUTION · OF · USN · CHAPLAINS · BY · RANK ACCORDING · TO · NAVY · REGISTERS · BY · YEARS · AS · INDICATED

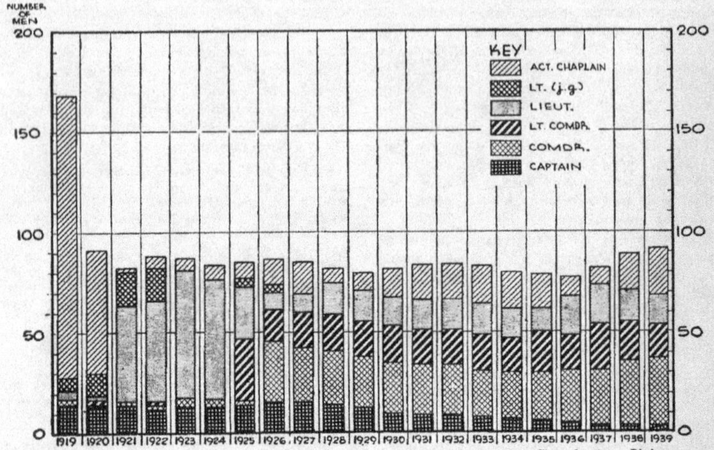

Chart by Anne Bird

Disciples of Christ, at its International Convention held in October 1936, voted to send no more of its ministers into the Army or Navy as chaplains and requested the Federal Council to "sever its official connection with the war system by dissolving the chaplaincy commission." It was not until May 1941 that this denomination took action to restore its relationships with the military chaplaincy. The movement was strongest in the Disciples, Methodist (North), Presbyterian (USA), Congregational-Christian, and Baptist (North) denominations. A greater ratio of the clergy accepted the pacificist than did the laity.

Chaplains were concerned about the injurious results of pacifist propaganda as it affected not only their own status as chaplains but also the youth of the land. The pacifists seemed to have a spiritual superiority complex. The implication was given in many circles that they had more conscience than non-pacifists. Young men raised in the Christian tradition who joined the Navy were sometimes shocked to find the reality of their Christian faith was doubted by some civilians simply because they wore the uniform. A definite line of cleavage had been drawn between Christianity and patriotism when the latter involved serving in the armed forces.

FEDERAL COUNCIL'S REPORT ON CHAPLAINCY

In January 1937, the Department of Research and Education of the Federal Council of Churches of Christ in America submitted a report on its study on the Army and Navy chaplaincy. The report went back to the Quadrennial Meeting of the Council in 1924 when a resolution was adopted which called for a study to be made of the advisability "of taking such steps as may be necessary and effective to place the service of ministers of religion to the men of both Army and Navy upon a non-military basis." The committee appointed to investigate made a report in December 1925 recommending that denominational committees and churches "make clear to the Chaplains their mind as to their attitude on this question."[39]

There the matter rested for several years. In the thirties several denominations took action. In 1934,

[39] CoC., Federal Council File. Additional material in this section is taken from the same source.

the Newark Conference of the Methodist Episcopal Church requested its bishops to refrain from appointing any of its members to the chaplaincy, and the New England Conference in the same year requested the Federal Council to perfect a plan whereby a Protestant chaplaincy for the armed forces could be "supervised" through a board or a department of the Federal Council.

In 1936, the General Synod of the Evangelical and Reformed Church refused to continue its participatin in the work of the General Committee on Army and Navy Chaplains. In May of the same year, the Methodist Episcopal Church passed a resolution requesting the Federal Council "to seek and find a method by which the spiritual ministry of the churches to the armed forces of the United States of America may be performed by ministers appointed and supported by and amenable to the churches." In October, the International Convention of the Disciples of Christ also adopted a resolution which called for the withdrawal of "official representation from the Chaplaincy Commission of the Federal Council of Churches," from "future participation therein," and

> that we request the Federal Council to sever its official connection with the war system by dissolving its chaplaincy commission and
>
> that the Federal Council be requested to provide a non-military ministry of religion to men in the armed services at the churches' own expense and under their own authority, without involving the church of Christ in any alliance whatsoever with the state or the military system.

Such actions by several denominations prompted the Department of Research and Education to investigate the attitude of Protestant chaplains in the Army and Navy on some of the issues then being debated. Questionnaires were mailed to the following:

	In service	Reserve*	Retired	Total
Army	88	128	49	265
Navy	66	30	15	111
				376

Replies were received from 139 Army or about fifty-two percent, and from fifty-two Navy or about forty-seven percent. The answers to some of the questions asked are given below:

1. Is the present chaplaincy system as a whole (check one) a. wholly satisfactory?
   b. fairly satisfactory?
   c. unsatisfactory?

Replies:	Wholly		Fairly		Unsatisfactory	
Army	57	41%	76	54.7%	1	.7%
Navy	20	38%	25	48%	1	1.9%

(Objections not on ethical grounds)

2. Is the wearing of the military uniform very important in relation to the chaplain's status?
   Yes ( ) No ( ) Please state why.

Replies:	Yes		No	
Army	133	95.6%	3	2%
Navy	43	82.7%	6	11.5%

3. Is the possession of military rank important to chaplains—
   a. in relations with the Commanding Officer?
   b. in relations with other officers?
   c. in relations with the men?

   Where the answer is "yes", please indicate in what way.

Replies:		Yes		No	
a. in relations with Commanding Officer	Army:	120	86%	14	10%
	Navy:	35	67%	13	25%
b. in relations with other officers	Army:	127	91%	8	5.7%
	Navy:	38	73%	7	13%
c. in relations with the men	Army:	126	90.6%	9	6%
	Navy:	38	73%	8	15%

Summarizing its findings, the committee reported:

> He [the chaplain] is very much a part of the military or naval establishment. He has, it is true, a strong sense of religious vocation. One cannot read the statements of these men about their calling without being impressed with the authentic pastoral consciousness which they reveal. But the chaplain has quite as definite a sense of "belonging" to the service and he believes that the wearing of a uniform is essential to his efficiency. He believes that as a civilian he would be an outsider and could not command the same measure of respect. This is due in part to the fact that the "service" consciousness in the Army and Navy is very strong and only the wearer of the uniform is regarded as measuring up to requirements.
>
> Further, and probably more significant, there is the intangible influence of rank throughout the services. They are built on rank. It is argued that without rank the chaplain would have little influence with the command and especially with the other officers, and could not effectually represent the men in getting attention for their needs. This intangible factor is evident at every point. The chaplains unconsciously reflect it in their characterization of their own office. They become assimilated to the establishment of which they are a part and they take the regimen as a matter of course. They are in and of the service; they believe in it profoundly and are happy in their belief.

CHANGING NATURE OF NAVAL PERSONNEL

There was a noticeable change in the type of men entering the service during the twenties and thirties. This was summarized by Chaplain T. P. Riddle in 1939.

> We have attracted a class of men who look at the Navy as a lifetime vocation, rather than as an adventure, as did the youngster 20 years ago. The devil-may-care lad who joined the service for a cruise of adventure gave little thought to a man's life. . . . The modern Navy man . . . joins the Navy to learn a vocation, and to make a career for himself. As he looks at his career, he naturally considers marriage as part of his scheme of life.[40]

In a speech delivered before the National Council of the YMCA at Detroit, Michigan, 20 October 1939, Chaplain Workman presented the following statistics:

> Of the 110,196 [men] in the naval service on June 30th [1939], 106,743 were native born. The ages of our enlisted men run as follows:

Under 20 Years	11,377
20 years	9,903
21–24 years	38,320
25–29 years	21,724
30–34 years	14,317
35–39 years	10,942

> Let us see the change in the character of this 1921 lad in the light of 1929 and 1939

	1921	1929	1939
Bad conduct discharges	4,000	2,193	965
Dishonorable discharges	6,000	632	106
Undesirable discharges		566	275
Desertions	11,000	2,055	338
Reenlistments		13,590	17,980
Men who had received good conduct medals		9,149	39,590

> In 1929 high school graduates 18%; in 1939 high school graduates 42%; high school work 85%; some college work 3%. In 1929 intelligence quotient was 85; in 1939 intelligence quotient 103. With the help of the Navy Department 61 men prepared themselves and entered the Naval Academy to become officers.[41]

The lean years of the depression in the thirties were partly responsible for this change. The Navy, having more applications for enlistment than could accepted, was able to select older and more mature men.

Another very significant change was the increase in the number of married men. Even men in the lower ratings were getting married and having children. Obviously, they could not support a family on their incomes without additional financial assistances. Sometimes the wife worked; sometimes she lived with her parents. Often there were financial problems which in turn affected the morale of the husband aboard ship. The situation grew until it attracted the attention of high ranking naval officials. Difficulties sometimes arose in regard to allotments. In 1933, the Bureau of Navigation published a pamphlet which set forth a guide for allotments to families when the question of support was in dispute. If a sailor were married, he was supposed to allot at least a third of his gross pay to his wife, and more if there were children.[42]

The Navy Relief was called upon increasingly, during these years, to assist in family difficulties. This involved a chaplain, for in many districts, he served as executive secretary or in a similar capacity of the local Navy Relief auxiliary. The relief load with its consequent need for investigation was particularly heavy in the Long Beach area where, according to Chaplain Riddle, some 9,000 families of enlisted men were living in 1936.[43]

Chaplain George S. Rentz in 1935 commented on the work of Navy Relief handled by his office. He stated:

> Last year there were more than 200 cases handled by the Chaplain that were outside the bounds of Long Beach and San Pedro. The total number of cases involving loans from the Navy Relief Society in 1934 was 959. There were also many cases handled by the Chaplain that did not involve loans.[44]

Rentz added that the work of the chaplain in this office was "almost wholly with the personnel of the Fleet and their families."

At the chaplains' conference held in November 1935 at Rancho Santa Fe, Mrs. T. P. Riddle, wife of Chaplain Riddle, presented a paper on "Personal Contacts with our Men's Families." This appeared as a chapter in the book, edited by her husband, entitled *Religion and Welfare in the United States Navy*, and had sound words of counsel for chaplains:

> About 50% of our men are married. Our appeal, spiritual or otherwise, can no longer be made to that 50%, as bachelors. It must be made to them as married men, or else we automatically eliminate them from the sphere of our influence. The greatest human appeal to the married man is through the wife and little ones.[45]

To combat the loneliness that many young Navy wives felt when their husbands were at sea, a Navy Wives' Club was organized in Long Beach which met

[40] Riddle, "Enlisted Men's Families," *Naval Institute Proceedings*, Jan. 1939, p. 83.
[41] CoC., History of Naval Chaplains File. It should be remembered that the number of naval personnel remained fairly constant through the years 1920–39.

[42] Directive, Nav. 5-K-Service, 28 April 1933.
[43] Riddle, *op. cit.*, p. 83.
[44] CoC., Rentz File, letter of 25 Apr.
[45] *Op. cit.*, p. 68.

at the local YWCA. Similar clubs became fairly common in port cities and often a chaplain and his wife sponsored them. Some of the clubs were named in honor of the wives of Navy chaplains, as were the Elizabeth Riddle Club of Long Beach and the June Glunt Club at Norfolk. Articles of Incorporation for the Navy Wives' Clubs of America were filed 3 June 1936 in the state of California.

A similar organization, the Navy Mothers' Club, secured its national charter in 1933. The first of these was organized at McAllen, Texas, in 1930. Navy chaplains and their wives have often served as sponsors for such organizations.

The 1920 edition of *Navy Regulations* encouraged pastoral calls, and required chaplains, in cases of home emergencies, to visit enlisted men's families in the vicinity of ports of call, greatly broadened the chaplains' ministry. Navy chaplains no longer were restricted in their duties to men in uniform; more and more they became chaplains to the families of naval personnel.

DIVINE SERVICES AND OTHER RELIGIOUS ACTIVITIES

The very heart of every Navy chaplain's ministry is found in his Divine Services. Even though he wears the uniform of a naval officer, the chaplain is first and foremost a minister, priest, or rabbi. While his duties often include many activities of a welfare or social nature, these activities never overshadow the special duties which only a properly trained and duly ordained clergyman performs.

On Sunday morning a regular routine is followed aboard ship which has been described by Chaplain E. W. Scott as follows:

> The decks being washed down and put in order during the early morning hours, breakfast served, and the crew mustered and dismissed, preparations for worship are initiated by the Officer-of-the-Deck, who gives the order, "Rig Church." If the weather is fair and mild, the service may be held out on the open deck under the awning, with canvas spread on the weather side for protection against the wind and spray. However, one of the crew's living spaces is generally used, the walls being covered and festooned with pennants and flags, which hide all trace of the fact that this compartment is utilized by the men for messing, sleeping, gun-drills and routine work. A movable desk is set up, the piano brought in, and backless benches and the available chairs serve as pews. "Church is rigged" by one of the divisions, the Chaplain putting on the finishing touches to suit his own fancy as to proper arrangement and to make the space serve to most advantage.[46]

[46] CoC., Scott File, *The Century Church Bulletin*, Mar. 1923.

The first reference to statistical summaries of attendance at Divine Services conducted by Navy chaplains was made by Secretary Daniels in 1920. He wrote that "there has been an average monthly attendance on the various religious services as conducted on board ships and at shore stations of 60,602."[47] At that rate, the total for the year was 727,224, or an average of five or six times a year for each person in the Navy and Marine Corps.

In 1925, attendance improved. "The statistics gathered from ships and stations," wrote the Secretary of the Navy, "indicate that the church attendance of officers and enlisted men has reached approximately 1,000,000 during the year, an average of nine per person per year."[48] The same average held for 1926.[49] On this latter year, the Secretary elaborated: "In addition to their other duties the chaplains have continued to stress the religious phase of their activities as paramount, particularly during the lenten and other religious festivals."

The picture in 1927 was even more favorable.

> Although attendance at these services is purely voluntary, the records of the several ships and stations indicate that the church attendance of officers and enlisted men for the year was approximately 1,027,000, an average attendance of nine and one-half times per person throughout the year. This is a high tribute to the efficiency and drawing power of the individual chaplains as well as an indication of the religious sentiment prevalent in the naval personnel.[50]

Affecting this ratio of attendance were these facts: many men were on watch during the hour for Divine Service, some were on leave or liberty, and others attended civilian churches.

An even higher percentage of naval personnel attended Divine Services in 1928. The Secretary reported a total of 1,222,000 present for the services held during the year by Navy chaplains, "an average attendance of 10.8 per person."[51] He added: "Also the attendance at communion services indicates the religious interest of the personnel of the Navy, 3,101 communion services having been held with a total attendance of 44,670." By 1929, the average attendance per person per year had increased to 11.5.

The annual report of Chief-of-Chaplains Sydney K. Evans, dated 10 July 1934, included a description of "Fleet Sunday" held in New York on 3 June of that year. Evans stated that "more than 5,500 officers

[47] *Report of Secretary of Navy*, 1920, p. 122.
[48] *Ibid.*, 1925, p. 168.
[49] *Ibid.*, 1926, p. 140.
[50] *Ibid.*, 1927, p. 157.
[51] *Ibid.*, 1928, p. 172.

and men attended Divine Service in St. Patrick's Cathedral and in the Cathedral of St. John the Divine." Cardinal Hayes spoke at the former and Bishop Manning at the latter service. The attendant publicity, not only in the press of New York City, but, throughout the nation, did much to emphasize the place the Navy was giving to religion.

In his annual report to the Chief of the Bureau of Naval Personnel, the Chief of Chaplains includes a summary of the statistical reports of individual chaplains. Such summaries are a tangible index of the variety and extent of the chaplains' activities. In an address delivered before the officers of the Bureau of Navigation on 28 February 1936, Chief-of-Chaplains Edward A. Duff explained the nature of these reports:[52]

Chaplains of the Navy made a monthly and annual report of all their activities, and through a careful checking of these reports, we are in position at all times, to give an accounting of the work of all chaplains. For example, the following statistics of the work for the year 1935, will give a pretty accurate idea of the manner of keeping check on the activities of the Corps.

Number of Divine Services held	5,433
Attendance at Divine Services	990,174
Number of Communion Services or Masses	2,782
Number making their Communion	34,773
Number of Military Services other than ship or station	632
Attendance at Military Services other than ship or station	75,435
Week-day religious activities	1,290
Attendance at week-day activities	67,883
Total attendance at all services, as above	1,168,245
SECULAR: Lectures, Song Fests, etc.	545
Attendance	22,661
Number of Sunday School Sessions or Bible Classes	937
Attendance at Sunday School Sessions or Bible Classes	44,975
Baptisms	400
Men Joining Church	187
Marriages	426
Funerals	686
Number of Civilian Services and Addresses	1,185
Attendance at Civilian Services and Addresses	320,861
School Sessions	4,754
School Sessions, Attendance	57,014
Letters Written, Re: Men	36,645
Number of books issued from libraries	746,448
Number of sight-seeing parties	246
Number of men in sight-seeing parties	16,200
Hospital or Sick Bay Visits	18,129
Prison or Brig Visits	3,442

[52] CoC., Duff File.

The Sunday Schools mentioned in the statistical summary refer to the Schools often conducted by Navy chaplains at shore installations for the children of naval personnel. In addition to their contacts with naval personnel, chaplains have many opportunities to appear before civilian churches, service clubs, and other groups. According to Chaplain Duff, Navy chaplains in 1935 addressed more than 320,000 people in these civilian gatherings.

To judge the effectiveness of a Navy chaplain's ministry by the same norm used to evaluate a civilian minister's does not give the full story. Navy life is not conducive to the organization of a church on board ship or at a shore base. Whenever a bluejacket became vitally interested in religion, he was encouraged by his chaplain to join the church of his choice in the nearest port or at his home community if not already a member. There are no statistics of such results. Moreover, many other factors peculiar to Navy life, already pointed out, enter into the picture.

A delightful and illuminating article on the work of a Navy chaplain entitled "My Parish", by Chaplain C. A. Neyman, appeared in the 15 April 1922 issue of *The Baptist.* Neyman compared the civilian parish he once had with the one on board the *Tennessee.* His former congregation was made up "of people ranging in size (and likewise in age) about as do the strings of a harp, but with a perceptible inclination to the feminine gender." These people lived apart and some were not seen by the pastor from Sunday to Sunday. On the other hand the parish on shipboard was compact.

But here, there are never less than 1,000 of us and we count each others' noses every morning at a given hour; and in a day's work not one of the 1,000 goes much farther than 400 feet from the chaplain's desk, and not a one but may be called to his elbow, if he wish, on five minutes' notice.

. . . .

Associations between a chaplain and officers and men in the Navy . . . are or may be delightfully direct. The propinquity is astounding. Instead of having to build up contacts, one finds them tumbling upon him, in plethora, willy-nilly. This is not a matter of standing behind a pulpit armed by cloistered preparation, or of going out, guided by a list, to pull at door bells and sit on parlor chairs. Here one's parishioners surprise him in his bath, or call him out of his bunk, or take him at mess, or at play, or walk in upon his meditation and study.

Chaplain Neyman had some pertinent remarks to make regarding the kind of preaching the sailor wanted.

Furthermore, they will listen to his preaching, these sailors. One never realizes what a scattering salvo he delivers at an audience which includes young and old of both sexes, nor how nearly impossible it is to minister adequately to the bill-peddler and the jurist at the same time, nor how near a miracle he attempts in breaking spiritual bread for both the mother, whose long-tried faith has a distinct pre-millenarian twist, and the high-school superintendent, who does not shudder at mention of higher criticism. One never realizes these things, I say—at least not completely—until he stands regularly before an audience of sailors who are much of an age and experience and who are never given to critically splitting doctrinal hairs. The discovery bobs up suddenly; and then one most likely flounders around for a while in an effort to deliver himself of some special message peculiarly adapted to these men. Yet, it dawns on him later that what is needed, yes, what is wanted, is a straightforward, unmincing "gospel" message. No frills, no tricks. But, for Heaven's sake, no shooting over heads, and no talking "down." Red-blooded presentation of red-blooded Christian principles.

Chaplain Neyman bore witness to a fact frequently observed by Navy chaplains. The men aboard ship love to sing.

In a small town once we could fill our church on Sunday evenings by announcing a song service, but city people turned up their noses at such garden-variety worship. The sailors, however, will fill a compartment on Sunday evenings and call for hymn after hymn—and sing them lustily—until the pianist is about to drop exhausted from the stool. They clamor, all at once after each last chorus, for their favorites, and he who leads the service must have a delicately-adjusted ear to select out of the bedlam's shout the number of the next song.

In his concluding paragraph Chaplain Neyman drew the following contrasts:

My institutional church is a battleship. My parsonage and my pulpit are on the same deck. Almost everything is here that I wanted back there, and numberless things beside. Certainly, more contacts than ever the clubs and gymnasiums would have secured. Better still than having the church stand on a strategic corner and invite the passer-by, is to have the whole institution, from keel to truck, move majestically out to sea and take its crew (my parishioners) bodily into new cities and far lands where there are great lessons to be learned and sights that inspire.[53]

Presidents of the United States have been known to conduct Divine services aboard United States naval vessels. Lovette in his *Naval Customs* makes the following claim:

The first commander in chief of the Army and Navy to hold divine service for Navy personnel was President Franklin D. Roosevelt. On Easter Sunday, 1 April, 1934, the President in the absence of a chaplain, stood on the quarter-deck of the *Nourmahal* and read the service from the *Episcopal Book of Common Prayer*. The officers and men of the U.S.S. *Ellis* were present. The flag of the President flew from the yacht.[54]

A number of printed aids appeared during these years to aid chaplains conducting Divine Services.

The first edition of *The Army and Navy Hymnal* appeared in 1920 under the joint sponsorship of the Chief of Navy Chaplains, J. B. Frazier, and Chief of Army Chaplains, J. E. Yates. The book was copyrighted by the Century Company. The first two paragraphs of the introduction tell how the book was compiled.

For a long time it has been felt that for the use of the Army and Navy, a hymnal should be published that would meet the demands of every occasion and of all the Chaplains. From time to time civilian organizations have attempted this work, and have produced books of merit. The fact remains, however, that the Chaplains should know, and do know, more about what the men want and what is appropriate for divine service as conducted in the Army and Navy, than those who have seen the situation only from the outside.

With a view of getting what we want in this book, representatives of the Chaplain Corps of the Army and Navy selected a number of popular hymns which were submitted to each Chaplain of both branches of the services, with the request that he suggest eliminations or material to be added; also that suggestions be made as to Orders of Service, Selections for responsive reading, et cetera.

This volume was divided into several sections. The first contained 242 hymns familiar to Protestants. Then came a section entitled "Orders of Worship" which was prepared by H. Augustine Smith, A.M., and had been copyrighted by The Century Company in 1919. This was followed by a series of "Responsive Readings" selected and arranged by Dr. Harry Emerson Fosdick. A Roman Catholic section of twenty-seven hymns and a few prayers came next. The last section contained eleven popular songs under the heading "Home and Patriotic Section." The book was supplied at Government expense and made available to chaplains. Both band and orchestra arrangements were separately published.

While the hymnal met a long-felt need, nevertheless, Navy chaplains found that the pitch of most of the hymns was too high for male voices. Many Army chaplains had mixed audiences and so were not faced with this problem. Consequently, a revised edition was brought out in 1925, sponsored by Chaplain Evan W. Scott of the Navy and Chaplain J. E.

[53] CoC., Neyman File.

[54] *Op. cit.*, pp. 289–90.

Yates of the Army. The fact that the two editions differed somewhat in pagination and choice of hymns, presented a difficulty when both were used in the same service. The pitch of many hymns was lowered in the new volume. In addition, four hymns, the words of which were written by Chaplain Henry van Dyke, were included in the revised edition as was a short Jewish section. There was no revision of the band and orchestra arrangement. On the whole, the second edition of the hymnal was preferred by Navy chaplains. A compact *Hymn Edition* of the *Hymnal* was issued in 1925 which contained the words but no music.

In June 1934, Chaplain T. P. Riddle relieved Chaplain J. J. Brady as Battle Force Chaplain. Being an Episcopalian, Chaplain Riddle was much interested in the forms of Divine Worship used by Protestant chaplains in the Fleet and the type of altar furnishings and fixtures used by all chaplains. In order to aid in beautifying Divine Services, Chaplain Riddle issued, in June 1935, an attractive fourteen page brochure entitled *Preparing for Divine Services on Board Ships of the United States Navy.* The pamphlet contained illustrations of altar furniture and of congregations at worship in well arranged compartments, and a set of six drawings of altar pieces that could easily be made by a ship's carpenter. With the illustrations were short articles on such subjects as: "The Church Compartment," "Rigging Church," and "A Setting Which Uses the Communion Table." The pamphlet, published on the *California* and delivered throughout the Fleet, was an aid to chaplains who wished to improve and dignify their Divine Services.

In 1935, Chaplain Riddle published his *Protestant Ritual, Adapted to Service Needs.* This was an eighty-four page booklet which contained the communion ritual of the Episcopal, Presbyterian, and Methodist churches, abbreviated somewhat for use at Navy Divine Services. Prayers and outlines of

Tent Chapel secured for the Philadelphia Navy Yard in 1927 by Chaplain C. H. Dickins.

The contrast between that which the Government provided for chaplains in the First World War and that of the Second is brought into sharp relief by comparing this picture with some of the chapels erected at naval bases during World War II. (See page 199.)

services suitable for funerals, baptisms, marriages, and special services of interest to the Navy were also included. On 17 February 1934, the Right Reverend Monsignor George J. Waring of the Military Ordinariate wrote to the Chief of Chaplains requesting:

> Inasmuch as Catholic Chaplains are not available at all ships and stations, may I suggest that you send an official message to all ships and stations, notifying them that "the Catholic Church grants a dispensation from abstinence from the use of flesh meat to all its members in the naval service and their families, on all days of the year, except the day before Christmas, Ash Wednesday, Good Friday and the morning of Holy Saturday."
>
> It is requested that all supply officers, commissary stewards and mess officers, carefully note these four days in order to prepare suitable menus. Good Friday, this year, occurs the 30th of March.[55]

A circular containing this quotation from Bishop Waring's letter was sent to all ships and stations by the Bureau. A similar circular was distributed in 1935.

CHAPELS SECURED

More and more attention was being paid by chaplains, during these years under review, to the necessity of having suitable and attractive places for Divine Services. In some localities, facilities were remodelled for religious services. In other places, funds donated by civilians made the building of a chapel or the purchase of a suitable building possible.

Following the First World War, the Red Cross building at the Naval Operating Base at Norfolk was changed into a Catholic chapel, another building at the Base being already used as a Protestant chapel. In 1919, an armory at the Great Lakes Naval Training Station was converted into a chapel with a seating capacity of 110. A frame steeple was added to the building to give it "atmosphere." At first, the chapel was used by the Roman Catholics but in 1925 it was declared by the Commanding Officer a "Chapel for all Faiths." Later, it again was used exclusively by the Catholics.

During Chaplain T. P. Riddle's tour of duty at the Naval Air Station at Newport, (May 1930–December 1933) the second deck of the gymnasium, large enough to seat 500, was converted into a permanent chapel. Writing in retrospect, Riddle summarized the extent of the renovations made and the nature of the gifts received:

> The old praecennium arch was removed and a chancel created. The authentic constellations with the north star above the Altar were charted by Lieutenant Commander Holton and reproduced in gold on a dark blue background as the ceiling of the chancel. The Altar, as well as the red brocade Altar hangings and Altar brasses were gifts from Mrs. Hamilton Fish Webster. The Reverend Charles E. Peck, vicar of St. Paul's Cathedral in Boston,

[55] CoC., Military Ordinariate File.

Interior and Exterior views of chapel, Naval Air Station, Lakehurst, N. J.

832159—49——16

arranged for the donation of two Chancel chairs and kneeling benches formerly used by Bishop Brooks.

During this period, a certain amount of publicity was given to the creation of this Chapel. This resulted in the Reverend Karl Reiland, Rector of St. George's Church, New York, donating an excellent used Aeolian pipe organ, a gift from the late J. Pierpont Morgan. The pews came from Appleton Chapel at Harvard, due to the interest of the late Bishop Lawrence of Massachusetts, and of his son Appleton, now Bishop of Western Massachusetts. The hand wrought lighting fixtures and stained glass windows at the entrance of the Chapel were donated by Father Hutchinson, the rector of St. John's Church, Newport. I understand that some twenty stained glass windows were installed due to the efforts of Chaplain Reuben Shrum while he was senior chaplain of the Newport Naval Training Station. During his time, a new facade was created for the building presenting a dignified and churchly appearance.[56]

Late in 1928, the American Legion Department of New Jersey inaugurated the plan to construct a nonsectarian chapel at the Naval Air Station at Lakehurst, "The Cathedral of the Air," as a tribute to the self-sacrifice and devoted service of flyers. The Act enabling construction was signed by Calvin Coolidge on 2 March 1929, one of his last official acts as President of the United States. An American Legion Memorial Chapel Association was incorporated and funds were solicited. Leadership in the project devolved mainly upon the Honorable Herbert H. Blizzard of Audubon, Dr. Samuel A. Loveman of Toms River, and Chaplain W. W. Edel, then on duty at the Lakehurst Air Station. The main financial campaign during which more than $71,000 in cash was raised, began in February 1930. Chaplain Edel was indefatigable in promoting the project, travelling in 1930 thousands of miles and gave more than 200 addresses in behalf of the financial campaign.[57]

Construction of the stone chapel on a modified Gothic plan on Government property was begun in the spring of 1932. The building, large enough to seat 300, was finished that year but no funds were available to purchase interior equipment. For this reason, the chapel was not used regularly until 16 February 1941, when, the Government having contributed $7,000 for furnishings, the Cathedral of the Air was opened for Divine Services. The station chaplain at the time was J. V. Claypool. The Cathedral of the Air has the distinction of being the only chapel constructed at a naval installation by popular subscription.[58] The chapel was dedicated on 26 October 1947.

In 1931, Chaplain M. H. Petzold arranged for the renovation of the building used as a chapel at the Marine Barracks at Parris Island, South Carolina. A special appropriation made possible such improvements as a new ceiling, interior walls, deck, wiring, electrical fixtures, and radiators, and the painting of the building throughout. On 4 May 1931, Petzold wrote: "Our new chapel was completed and reopened on Easter Sunday and it is certainly a pleasure to conduct services there now."[59] In the same letter he stated: "One hundred and twenty-two persons have been baptized and two hundred and thirteen have been received into the church." Here again is evidence of the importance of suitable physical equipment for an effective spiritual ministry. The chapel, including the balcony, was large enough to seat approximately 500.

During Chaplain E. L. Ackiss' tour of duty at the Naval Training Station at San Diego (June 1930–July 1934), two new mess halls at Camp Ingram (later known as the South Unit) were converted into Protestant and Catholic chapels. One-half of the building which contained the Catholic chapel was used as a library and recreational room for the incoming recruits. The alterations were made by the station, and the equipment—a suitable altar, communion rail, and pews—was purchased with station funds. ~~Imitation stained glass~~ windows were placed in both chapels to contribute a church-like "atmosphere."

When Chaplain R. L. Lewis reported for duty at the Marine Base at Quantico in the spring of 1934, the Marines were using as a chapel a World War I barracks, which had "suffered" some repairs. One end of the building was rigged with an altar; a small Catholic chapel occupied the other. During his tour of duty there, the building was moved to a new location and enlarged by the addition of a wing at one end. This permitted an arrangement of two chapels, one for the Protestants and one for the Catholics, each seating from 150 to 200 people. By means of folding doors, each chapel was able to accommodate overflow audiences.

When Chaplains Stanton Salisbury and John Robinson were on duty at Quantico in 1936–1937, they found it necessary to make certain changes in the chapel to permit the maximum use of facilities. A

---

[56] CoC., Riddle File, letter of 2 Feb. 1946.

[57] CoC., Edel File, letter of H. H. Blizzard, 25 Feb. 1931. Chaplain Edel was transferred to new duty in May 1931 before work on the new building was started.

[58] CoC., Edel File, letter 9 Jan. 1946.

[59] CoC., Petzold File.

revolving platform was designed which, by turning halfway around, brought into place either a Protestant or a Catholic altar fully rigged. Thus, both chaplains were able to use the same auditorium and quickly rig for church when one service immediately followed another. This was the first of a number of revolving altars used in Navy chapels.

Chaplain T. P. Riddle was instrumental in securing the erection of a beautiful small chapel at the Navy Yard at Bremerton in 1938. The new chapel was built from the salvaged material of an old barracks. Several Navy chaplains contributed to the cost of an exquisite window, made of antique cathedral glass and depicting the crucifixion, which was placed over the altar. Officers, attached to the Yard, donated a pipe organ, and a baptismal font was given by the enlisted men stationed in the Yard and aboard the *Patterson* and *Jarvis*.

Another pioneer chapel appeared during these years at the Norfolk Navy Yard, Portsmouth, Virginia. After the new transportation building was completed in 1939, the fire-fighting equipment was transferred to it from the old fire house. One day Chaplain W. W. Elder, then on duty there, called on Admiral M. H. Simons, the Commandant, and asked if the empty fire house could be used as a chapel. Admiral Simons looking from his window noted the steeple on the building and said: "We already have a steeple on it. We can put a bell in the steeple and have a good start toward a chapel. Go ahead and see what you can do."

There were no available appropriations to make the necessary alterations, but Chaplain Elder found whole-hearted interest and cooperation from officers and men. The Navy Yard Public Works put in flooring and lighting. The Navy Yard Civilian Cooperative Association donated $1,000 to pay for an altar, pulpit, the window back of the altar, and other accessories. The Navy Yard Automobile Tag Fund gave $1,500 for pews and, later, enough additional funds to purchase a Hammond organ. One of the contracting companies in Portsmouth gave the cement for the floor of the vestibule and the entrance. Thus, another chapel became available. Dedication services were held on 27 October 1939.

THE CHURCH PENNANT

A conference was held in Washington in May 1923, sponsored by several patriotic bodies for the purpose of codifying rules for the proper display of the national emblem. This conference looked with disfavor upon the Navy custom of hoisting the church pennant over the national emblem as a signal that Divine Services were in progress. This disapproval of an old Navy custom was also found among members of Congress. On 28 January 1924, Commander J. H. Sypher, Superintendent of Naval Records and Library, wrote to the Chief of Naval Operations on the subject.[60]

The church pennant in use.

> The recent National Convention held in this city on Flag Day to draw up rules for the uniform observance of proper flag etiquette throughout the country, made a rule that no flag *or pennant* should be hoisted above the National Ensign. I am informed that this rule was aimed particularly at the Navy custom of half-masting the colors and hoisting a pennant over them whenever the crew is at church.
>
> Some years ago, this matter was taken up in Congress and the following resolution was passed by the United States Senate:
>
> "That the Secretary of the Navy be, and he is hereby, directed to inform the Senate whether or not at any time or under any circumstances any flag, emblem or banner is raised above the Stars and Stripes on any vessel, building, or ground under the jurisdiction of the Navy Department of the United States."
>
> In the discussion of this subject, the Senator from Idaho stated that his object in speaking was that the Senate might know that, "It is claimed to be the practice of the United States Navy to raise a flag above the Stars and Stripes. I do not believe that any considerable portion of the people understand that this is the practice in the Navy of the United States, but I do

[60] Nav. Rec. Coll., Church Pennant File.

speak with confidence when I say that no senator will approve of any such act. No flag, church or state, has a right to be placed above the Stars and Stripes. We do not recognize the precedence or prominence of the flag of any church above the Stars and Stripes."

Commander Sypher recommended that in view of the "dislike of this practice both in the Navy and throughout the country, as is indicated in the last Flag Convention . . . the Navy practice be modified so that it may show honor to God without at the same time degrading our National Emblem." There is, however, little evidence to indicate that there was much "dislike" of the custom within the Navy. The old tradition continued to be followed both on ships and at many shore installations during these years.

UNIFORM CHANGES

Not many changes in the chaplain's uniform occurred during these years. By 1 July 1921, the old type blue uniform with the high collar had been replaced by the present double-breasted service coat with the roll collar. The 1922 edition of *Uniform Regulations* (pp. 16, 19) authorized slightly different dimensions for the chaplain's cross worn on the sleeve. The new regulation read in part:

> In addition to the insignia of rank [sleeve stripes] . . . officers of the several corps and warrant officers shall wear 1/4 inch above the sleeve stripes, except on overcoats, the following insignia of corps, all to be embroidered and of a size to be inscribed in a circle of 1 1/4 inches in diameter.
>
> *Chaplains*—A Latin cross, embroidered in gold, the long arm 1 inch long, the short arm 9/16 inch long, and each arm 3/16 inch wide; to be set inclined toward the rear, the longer arm making an angle of 60° with the upper stripe of lace.
>
> The full-dress belt shall be worn by all commissioned officers, except chaplains and chief warrant officers. . . .

Although the shepherd's crook had been authorized for chaplains of the Jewish faith in 1918, nevertheless, this provision did not appear in the 1922 edition of *Uniform Regulations*. This omission was, no doubt, due to the fact that there were no Jewish chaplains on active duty at that time. Chaplain David Goldberg, the only chaplain to wear the shepherd's crook, resigned from the Regular Navy in 1919 and immediately joined the Reserve. While he remained in the Reserve, until 1941, he was never again on active duty. However, it may be assumed that he would have worn the shepherd's crook if he had been.

When H. Cerf Straus, of the Jewish faith, was sworn in as a Reserve chaplain in December 1931, the question of insignia was raised. Since the most recent uniform regulations (1922) specified only the cross for chaplains, it was natural that he should raise the question as to appropriate insignia. Straus was notified that action would be taken to approve suitable Jewish insignia as a substitute for the Latin cross.

Chief-of-Chaplains Evans, on 25 January 1932, wrote to Admiral T. J. Senn, Commandant of the Eleventh Naval District where Straus resided, as follows: "The present Regulations make no provision whatever for any Chaplain's uniform which does not carry the Cross on it."[61] Chaplain Evans recommended that Rabbi Straus submit an official request for permission to wear the shepherd's crook in lieu of the Latin cross.

On 1 February 1932, Straus submitted his request for permission to wear the suggested Jewish insignia. On 8 February, the Bureau of Navigation added its endorsement in sending on the request to the Secretary of the Navy. The endorsement included the following sentence which indicates that Chaplain David Goldberg had been completely forgotten: "He is the only representative of the Jewish faith in the Chaplain Corps of the Navy or Naval Reserve." The Secretary of the Navy, C. F. Adams, approved the request.

The 1922 edition of *Uniform Regulations* contained the old restriction "except the Chaplain" when referring to the full dress uniform. It should be noted, however, that dress and evening dress uniforms which had been authorized for chaplains differed from full dress only in the accouterments and the hat.

Before World War II began, the President of the United States held an annual reception at the White House for the ranking Army and Navy officers of Washington On one such occasion, Chaplain R. D. Workman, as Chief of the Chaplains' Division, attended. The fact that he could not wear the full dress hat or trousers and that his coat had no epaulets made him conspicuous. The next day, a report appeared in a Washington paper which called attention to the fact that one of the officers had carelessly forgotten to dress properly for the occasion.

On 14 April 1938, Chaplain Workman officially requested that the *Uniform Regulations* be modified so that uniforms for chaplains conform to uniforms specified for other naval officers in every respect "except that Chaplains be not required to wear

[61] CoC., Straus File.

swords." This proposal was submitted to high ranking naval officers for their reaction. The majority was opposed to this change so nothing was done.[62]

"ALL MANNER OF THINGS THEREUNTO BELONGING"

When William Balch received his commission, dated 30 October 1799, making him a chaplain in the United States Navy, he read thereon the following description of his new responsibilities: "He is therefore carefully and diligently to discharge the duty of a Chaplain *by doing and performing all manner of Things thereunto belonging.*"[63] These words, which appear also on other naval officers' commissions suggest particularly the limitless possibilities of a chaplain's collateral duties.

A good description of the expanding duties of chaplains during the period may be found in their annual reports. Here are human interest stories which reveal the length, breadth, and depth of the influence of chaplains. In the annual report submitted by Chaplain Duff for 1935 is the item: "Letters Written, Re: Men . . . 36,645." All the emotions that can stir the human heart are involved in those letters. Men worried about domestic difficulties such as sickness, debt, and marital trouble; some were maladjusted to naval service; boys were often homesick. These and many other problems were presented to the chaplain. Chaplain W. P. Williams, in his paper "Personal Contacts with My Men," which he read before the Chaplains' Conference of November 1935, declared:

> A chaplain must be sympathetic. There is an old "gag" about "sympathy chits." Yet the persistence of that time-worn joke is indicative of the interest our men have in the subject. Lack of sincerity is the Navy's unforgivable sin. A chaplain who is not sympathetic is not a sincere clergyman.[64]

Out of the chaplains' conferences with bluejackets poured a stream of letters which cleared away a multitude of difficulties and contributed to their contentment and efficiency. The wise chaplain often discerned a spiritual problem at the basis of the difficulty.

While Chaplain M. M. Witherspoon served at the Naval Training Station in San Francisco during 1920-1922, he not only gave a series of five lectures to the new recruits, he also "sent a personal letter to the mother of each new man" Out of this grew a correspondance that lasted many years.

Chaplain James S. Day, in his annual report for 1929, told of his activities as a transport chaplain on the *Chaumont* which had made three round trips that year between San Francisco, Guam, and Manila. Approximately, 4,000 men had been transported, many of whom were recruits going on their first tour of duty. Day comments in his letter:

> It has been my constant endeavor to emphasize "keeping in touch with the home folks," and it has been my custom to send a circular letter to the parents of these new men telling them something of the new life their boys are entering. The response received from these circular letters has been very gratifying.[65]

Chaplain C. A. Neyman, serving aboard the *Indianapolis,* made a study of the books bluejackets borrowed from his library for the period 15 November 1932 to 30 June 1934. Of the 1,323 books catalogued, 690 were classified as non-fiction and 633 as fiction. The total loans during the period studied were 3,012 non-fiction and 14,476 fiction. In the first classification, history was most popular with 1,057 books drawn out. Forty-three books on religion were borrowed. The most popular fiction authors were Zane Grey, Peter B. Kyne, William MacLeod Raine, and Clarence E. Mulford. Gray's book, *Arizona Ames,* was loaned sixty times in one year. The list of the fifty most popular books did not include one sea story. The favorites were those that dealt with the great open spaces of the West with accounts of Indian fighters, western pioneers, and cattle thieves.[66]

A consideration of the number of books loaned by the libraries under chaplains' care during 1935 will suggest the importance of this phase of their activity. They totaled 746,448! Only those who have spent many leisure hours at sea can appreciate the importance of this work to morale.

"All manner of things thereunto belonging" included greater solicitude for the financial welfare of the families of naval personnel. Chaplain H. S. Dyer reported, in December 1935, that during the sixteen preceding months, 6,920 government life insurance policies had been issued in his San Diego Training Station office for a total of $16,495,500. He added: "The possibility of relieving human distress simply staggers the imagination!"[67]

---

[62] *Ibid.,* Uniform File.

[63] Italics the author's.

[64] Riddle, *Religion and Welfare,* p. 6.

[65] Unless listed to the contrary, the quotations from chaplains in this section come from their annual reports for the years indicated. These are on file in CoC.

[66] From Chaplain Neyman's private notes.

[67] Riddle, *Religion and Welfare,* p. 59.

Chaplain L. N. Taylor, reporting his 1926 actvities wrote:

> From 1 January 1926 to 21 September 1926, I served as District Chaplain of the Eleventh Naval District. My chief duties in that District were those of Secretary and Treasurer of the San Diego Naval Aid Society and Chairman of the Investigating Committee of the Navy Relief Society. In these connections, I handled an average of ten cases per day at the office and was frequently called upon to visit homes and places of business in ascertaining the details of the cases of those requesting aid from these Societies.

The Navy Relief Society referred to was incorporated early in the century. It was a semi-official organization with headquarters in Washington and was officered largely by active and retired naval officers and members of their families. Funds were derived from Navy and Marine Corps personnel contributions. Arrangements were later made whereby the recipients of loans could repay them by monthly allotments from their pay.

The Society operated through local auxiliaries in which chaplains often served as executive secretaries. Emergency loans without interest to tide over periods of sickness, to provide funds when pay or allotments were delayed, and even to furnish educational scholarships were available to Navy and Marine Corps personnel and to their dependents. In exceptional cases, gratuities were granted. Extending financial aid, however, was often the least of the services rendered. Chaplains calling at the homes of families in financial distress offered advice and assistance. The Eleventh Naval District, for example, found that its Navy Relief work was expanding in the years following the First World War because so many families of enlisted men settled in San Diego and Long Beach.

Sometimes situations arose which called for financial assistance beyond that which the usual relief agencies could handle. Chaplain F. L. Albert, in his annual report of 1921, tells of sponsoring a relief fund for the destitute widow and child of a chief boatswain's mate. The liberal hearted bluejackets, out of their modest pay, responded with $2,124.79. "With this sum," wrote Albert, "a modest home was purchased, and, in other practical ways, the bereaved family was assisted in its readjustment."

Often chaplains assisted in raising Navy Relief funds. Chaplain M. M. Witherspoon, who was responsible for most of the relief work at Mare Island during 1933 and 1934, planned a pretentious carnival to procure funds. He had the approval of the Commandant, Admiral Yancey Williams, and secured such talent as Mary Pickford, Max Baer, and the original cast of "The Drunkard," then playing in San Francisco, for his project. Witherspoon, in his "Naval Autobiography" wrote:

> Publicity for the carnival reached gigantic proportions; we spent over 6 weeks on the ground work. But triumph was ours; 6000 people the first evening, 7500 the second and 8500 the third—the largest crowd Mare Island ever had. And the amount cleared—$14,500.[68]

In addition to the distinctively religious festivals as Christmas and Easter, chaplains were paying increasing attention to such special days as Mother's Day, Thanksgiving, and Memorial Day. Christmas parties for children from orphanages or underprivileged homes became events long to be remembered by the children as well as the sailors themselves.

Chaplains cooperated with clergymen and with chaplains of faiths other than their own. Chaplain M. A. Hally's comment, on 3 January 1934, exemplifies this spirit. "In connection with Jewish feast days, I have always been ready to assist Jewish men of the Service to be present at rites of their belief."

Chaplains continued to spend much time in such collateral activities as arranging smokers, dances, athletic events, and moving picture programs; conducting sightseeing parties; coaching men for the Naval Academy examinations; editing the ship's paper; serving as wardroom mess treasurer; and carrying on Navy Relief work. Chaplain W. W. Edel trained a choir of thirty-five voices on the *Wright.*

For two years, 1927–1929, M. M. Witherspoon served as chaplain for the Marines at Quantico. Describing one aspect of his work, Witherspoon in his "Autobiography" wrote:

> For the first time, I undertook educational work, serving as superintendent of the post school. Believing that the pupils should have attractive teachers, I went to the State Teachers College at Harrisonburg and hand picked six girls who had, not only high scholastic standing, but, charm and good looks. When I returned with them, the officers were astounded with their pulchritude and accused me of selecting them for their beauty, not their brains. But, time proved they had both; the school reached a new high and, eventually, the teachers all married Marine officers and are now lending prestige to the Corps.[69]

Chaplains were given special responsibility for educational work in Navy prisons. The following sections are from the *Manual for the Government of United States Naval Prisons:*

---

[68] *Op. cit.*, p. 55.
[69] *Op. cit.*, p. 34.

77. (b) The chaplain shall be the educational officer of the prison. Under the direction of the commanding officer, he shall exercise general supervision over all educational instructions. He shall recommend to the commanding officer, from among the prisoners, competent instructors to give instructions to prisoners in elementary and special subjects.

. . . .

78. The chaplain shall have charge of the library and shall see that no improper books, periodicals, etc., are placed in the possession of the prisoners. He shall also be charged with the custody of all textbooks which are issued for the use of prisoners.

An unusual type of duty came to chaplains who were assigned to the Sixth Division of the Bureau of Navigation, later called the Morale or Training Division. This duty, which was inaugurated in 1919, entailed the editing and publishing of training or rating courses for enlisted personnel. The actual writing of these publications was done by line officers and chief petty officers, experts in various fields, who were ordered to the Bureau for this purpose. J. F. B. Carruthers was the first chaplain detailed to this duty and E. L. Ackiss, who was detached from this assignment in 1927, was the last. During Ackiss' tour of duty, the four Navy manuals—*Bluejacket's Manual, Landing Force Manual, Ship and Gunnery Drills,* and *Boat Book* were revised, enlarged, and reprinted.

On the whole, Navy chaplains accepted their collateral duties in good spirit, believing that the faithful performance of them increased their effectiveness as religious leaders. Friendships that arise out of the personal contacts of chaplains with the men in activities that are not strictly religious often lead to a deeper religious experience for the individuals.

CHAPLAINS AT WORK OVERSEAS

Navy chaplains on duty with the United States Asiatic Fleet were deeply concerned over the deporable moral conditions in such places as Chefoo and Shanghai, where the Fleet was seasonally based. Writing to Chaplain E. W. Scott from Chefoo, on 15 July 1924, Chaplain T. P. Riddle, then Fleet Chaplain, described conditions:

I found out that Chefoo had been opened as wide as a Western mining camp to all the undesirables who wished to come in from any of the nearby ports or even Russia, (over fifty new "joints" had opened up—some fairly respectable and others with narcotics, gambling and prostitutes of a low order).[70]

Some remedial measures were taken by the Tao Yin (Chinese Mayor), as the result of pressure from Navy chaplains, foreign residents of Chefoo, and certain leading Chinese citizens. Much more was needed, however. In an attempt to counteract the bad conditions, chaplains instituted an active program to occupy the leisure hours of the men of the Fleet. In this, they had the full cooperation of their ranking officers and the officials of the local branch of the YMCA.

An important part of the Fleet was the destroyer unit with its 2,400 men. It was difficult to reach these men with Divine Services because they were scattered. Writing to Chaplain Scott, then Chief of Navy Chaplains, on 12 September 1924, Chaplain Riddle summarized what was done.

I was convinced from the start that church should be held ashore for all hands, provided I could get the cooperation of the Commander in Chief, Y. M. C. A., and Catholics. Everybody cooperated. . . . The "Y" gave me everything I wanted turning their godown [ie., warehouse] into a dignified church in the morning and their boxing ring into a Billy Sunday set up in the evening. The Catholic Bishop did his bit by having one of his English speaking priests celebrate at a special Naval Mass each Sunday at ten.

The plan proved successful. Attendance at the morning services, including some civilians, averaged between four and five hundred; the evening meetings averaged seven hundred. Each Protestant chaplain took his turn speaking. Y secretaries also had a part in the services.

From such gatherings came the term Fleet Church used to describe Divine Services for all Protestants of the Fleet. The term occurs increasingly throughout 1925 and 1926. Chaplain T. L. Kirkpatrick relieved Riddle as Fleet Chaplain in September 1925 and carried on the work. The service held for all Catholics was referred to as Fleet Mass.

Kirkpatrick, in a letter to Chaplain Dickins, written from Chefoo on 7 July 1926, declared that

Chefoo is a moral cesspool in spite of all. Officers are well off here, with plenty of tennis courts, a good club with bowling alleys and all conveniences, and plenty of good beach available. Enlisted men have only the Y.M.C.A. as a wholesome recreation center. Five tennis courts are available, and an outdoor basketball court.[71]

The enlisted men also had two baseball diamonds. No beach was available because of the lack of public bath houses. Chaplain Kirkpatrick complained that so little entertainment was available for liberty parties which sometimes numbered almost 2,000 men.

[70] CoC., Riddle File.

[71] *Ibid.*, Kirkpatrick File.

"So you see," wrote Kirkpatrick, "it's a great fight we have here."

When Chaplain W. A. Maguire, a Catholic, was assigned in August 1929 to the *Black Hawk,* mother ship of the destroyers based at Chefoo, he had to take over the responsibilities of the Fleet Church for Protestants. Maguire, in his book, *Rig For Church,* tells how a committee of Protestant missionaries met with him before the Fleet went north in the summer of 1930 and besought him to carry on the fine work that had been started in the Fleet Church. Regarding this experience, Maguire wrote:

> Chefoo, that summer and the summer following, experienced the unusual. Protestants of various denominations listened week in and week out to Catholic sermons on the Gospel of the day. They listened to the same words that were spoken each Sunday from the altar of St. Mary's.
>
> The experience was a trying one but I believe, for me, it was highly beneficial. Never have I worked so hard on my sermons. It was encouraging to have the congregation come to me after the services and say, 'I so enjoyed your message.'[72]

Letters from chaplains stationed in Asiatic waters show their continued concern over the moral conditions in the Chinese port cities. Again and again, because of some chaplain's intercession, remedial measures were taken to reduce the venereal disease rate or to provide wholesome recreational activities to offset the downward drag of shore conditions.

In the early part of 1927, the Fourth and Sixth Regiments of United States Marines, with Chaplains H. M. Peterson and W. W. Edel, respectively attached, were sent to Shanghai to share in the defense of American lives and property threatened by the unrest then developing in China. Chaplain Peterson remained with the Fourth Regiment until October when he was sent to Guam. Chaplain Edel accompanied his Marines when they were transferred to Tientsin in the summer of 1927. In May Chaplains Kirkpatrick, Edel, and Peterson, with Mr. Arthur Clark of the Navy Y in Shanghai, met in the Y and organized "the Asiatic Fleet Church" with three components, the USS *Pittsburg* Church, the Sixth Regiment Church, and the Fourth Regiment Church.

Chaplain W. R. Hall reported for duty with the Marines in Shanghai in May 1928 and soon thereafter revived the Fourth Marines Church. There has been nothing comparable to this unique church in the history of the Chaplain Corps of the Navy. Prior to the time Hall reported for duty, Divine Services for the Fourth Marines were held in mess halls and barracks. Men were scattered in a dozen billets, some more than a mile from headquarters, a situation which made it extremely difficult to assemble them for Divine Services. After trying the experiment of holding several services every Sunday morning, Hall secured the use of the Embassy Theater on Bubbling Well Road. Colonel F. D. Kilgore, Commanding Officer of the Fourth Marines, arranged for the Marine Band to give a concert preceding Divine Service. Civilians were invited to join with service personnel. The first service was held on 9 September 1928 and attracted an audience of more than 600. Many tourists, consular officials, and Chinese attended the services. Noted clergymen, passing through Shanghai, were often invited to speak. The happy combination of location, type of service, and leadership soon make the Fourth Marines Church an outstanding unit in the religious life of the great oriental city.[73]

In February 1930, Chaplain R. W. Truitt relieved Chaplain Hall and was installed with appropriate ceremonies as pastor of the Church. Shortly after taking over, Chaplain Truitt found it necessary to change its location to the Italian Gardens of the Majestic Hotel. Since this was four blocks closer to the Marines' billets, the move proved advantageous. In October, the Church moved again, this time to the Carlton Theater where services were held during two winters. In the summer of 1931, however, the Church returned to the Italian Gardens which afforded more comfortable out-of-door arrangements.

Chaplain Truitt was relieved by Chaplain J. H. Brooks in April 1932. In June, the services were transferred to the Cathay Theater, which had recently installed air conditioning, thus making it an ideal meeting place, even during the hot summer months. The following extracts from Chaplain Brooks' book *Distinguished Service* give further details about the activities of this church:

> On July 3rd, the first service in the Cathay Theatre was held. Both the church program and the band concert were commemorative of the anniversary of American Independence.
>
> This service was remarkable also because it marked the entry of the Fourth Marines Church into another field of usefulness. Mr. Jimmy James, owner of Radio Station RUOK, offered to broadcast the church service and band

[72] *Op. cit.,* p. 205. Chaplain Maguire in his book relates many experiences as a naval chaplain from his entry into the Corps in 1917 to Pearl Harbor.

[73] An account of this unique church down to 1933 is found in the first chapter of the book *Distinguished Service* by Chaplain Joseph H. Brooks.

concert from his Station, paying the expenses of the broadcast himself. This kind offer was accepted, and on July 3rd, radio listeners were surprised to hear a new Sunday morning greeting: "You are listening to Radio Station RUOK broadcasting by remote control, the program of the Fourth Marines Church in the Cathay Theatre, Shanghai". The broadcast has continued every Sunday since that time except for occasional interruptions due to technical difficulties.

. . . .

The services held in the Fourth Marines Church have been reported in all of Shanghai's morning English language newspapers. In two of the papers, the sermons have been printed in full. During the year ending July 1st, 1933, a total of 3,350 column inches of space was given to the activities of the Church.

. . . .

On Christmas Day, 1932, the Fourth Marines Church remembered the children in the Russian Refugee School and in the China Christian Day Nursery. More than one thousand presents were brought to the church by Marines and their friends. Two Christmas trees stood in the lobby of the Cathay Theatre, aglow with lights and resplendent with toys, dolls and vari-colored decorations. Around the trees were placed the presents as they came in. After the church service two trucks, provided by Mr. Honigsberg, were loaded with trees and presents and sent to the schools.[74]

Writing to Chaplain Workman on 28 July 1933, Brooks stated:

Our attendance is holding up very well. It averages 400 to 600 a Sunday and on special occasions we have more. On Mother's Day, the attendance was over 1200. . . . I think the Fourth Marines Church fills a real need in Shanghai. It is attended regularly by many English speaking Chinese teachers, students, business men, and by a large number of Americans and British. While we have no membership roll, we have two or three hundred civilians who attend as regularly as if their names were on a Church roster.[75]

During Chaplain M. M. Witherspoon's tour of duty in Shanghai, December 1934 to October 1936, the attendance at the Regimental Church continued to grow. In a letter to Chaplain E. A. Duff, then Chief of Chaplains, Bishop Herbert Welch of the Methodist Church writing from Shanghai on 17 December 1935 commented on the effective service Chaplain Witherspoon was rendering:

For some years, a church service has been conducted for the Marines which has been attended by a certain number of people from the community. But, under Chaplain Witherspoon's vigorous efforts, this service has been built up until it has become a notable contribution to our American life and to the community at large. It is held at 10 o'clock on Sunday mornings in the Grand Theatre, which, I believe, seats about two thousand people, and the theatre is actually full. . . . Numerous Chinese, especially of the younger generation, are to be found there. While the music of the band and soloists is, undoubtedly, an attraction, there is always a stimulating message given by the Chaplain himself.

Writing to Chaplain Dickins from Tientsin on 18 July 1927, Chaplain Edel enthusiastically commented on the success of the Regimental Church plan. Edel wrote:

It proved to be exactly the thing that was needed in my outfit at least, and the Regimental Church of the Sixth Regiment at once became the focus for all the interests of the Chaplain and of men who were at all inclined to live the Christian life. It formed a definite organization to which men could give their loyalty and for which men were willing to work, and in two weeks there were nearly sixty men in the Sixth Regiment who had signed up for the Regimental Church.

The unique "Membership Covenant" which provided dual membership was signed by all who wished to be considered active members in the Regimental Church. It follows:

I believe in Jesus Christ; and pledge my loyalty to Him as He is made known to me thru prayerful study of the New Testament, and to the manner of life He taught men as it is found in the same pages. It is understood that this membership is not intended to take the place of membership in any other church organization, or to invalidate such membership, but, is to supplement such membership while I am in the military service.[76]

On 9 September 1927, a memorandum was issued by General Smedley Butler, from Third Brigade Headquarters in Tientsin, which further defined the organization of the church. It read:

1. The Third Brigade Church is hereby organized, effective this date.

2. The Third Brigade Church will for the present be affiliated with the Asiatic Fleet Church, and will elect representatives to the Asiatic Fleet Church Council.

3. The Third Brigade Church will consist of the following units:

(a) Brigade Special Troops Church
(b) Fourth Regiment Church
(c) Sixth Regiment Church
(d) Tenth Regiment Church
(e) Provisional Regiment Church.

4. The Chaplains in charge of the Third Brigade Church will be as follows:

Commander William W. Edel, ChC., U. S. Navy
Lieutenant-Commander H. M. Peterson, ChC., U. S. Navy

5. The Third Brigade Church is Inter-Denominational and all officers and men who are Church Members at

[74] *Op. cit.*, pp. 3–5.
[75] CoC., Brooks File.
[76] CoC., Edel File.

home are urged to become members of the Third Brigade Church. Such membership does not invalidate membership in any other church and no letter of transfer is necessary from another church. On transfer to the United States, if not a member of another church, a transfer of church membership will be issued on request.[77]

Experience proved that the idea was practical and helpful. On 19 September, Edel noted that the Brigade Church, exclusive of the Fourth Regiment in Shanghai, had 492 members and was still growing. "The Sixth Regiment Church has 25% of the entire complement in it," wrote Edel, "and the Tenth Regiment Church has nearly 40%. . . ." Chaplain Edel organized an auxiliary group of the strongest Christians and called it the "Century of Cornelius." Membership in September numbered fifty-six. By the use of initiations, rituals, and insignia, the men were bound together in a close Christian fellowship. Chaplain Edel reported: "It is a tower of strength to the Chaplain, and its members can be called upon at any time to do any service that may be needed." By 11 November, the membership of the Brigade Church passed the 1,000 mark. It was then holding its meetings in the Empire Theater in Tientsin. Chaplain Edel's fine work with the Marines at Tientsin was terminated in December 1927 when he received orders to return to the States.

Navy chaplains on duty in the Virgin Islands were included on the staff of the Governor as Aid for Public Welfare. They administered the "Poor List" and also served as Assistant Federal Prohibition Director and Prohibition Commissioner for the Virgin Islands. When Chaplain E. L. Ackiss was stationed there, June 1923 to June 1925, he organized and became Director of the Department of Agriculture, Commerce and Labor.

Chaplain Herbert Dumstrey reported for duty in Samoa in the spring of 1919. During the previous three years, there had been no chaplain on duty and a missionary from the Church of England had conducted Divine Services once a month for naval personnel. Chaplain Dumstrey stepped into a difficult moral situation at Samoa which was characterized by excessive drinking and immorality. His written protests led to an official investigation in which he was able to substantiate his charges. As a result the Governor took a firm stand and the abuses were corrected. This in turn made Dumstrey unpopular with a certain element of the naval personnel. These individuals, however, eventually received orders to other duty. Dumstrey kept Chaplain Frazier informed of the progress of the investigation and his Chief wrote back words of encouragement. "I am a firm believer in the final triumph of the Golden Rule," wrote Frazier on 23 March 1920. "Now that you have the Commanding Officer back of you, I feel that you will find it much pleasanter, and that his position will influence the attitude of others.[78]

Dumstrey was the first chaplain to be appointed Superintendent of Education of the Island Government. His appointment in February 1921 marked the beginning of the notable services chaplains have rendered the natives of Samoa in the field of education. Chaplain Edel was engaged in similar activity at Pago Pago and on 5 August 1924 reported that he had twenty schools to inspect each quarter. Great difficulty was being experienced in securing a sufficient number of trained teachers. In December, Edel reported that a Teachers' Institute was being conducted, and that the School budget for 1925 was set for $17,918. In August 1925, Edel stated that the First Annual School Field and Track Meet for American Samoa was scheduled for Labor Day. Poyer High School then had an enrollment of 248, an increase of 100 percent during Edel's administration.

Writing on 25 March 1927, Chaplain James M. Hester, then stationed at Samoa, reported that the work for the Department of Education required practically all of his time. He also stated that ninety percent of his evening congregations was composed by Samoans and half-castes. The work continued to grow in importance with the passing of the years. Chaplain J. W. Moore added his testimony to that of his predecessors when he wrote to Chaplain Evans on 28 March 1933.

The position of the Chaplain here is one of unusual power and prestige. He enjoys said prestige among the Samoans, primarily because he is a Christian Minister. For over one hundred years, Christian Missions have been at work in Samoa and now seem to be more influential than the Government itself. The Chaplain, moreover, is clothed by the Government with great dignity as the head of the Department of Education. This dignity, incidentally has been enhanced recently in my being appointed "President of the Board of Education", an office until March 6, 1933, held by the Chief Justice of the Government of American Samoa. The duties in Education are no sinecure; and, at present are of more importance than ever. . . .

The educational activities of chaplains stationed on Guam was analogous to those on Samoa. The

[77] *Ibid.*

[78] CoC., Dumstrey File. Unless otherwise noted, additional information in this section has been taken from the file of the chaplain concerned.

problems faced by chaplains on these islands were quite similar. There was a lack of trained teachers and an ever expanding program.

Chaplain Stanton W. Salisbury, in his annual report submitted to the Governor of Guam on 14 July 1925, reported that there were then twenty-three schools under the Department of Education; that a staff of ninety-eight, including seventy-nine native teachers, was employed and that the "curriculum is that of the State of California and the high standard of that State was rigidly enforced through frequent supervision and constant checking." "Every child in the schools of Guam," wrote Salisbury, "receives industrial training." An evening High School was begun in January 1925 for the special benefit of native teachers who wished to continue their education. The expenditures for the year amounted to $36,-242.41. A total of 2,720 students were registered in the schools with an average daily attendance of 89.8 percent.

During the years 1920–1931, Governor Willis W. Bradley, Jr., inaugurated a school-naming program which resulted in designating the public school at Sinajana the Chaplain Stanton W. Salisbury School in honor of the chaplain who headed the Department of Education with such distinction during the years 1924–1926. Another chaplain so honored was Harry M. Peterson. The first industrial school on the Island was named after him in appreciation of his leadership in practical education during his duty in Guam in 1927–1929. When Chaplain Francis L. Albert completed his tour of duty in December 1932, he was also honored by the appreciative natives of Guam. On request of the Guam Teachers' Association and with the unanimous approval of the joint Houses of the Guam Congress, Governor Edmund S. Root published an order naming the public park adjacent to the Seaton Schroeder High School in Agana the Chaplain Albert Plaza.

Chaplain Albert laid great emphasis upon vocational education during his service as Head of the Department of Education in Guam. The following extracts from his report for 1930–1931 tell their own story:

> There is a growing conviction in Guam that, while academic education should not be neglected, there should be increasing emphasis placed upon practical vocational guidance. To carry out this program, an industrial group has been built on the edge of the Government House grounds facing San Ramon Street. In addition to the School Carpentry Shop, there is now a new Weaving School, of frame construction, completed at a cost of $730.16, and opened to the weaving classes on October 27, 1930. A companion building, to house the Sewing School, was completed on November 17, 1930, at a cost of $658.07. Three new "Singer" machines, costing $225.00, have been added to the equipment of the Sewing School. . . . Plans are underway for the erection of a new Cooking School, to be located just beyond the Sewing School, and to be of corresponding design. It is intended that these four schools shall take care of the industrial training of boys and girls from all the Agana schools, from the fourth to the ninth grades inclusive.
>
> . . . .
>
> The United States Department of Agriculture gave the Department of Education, through the Island Government, enough blooded live-stock to start an Agricultural School Farm in Guam. The United States Navy Department furnished transportation from the States to Guam via the U.S.S. NITRO, arriving on 22 February. The animals, consisting of two pure-bred Ayrshire bulls and one pure-bred, dual-purpose shorthorn bull, one registered Duroc Jersey boar and two Duroc Jersey sows; ten Rhode Island Red hens and two cocks, nine White Leghorn hens and two cocks; came through in excellent condition.

Naturally, chaplains during their tours of duty in Samoa and Guam, carried on the unique activities referred to in this section in addition to the customary duties associated with the chaplain's work.

During the years in between the two great World Wars, chaplains aboard ships often conducted sight-seeing parties for members of the ship's company while visiting foreign ports. These were frequently of great educational value to the men as they visited the historical shrines, temples, churches, muesums, and art galleries in such great cities as Rome, Paris, Cairo, Peking, Shanghai, and London. Occasionally, pilgrimages were even arranged to the Holy Land.

In the spring of 1929, Chaplain W. N. Thomas sponsored a series of such tours in several European cities. He took a party of fifty-four from the *Raleigh* to see Paris. Accommodations for the entire party were secured in Lafayette Hotel. One evening, shortly after the arrival of the party, the manager of the hotel approached the Chaplain and in his characteristic excitable manner said that he had no maids to clean the rooms and make the beds and carry on the other duties of the hotel.

"Well," replied Chaplain Thomas, "where are your maids? You had them here this morning."

"Ah, M'sieu," the manager explained, "but ze sailors 'ave taken all ze maid to ze théatre!"

"In that case," said the Chaplain, "let the men clean their own rooms!"

Navy Sightseeing Party, Cairo, Egypt, about 1924.

WELFARE AGENCIES BETWEEN WARS

From their surpluses at the close of the war, many welfare agencies granted ex-servicemen educational scholarships. The YMCA expended $5,250,688 in this way, reaching 106,946 individuals, and the K of C distributed $1,250,000 for the same purpose.[79]

During these years, between the two great World Wars, every Catholic chaplain who entered the service received from the Chaplain's Aid Association of the Catholic Church a Mass kit, vestments, altar linens, altar wines, and other ecclesiastical items. All Catholic chaplains in service were also supplied with rosaries, religious literature, and other religious items needed in their work.

The Luthern Service Center in Bremerton, Washinton, was one of the few service men's centers to carry on in the years between the wars. There the Reverend and Mrs. Theodore Hokenstadt endeared themselves to thousands of bluejackets. Motherly Mrs. Hokenstadt has a phenomenal memory for voices and often recognized bluejackets by this means long after the face had been forgotten.

[79] NRSO: 28234-25. Letter dated 21 Mar. 1923, from George R. Landis of the YMCA to Secretary of the Navy Edwin Denby; Egan and Kennedy, *The Knights of Columbus*, p. 400.

"Do you remember me?" a sailor might ask after being away for five or even ten years.

"Talk," Mrs. Hokenstadt would say as she turned her gaze from his face. "I must hear your voice." And then as he talked, her face would light up with recognition. "I remember. You are . . .," and she would mention his name.

The years 1920–1929 were years of rapid expansion for the Y both in the United States and in its overseas activities. Likewise, Y work for service personnel, in particular naval personnel, broadened tremendously in scope.

The following is a list of the larger building acquired in the United States during the period under review with the approximate cost and year of dedication: $200,000 annex, Norolk, 1923; $275,000 main building, Bremerton, Washington, 1924; $800,000 building, San Diego, 1924; $900,000 building on the Embarcadero in San Francisco, 1926; $750,000 center at San Pedro, 1926; $850,000 building, Philadelphia, 1929; $35,000 annex, Vallejo, 1931; $260,000 building, Seattle, 1931; $12,500 annex for San Pedro, 1933; and a $70,000 building at Long Beach, 1936.[80]

[80] See Pond, MS. "History of Army and Navy YM A." The building in Seattle was a general building with a section set apart for Navy use.

Two new buildings were opened in the Canal Zone, one at Cristobal in December 1923 and the other at Balboa a year later. A commodious new building was erected in Manila at a cost of $325,000 which was dedicated 19 June 1926. An $830,000 center was dedicated in Honolulu, 16 March 1928.

The work in China was especially noteworthy because the Y there filled such a desperate need. Moral conditions in the large port cities in the Orient visited by the Fleet were such as to give grave concern, not only to chaplains, but, to line officers as well. Naval personnel from time to time had shore liberty in strange cities where temptations abounded. The Y work which had been started in Shanghai in 1912 was greatly expanded to keep pace with the increased number of United States bluejackets on duty in Far Eastern waters.

A fine new Y building costing more than $330,000 was opened on North Szechuan Road in Shanghai in 1923. This had the usual provisions of a dormitory, cafe, gymnasium, swimming pool, reading and writing rooms, canteen, and other facilities. An extension of the Shanghai work was conducted at Chefoo where the Fleet spent much of the year. Here property worth about $100,000 which included "three tennis courts, a basketball court, and a boxing pavilion in addition to nine substantial buildings of various sizes had been acquired by 1934.[81] The Y, "farthest from home," was located at Hankow far in the interior of China to serve the men on the Yangtze patrol. Here a building was first rented in 1924 for Y activities. Six years later property was purchased. The Y resumed work at Peking in 1922 for the Marines assigned to the Legation Guard. Other work in China was located at Tientsin and Hong Kong.

[81] *Ibid,* p. 140.

Navy Sightseeing Party, Soochow, China, June 1922.

Navy chaplains conference at Rancho Santa Fe, near Long Beach California, 25–26 November 1935.

Thirty-two chaplains of the Battle Fleet and West Coast attended. The total number of USN chaplains in 1935 was only seventy-nine.

Left to right (men only): 1st row; Wilfred Hall, L. P. Johnson, M. J. Hallyn, G. S. Rentz, V. J. Gorski, C. H. Mansfield, E. W. Davis, T. P. Riddle, R. R. Marken, F. W. Meehling, W. W. Edel, D. L. Quinn, A. L. Park, W. P. Williams, H. Rountree.

2nd row; Richard Ziegler, Ralston Holmes, T. A. Ryner, Bishop Cantwell, Adm. Harris Laning, E. A. Duff, Bowen McCoy, Monsignor B. J. Dolan, R. D. Workman

3rd row; H. G. Glunt, J. P. Forsander, J. F. Hugues, E. J. Robbins, W. F. Cuthriell, Harry Peterson, M. H. Petzold, W. W. Peirce, T. C. Miller, F. L. Albert, E. L. Ackiss, A. deG. Volger, H. S. Dyer, W. H. Rafferty.

Top row; James Gainfort, C. A. Dittmar.

(See page 216)

The Y not only provided a clean wholesome place for bluejackets to spend their leisure hours, but was also active in promoting recreational and social events. Many sailor boys astride diminutive donkeys participated in Y sightseeing tours to old temples and pagodas.

A typical Sunday at the Shanghai Y started with a "Java Club" meeting around the breakfast table with an open forum on some religious subject. Church parties were sponsored for the usual Sunday morning services held in Community Church (American) or in one of the English churches. In the evening, the Y gave a sukiyaki dinner in the gymnasim for sailors and for families from Community Church. Here, perhaps for the only time in months, American sailor boys had the opportunity of fellowship with civilians of their own country and especially with American girls. Following the dinner, the low tables and mats were removed, folding chairs unstacked and put into position, and the gymnasium converted into a church. The evening service was usually informal, with the emphasis on singing.

Thomas A. Rymer, Senior Secretary of the Army and Navy Department of the National Council, presented a summary of the activities of the Y before a meeting of chaplains near Long Beach in November 1935. He stated:

> Today there are 34 Army and Navy Associations, 24 of them in continental United States, 2 in Canal Zone, 2 in Hawaii, 2 in the Philippine Islands and 4 in China. They have property and endowments amounting to approximately $10,000,000. These Associations spend for the men of the Service in excess of a million dollars a year.[82]

Thousands upon thousands of American boys of all creeds at home and abroad have benefitted by the facilities provided and services rendered by this institution. In all this, the Y has had the cooperation and assistance of Navy chaplains.

In the absence of Jewish chaplains on active duty, the Jewish Welfare Board continued to provide for the religious needs of men of that faith to the best of its ability. The Bureau of Navigation issued annual directives after 1921 which made it easier for Jewish personnel to obtain special leave to attend the Passover rites. The following was issued on 16 February 1921 to all ships and stations:

> Subject: Jewish Holiday.
>
> 1. Commanding Officers are authorized to grant leave at discretion to men of Jewish Faith from noon Friday, 22 April, until midnight, 24 April, 1921, with such additional travel time as may be practicable.
>
> 2. The Jewish Welfare Board will furnish special prayer books for the Seder services and unleavened bread to men of the Jewish Faith for the eight days of Passover; and Commanding Officers are directed to arrange for distribution to men of Jewish Faith under their command.

A similar directive appeared each year up to and including 1938. Beginning with 1925 and continuing through 1938, another series of directives granted similar privileges for the Jewish High Holy Days. The following directive which appeared on 16 July 1925 is typical:

> Subject: Jewish High Holy Days.
>
> In order to permit observance of the High Holy Days, including the New Year (Rosh Hashana) and the Day of Atonement (Yom Kippur), Commanding Officers will grant leave, when the exigencies of the Service permit, to men of the Jewish Faith from noon of September 17th until noon of September 22nd, and from noon of September 26th until noon of September 30th, 1925.

Thus, consideration was given to a minority religious group to observe their religious festivals in the traditional manner.

For years, the American Seamen's Friend Society had presented Bibles to the members of the graduating class at the Naval Academy. In 1933, this project was taken over by the Maryland Bible Society, a branch of the American Bible Society, and has been continued without interruption. In appropriate services at each graduation, the Maryland Society has presented to each graduate a beautiful bound copy of the New Testament and Psalms with the Naval Academy seal stamped in gold on the cover.

RELIGIOUS ACTIVITIES AT THE NAVAL ACADEMY

The four Chaplains who saw duty at the Naval Academy during these years under review were:

- Bower R. Patrick, 26 Dec. 1919–5 July 1921.
- Sydney K. Evans, 5 July 1921–10 June 1929.
- Frank H. Lash, 1 June 1929–3 June 1933.
- William N. Thomas, 3 June 1933–.

In addition, there were assistant chaplains, each of whom usually stayed approximately two years.

Various customs and traditions pertaining to religious life of midshipmen were carried on and added to by the chaplains on duty at the Academy during these years. The Naval Academy Christian Association, an independent organization which grew out of the YMCA of earlier years, continued to meet each Sunday evening. In his annual report for 1925, Chaplain Evans wrote: "This is a voluntary meeting conducted by the midshipmen themselves, the attendance varying from 300 to 1,000, and on special occasions

[82] Riddle, *Religion and Welfare*, p. 55.

Present Chapel at Naval Academy, 1940–

it is higher." Morning prayers continued to be offered in the mess hall after breakfast each week-day morning for the whole Regiment. A Sunday School was conducted for the children of the officers and civilian instructors on duty at the Academy.

The Naval Academy Chapel contains many interesting treasures and memorials. A Communion Rail was placed in the Chapel and dedicated to the memory of Chaplain Henry Howard Clark on Sunday, 22 February 1925. Among the memorials is the $7,500 window, "The Commission Invisible," presented by the class of 1927. This stained glass picture, the idea of which was suggested by Chaplain Evans, portrays a graduate in the white ensign's uniform holding his commission. The young man appears just to have finished reading his commission and has become consicious of another and co-extensive call. The figure of Christ appears in the sky at the right pointing with one hand in benediction towards the American flag and with the other towards the young officer. Thus, the Annapolis graduate standing at the beginning of his professional career is reminded of his double duty, in no wise contradictory, to God and country. This striking window, so original in design, has attracted much attention.

In 1934, the Regiment had so increased that two Sunday morning services became necessary in order to accommodate all who wished to attend. One service was held at 0900 and the other at 1030. Chaplain Thomas, then on duty, made the first service more informal than the second and gave a different sermon. Thomas further enriched the services by introducing prayers written especially for the midshipmen and for the Navy. He wrote "The Midshipman's Prayer" in 1938. All plebes at the Academy are urged to memorize this prayer, which is repeated in unison every Sunday morning at Divine Service. It has been printed on the last page of the 1942 edition of *Song and Service Book for Ship and Field*. This prayer, widely known throughout the Navy, reads:

> Almighty Father, whose way is in the sea and whose paths are in the great waters, whose command is over all and whose love never faileth: Let me be aware of Thy presence and obedient to Thy will. Keep me true to my best self, guarding me against dishonesty in purpose and in deed, and helping me so to live that I can stand un-

ashamed and unafraid before my shipmates, my loved ones, and Thee. Protect those in whose love I live. Give me the will to do the work of a man and to accept my share of responsibilities with a strong heart and a cheerful mind. Make me considerate of those intrusted to my leadership and faithful to the duties my country has intrusted to me. Let my uniform remind me daily of the traditions of the Service of which I am a part. If I am inclined to doubt, steady my faith; if I am tempted, make me strong to resist; if I should miss the mark, give me courage to try again. Guide me with the light of truth and keep before me the life of Him by whose example and help I trust to obtain the answer to my prayer, Jesus Christ our Lord. Amen.

The October 1945 issue of *Naval Institute Proceedings* contains an interesting article by Sarah Corbin Robert on "The Naval Academy Chapel-Cathedral of the Navy" in which she makes the following comment on the form of service used at the Academy:

> It is true that the Prayer Book of the Protestant Episcopal Church is used. This practice came about in this way. On ships of the Navy which have no chaplain, the captain, or other officer, conducts religious worship. The psalms, lessons, and prayers for all occasions contained within this one volume made it easy for a captain unaccustomed to such duty to conduct a dignified service. The Chapel uses this Prayer Book, but it has gradually worked out its own order of worship for a service little longer than an hour. The special prayers are "For the Navy," "For All in the Armed Forces," including prisoners of war, "For the Naval Academy," and "The Prayer of a Midshipman."
>
> Captains of ships who have known the Chapel service write back for the order of worship, and chaplains unfamiliar with this service often write, when the Captain asks that the prayers which became meaningful to him at the Naval Academy be used aboard his ship. As in the past when the churches looked to the Cathedral for the liturgy, so the Navy Academy Chapel is setting the pattern for religious services throughout the Navy.

Increasingly, the chaplain at the Academy was called upon to perform marriages for the graduates, to baptise their children, and to officiate at funerals. The following records, taken from the Chapel log, during Chaplain Thomas' first six and a half years of duty there, is indicative of the ever-expanding importance of the chaplain's ministry at the Academy:

	Baptisms	*Weddings*	*Funerals*
1933	9	8	1
1934	37	17	14
1935	36	23	16
1936	44	26	27
1937	63	23	18
1938	53	9	11
1939	67	22	16
	309	128	103

832159—49——17

In January 1939, alterations to lengthen the nave were initiated which were to cost $400,000 or approximately the original cost of the Chapel. It was possible to hold services in the main part of the Chapel during most of the reconstruction. When completed, the alterations permitted a seating capacity of 2,500. Upon Chaplain Thomas' suggestion, the naval guns which formerly decorated the main approach to the Chapel were removed and two large old-fashioned anchors, made for the Navy's first armored cruiser *New York,* were substituted as being more symbolic of the Christian tradition. The enlarged Chapel was rededicated on 28 April 1940.

OF GENERAL INTEREST

On 5 October 1920, the Federal Council of the Churches of Christ in America, through its Committee on Army and Navy Chaplains, presented to Secretary Daniels and Chaplain Frazier handsome war medals for each Navy Protestant chaplain who served in World War I. The members of the presentation committee "expressed their high appreciation of the devoted services rendered by those who in the time that tried men's souls ministered to the moral and spiritual needs of those who faced danger and death for their country on the seas." The President of the United States and the Secretaries of War and of the Navy were also recipients of one of these medals "given in recognition of their deep interest in the more important work of the chaplains."[83]

Following World War I, several of the larger denominations continued their interest in the work of their representatives in the Navy chaplaincy by granting small annual subsidies of $25 to $100 to be used by the chaplain in his work. However, by 1930, practically all such subsidies had been discontinued.

In June 1921, an effort was made to guarantee the publication of the history of the Chaplain Corps which had been prepared by Chaplain Hoes. As early as 1908, Chaplain Hoes was at work on his manuscript. Writing to Chaplain Helms on 10 October of that year he said:

> I am now engaged upon its final revision, preparatory to sending my manuscript to the publisher, by whom it will be issued in sumptuous form—probably in two volumes, with portrait pictures of all the Chaplains excepting a very few of the earlier ones whose pictures either

[83] *Report of Secretary of Navy,* 1920, p. 122. The members of the presentation committee included Bishop W. F. McDowell, Mr. W. K. Cooper, and the following ministers: E. O. Watson, A. R. Bird, and C. F. Steck.

do not exist or I have not been able to discover. I have succeeded, however, in obtaining most of the early ones.[84]

To Chaplain Hoes goes the credit of collecting and preserving invaluable data relating to the history of the Corps. From his collection of photographs and copies of paintings of the earlier chaplains came most of the pictures which are today in the office of the Chief of Chaplains. For some reason, Chaplain Hoes was never able to get his history published. Perhaps, it was because he never completed writing his notes in presentable form for a publisher. In May 1921, when Hoes was in his seventy-second year, he met with an accident which sent him to the Naval Hospital in Washington.

Chaplain Frazier, appreciating the value of Hoes's researches, took the lead in persuading him to turn over all his notes and materials to Chaplain Edel. On 9 June 1921, Chaplain Frazier sent out letters to other chaplains requesting five dollar contributions to subsidize the publication of a book. Chaplain Frazier explained:

> Chaplain Hoes has been collecting for a great many years, data and photographs concerning all men who have served as Chaplains in the Navy since we first had a Navy. This data is extremely interesting and instructive and should be put in permanent shape. Chaplain Hoes has refused, up to the last week, to let this matter get out of his own hands. He is an old man, however, and has just had an accident that will make him an invalid for the rest of his life. At the earnest solicitation of Chaplains Tribou, Wright, and myself he has agreed to turn this matter over to us on condition that we prepare it for publication in book form. Chaplain Edel, who had just written a long thesis for his degree at Boston University . . . has agreed to undertake with us the writing of this history.[85]

The financial response to the appeal was disappointing. Edel's interest in the history of the Corps antedates the receipt of the material from Hoes. Edel had completed a 175 typewritten page thesis on the history of the Corps in 1921 from notes gathered by Chaplain Tribou and from assistance rendered by other veteran chaplains as Frazier, Clark, Gill, and Dickins. Edel found that Hoes had so much additional material it became necessary for him to rewrite his manuscript. In 1942, Edel prepared the twenty-seven page mimeographed pamphlet entitled "Navy Chaplains from 1775 to 1917" which has been widely circulated among Navy chaplains in World War II as the best available brief history of the Corps. Chaplain Edel has also published a number of articles on various aspects of the history of the Corps in several periodicals and has used interesting biographical material about Navy chaplains on his unique Christmas greetings.

The *Manual* which Chaplain Frazier issued during the First World War was not inclusive enough to meet all conditions faced in the thirties. The widespread need for some adequate guide was met by a manual issued by Chaplain J. J. Brady when he was serving as Battle Force Chaplain of the Pacific Fleet. During his tour of duty, which began in March 1930, he compiled a Manual for the benefit of the chaplains under his jurisdiction. This loose-leaf compilation consisted of printed and mimeographed material. Included were an outline for a burial service at sea, suggestions for conducting the annual Christmas parties for children, and ideas for conducting sightseeing excursions. This Manual served a very useful purpose.

The second Christian Scientist to serve as chaplain in the Navy was Joel H. Benson who was appointed acting chaplain on 10 March 1920. Chaplain Richard J. Davis, the first to serve from this denomination, resigned on 31 January 1920. Since the Christian Science Church had secured recognition by having one of its number included in the Corps, the effort was made to have a successor to Davis appointed as soon as possible. Chaplain Frazier, knowing the difficulties he had in placing Chaplain Davis, objected to the appointment of others from this church. These objections were overruled, however. Benson had enrolled in the Naval Reserve Force, Class 4, on 13 December 1918 in the grade of assistant paymaster with the rank of ensign. After he resigned this commission in the Naval Reserve, he was appointed chaplain in the Regular Navy.

Chaplain Benson was sent first to the Naval Training Station at Newport and then attached to the *Bridgeport*. His presence as chaplain aboard that ship soon raised difficult problems both for his Commanding Officer and for Chaplain Frazier. Chaplain Benson was a layman who, although an acceptable religious leader in his own faith, had never received the seminary training other chaplains had taken. While competent as officer-in-charge of recreation, athletics, and moving pictures, he was unable to serve acceptably as the leader of the religious life of the ship. According to a letter written by the Commanding Officer dated 9 May 1922, Chaplain Benson

[84] Edel Coll., Hoes File.

[85] CoC., Tribou File.

was the only member of the Christian Science Church aboard.

Chaplain Benson was transferred to the *Denebola* on 13 June 1922 where he remained until 16 May 1924. For several months this ship was in Near Eastern waters where Chaplain Benson, provided with funds from his church, carried on extensive relief work with refugees in Constantinople. For these outstanding services Chaplain Benson received the Croix de Chevalier de l'Ordre de Sauveur from the Greek Government. He returned to the States in January 1924 and was ordered in May to the Navy Yard Charleston, South Carolina, where he remained for more than two years. Here the same difficulties arose as to his ability to minister acceptably to the religious needs of other Protestant personnel. The Commandant of the Yard wrote to Chaplain Scott, then Chief of Chaplains, on 22 September 1925, pointing out some of the problems involved. Clergymen had to be called in from Charleston to conduct funerals and weddings. Protestant parents in the Yard and in the immediate vicinity objected to sending their children to a Sunday School conducted by a Christian Scientist. The Commandant mentioned another problem:

> Paragraph 1245 (6). U. S. Navy Regulations, provides that one of the duties of a Chaplain shall be to "visit the sick . . ." It is not clear what benefit or comfort a sick person could derive from the visits of a Chaplain of a faith which denies the existence of disease.

Since no family attached to the station belonged to the Christian Science church, a change of chaplains was requested.

It was Chaplain Scott's opinion, which he stated was free of prejudice against either Chaplain Benson or the Christian Science Church, that the Chaplain was not professionally qualified to perform the work naturally expected of a naval chaplain. Chaplain G. H. Dickins, the third Chief of Chaplains, was of the same opinion as his two predecessors.

Chaplain Benson resigned on 15 September 1926. The Christian Science Church, notified of this action, did not request the appointment of a successor. No other Christian Science chaplain was appointed before World War II.[86]

Salt water humor has always been strong in the Navy. During the twenties a clever cartoon entitled "Spiritual Overhaul Ship SOS No. 1," which lampooned chaplains, appeared in Navy circles. It is said that it was drawn by the first lieutenant of a ship stationed in the Philippines who was thoroughly annoyed by the incessant requests of the ship's chaplain for special boat trips. The artist depicted a cross-section of the ship, labeling compartments with subtle humor: a Baptist Immersion Chamber, a Pork Hole in the Rabbi's Galley, a Fish Locker. The cartoon has acquired additional embellishments during the intervening years. The illustration on page 246 incorporates the most recent additions to the cartoon, which include the "bright and shining light" on the steeple.

An example of the concern that many chaplains felt for the physical welfare of naval personnel is illustrated in the gift of a commodious swimming pool to the Naval Base at Cavite, Philippine Islands, by Chaplain Truman P. Riddle. This pool, costing approximately $5,000 was constructed during Riddle's tour of duty there from July 1927 to February 1930, and has been much used and greatly appreciated by all hands.

The United States Coast Guard, originally known as the Revenue Marine, was created by Act of Congress on 4 August 1790 and was placed under the Treasury Department. The Act of 2 March 1799 provided that this service "shall, whenever the President of the United States shall so direct, cooperate with the Navy of the United States."[87]

The Coast Guard has no Chaplain Corps. During the First World War, when the Coast Guard came under the jurisdiction of the Navy, its personnel automatically fell within the scope of the Navy chaplains' responsibilities. The ministry of naval chaplains to the Coast Guard began in 1932 with the opening of the new Coast Guard Academy at New London, Connecticut. The chaplain assigned to the Submarine Base at New London had additional duty orders to the Academy. The first to serve in this two-fold capacity was Chaplain Roy L. Lewis. In 1932, he wrote to the Chief of Chaplains Evans describing his extra duties:

> I have arranged to conduct morning prayers at the Academy during the period of Lent; also to conduct a Bible study class during this season. I am convinced that the value of these services is worth all the extra effort it costs me.[88]

As early as 14 January of that year, Admiral F. C. Billard, Commandant of the Coast Guard, officially requested the Chief of the Bureau of Navigation to assign a full time Navy chaplain to the Academy. The Admiral asserted that the detail of Chaplain

[86] CoC., Benson File.

[87] Lovette, *Naval Customs*, p. 297.

[88] CoC., Lewis File, letter of 18 Feb.

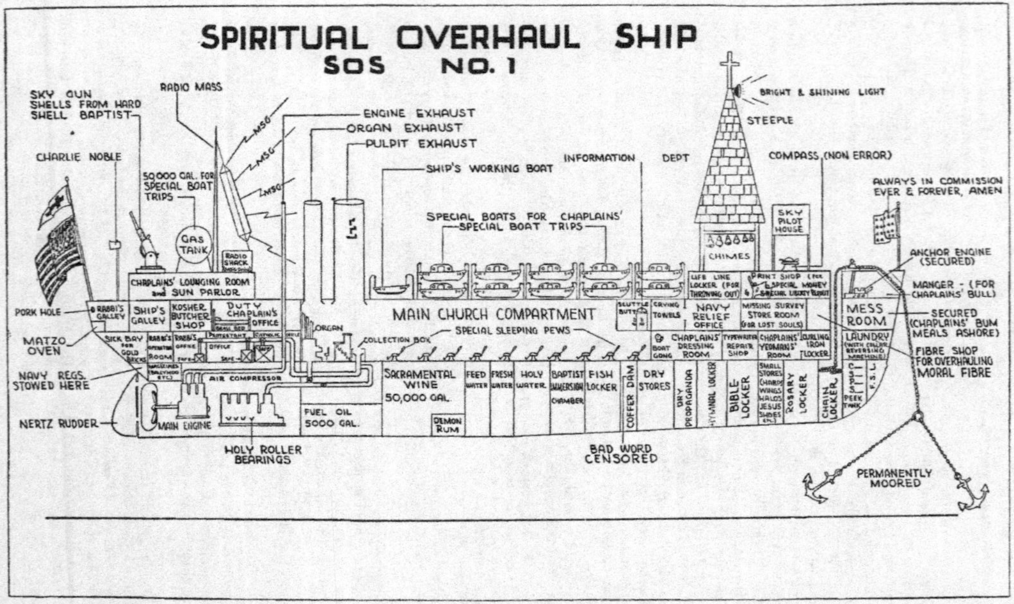

EXPLANATION

Since repair ships carry various special mechanics and hospital ships carry assorted brands of Medical Officers, it is suggested that the spiritual activities of the service be concentrated in one ship, specially designed for the purpose. The above plan was conceived for the benefit of the Chaplain Corps and the proposed name of this vessel is U.S.S. PADRE (S. O. S. No. 1)

It is to be permanently moored for the convenience of the chaplains going ashore. The Special boats are shown hoisted for the purposes of illustration only, for naturally they will be in the water at all times ready to make trips for chaplains five minutes after the regular boat has shoved off.

When force commanders consider that the moral tone of one of their ships is low, or at any rate, lower than usual, they will order her along side the U.S.S. PADRE for spiritual overhaul. The Commanding Officer will submit a work list of all souls needing and warranting repair. The chaplains will then proceed to do their stuff.

(Cartoonist and author of explanation unknown)

Lewis to such duty would be most agreeable to the Superintendent of the Academy and to him.

While budgetary technicalities prevented the Navy Department from acceding to this request, nevertheless similar additional duty orders were issued to Lewis's successors.[89] Chaplains after Lewis who ministered to the Coast Guard Academy while serving at the Submarine Base were: R. W. Shrum, September 1932 to 1934; G. L. Markle until 1936; and J. W. Moore, still on duty when the period under review closed.

Three Naval Reserve chaplains served with the Civilian Conservation Corps in the thirties. Chaplain A. O. Martin reported for duty at Fort Sheridan, Illinois, on 3 June 1935 and was sent to Fort Brady, Michigan, where he served the CCC camps in that vicinity. On 24 January 1936, Martin began duty at Trout Lake, Michigan. He continued this type of duty until June 1940.

Chaplain H. F. Hanson served with the CCC from October 1935 to April 1946 with headquarters at Fort Thomas, Kentucky. Chaplain C. M. Lever began duty in a group of camps near Superior, Wisconsin, in May 1936. Writing to Chaplain Workman on 2 February 1940, Lever stated:

> For three years and eight months, I have had the same CCC Camps as Chaplain most of that time as a Naval Reserve Officer on Active Duty. I have averaged about 28 services per month during that period, travelling about 300 miles per week in all kinds of weather and I have enjoyed all of it.[90]

Army officers were in command of these camps. The Naval Reserve Officers who served under them wore naval uniforms. Several Army Reserves who served as CCC chaplains later entered either the Regular or the Naval Reserve. Among those entering the Regular Navy were chaplains E. P. Wuebbens, W. G. Irwin, and Otto D. F. Herrmann.

The number of chaplains on sea and shore duty overseas fluctuated between fifty-two percent and fifty-eight percent of the Corps.

During the twenties the Catholics, Methodists, and Presbyterians each had about seventeen chaplains in the Corps. During the years 1929–1932, the Presbyterians had nineteen. An analysis of the denominational affiliations of chaplains who entered the Corps during the years 1920–1939 is shown in Column A. The affiliations of chaplains who were on duty 8 September 1939 is indicated in column B.

	A	B
Baptist	12	13
Catholic	27	20
Christian Science	1	
Congregational	3	4
Cong.-Christian	1	1
Disciples	3	5
Lutheran	7	9
Methodist	9	18
Presbyterian	11	17
Prot. Episcopal	11	5
Reformed	2	2
United Brethren	1	
	88	94

IN RETROSPECT

Tremendous changes had taken place in the life and organization of the Navy since 1778 when Benjamin Balch, the first known chaplain to serve in the Continental Navy, walked the decks of the small frigate *Boston.* One hundred and sixty-one years later, or on 8 September 1939, there were ninety-two chaplains on duty with the mighty ships of the modern fleet as well as on shore at home and abroad. During these years, a total of 471, counting the sixty who served in the early days of the Navy without warrant or commission, were listed as chaplains.

At first, the Continental Navy was copied after the British Navy. As in other matters, the Colonies borrowed freely of the customs and traditions of the Mother Country. Gradually, through the years following the official establishment of the Navy Department in 1798, the United States Navy developed its own regulations, customs, and traditions, which, while indebted to the British for basic ideas, were yet indigenous to America.

No Corps of the Navy developed under such handicaps and difficulties as did the Chaplain Corps. For decades, Navy chaplains were the poorest paid officers in the Navy. For about twenty years, their number was limited to nine, and then for seventy-two years the total on duty could not exceed twenty-four. During most of the Navy's history, chaplains labored under discriminatory regulations regarding status, uniform, and rank. These in the main had been rectified by the eve of the First World War.

This great world struggle brought the Chaplain Corps of the United States Navy into its own. The office of the Chief of Chaplains, established under the pressure of war conditions, demonstrated its value in the efficiency and effectiveness of the work of the

[89] *Ibid.*, Coast Guard Academy File.
[90] *Ibid.*, Lever File

Corps. With the ever rising educational and professional standards required of new appointees to the Corps, the chaplains won increasingly the respect and confidence of both enlisted men and officers. By 1939, the chaplains' place in the life and organization of the Navy was accepted and secured.

Chaplains share with other naval officers all the difficulties and compensations of seafaring life. Normal home relationships are disrupted. When a Navy man is away on an extended tour of sea or foreign duty, extra burdens are laid upon the shoulders of the wife and mother. If a chaplain is married, he faces the problem of finding suitable living quarters for his family in the vicinity of the home port of his ship. With a change of duty every few years, chaplains are unable to enjoy the accumulative benefits which come from long residence in the same locality. It is difficult to move one's library and possessions from one place to another. These are some of the disadvantages in a naval chaplain's life.

On the other hand, the compensations which come with long service in the Navy have a way of balancing these difficulties. Even though the children of naval officers sometimes find their education interrupted, yet through their extensive travels, often abroad, they secure a broader education than other young people of their age usually receive. They obtain a knowledge and experience of life which does not come from school books. A chaplain and his family soon discover that the Navy has given them an opportunity to make a host of friends all over the world. Navy chaplains are at home everywhere; there is nothing of the provincial about them.

In an article entitled "Sea-Going Clergymen," which appeared in the Christmas 1942 issue of *The Trident,* Chaplain William N. Thomas, then chaplain at the Naval Academy, gave the following testimony which sums up the convictions of his colleagues in the naval chaplaincy:

> Out of a quarter of a century of service, the writer of these lines can say, without reservations, that no clergyman could ask for a happier environment and more pleasant associates than those furnished by the officers and men of the Navy. It is a great privilege to be a "padre" in the American Navy.[91]

[91] *Op. cit.*, p. 45.

APPENDIX I

UNITED STATES NAVY CHAPLAINS, 1778–1939[1]

Arranged alphabetically by years

Figures in parentheses indicate total for year.

* Indicates non-commissioned chaplain chosen by the commanding officer from the ship's company to serve for the duration of the cruise. Such names do not appear as chaplains in *Navy Registers.*

A year date after a name indicates first year of previous service.

ABBREVIATIONS

Bap (N)	Baptist (Northern)	PE	Protestant Episcopal
(S)	(Southern)	Presby (USA)	Presbyterian (Northern)
Breth	Brethren, Church of	(US)	(Southern)
Chris	Christian Church	RC	Roman Catholic
Cong	Congregational	Ref	Reformed Church of America
Cong-Chris	Congregational-Christian	UBreth	United Brethren
Disc	Disciples of Christ	Unit	Unitarian
Luth	Lutheran	Univ	Universalist
Meth (N)	Methodist (Northern)	UPresby	United Presbyterian
(S)	(Southern)		

Where no denomination is indicated, none is known.

1778
*Balch, Benjamin — Cong
1782
*Geagan, Timothy
1798
*Austin, William
1799
Balch, William — Cong
1800 (5)
Chandler, Samuel — PE (?)
*Flannery, Matthew
Morris, Noadiah
Thompson, Robert
Vallette, Elie
1802
McFarlan, Alexander — PE
1803
Leonard, Peter
1804
Dennison, Robert
*Schnetter, John
1805 (6)
*Budd, Charles A.
*Cerdona, Batta
*Cruize, Lawrence
*Lyde, Nathaniel
*Ormsby, Robert
*Ravara, Joseph
1807
*Fairchild, Lewis
Petty, William
1808
*Turner, Edward W.
1809 (6)
Barry, Garrett
Briscoe, William H.
Moore, Andrew L.
*Morris, Lemuel
Robinson, William
*Shaw, Charles
1810
*Denison, Henry
1811
Adams, David Phineas
Hunter, Andrew — Presby
*Smith, Nathaniel R.
1812 (7)
*Anderson, Samuel T.
*Carlton, John
Cook, John — PE
*Kelso, James
*Morris, George N.

[1] Totalling 477, including six names which appear twice. The biographical and service-record sketches of these 471 chaplains were included in volume III of this history issued by the Government Printing Office in 1948.

*Waine, Thomas	
*West, David	
1813 (7)	
*Bancroft, Henry	
*Breese, Thomas	
*Davis, George	
*Livermore, Samuel	
*Penny, Francis H.	
*Sands, Richard A.	
*Sands, William M.	
1814	
*Gatewood, Richard	
Humphreys, A. Y.	
*Lloyd, Richard	
1815 (9)	
Cooper, Colden	
Felch, Cheever	PE
*Handy, Charles O.	
*Hanlon, Alexander I.	
*Morse, E. B.	
Norton, Richard C.	
*Thomas, Gardner	
Watson, Joseph	
*Whitney, A. G.	
1816	
Andrews, Nathaniel	PE
Folsom, Charles	Unit
Ireland, John	PE
1817	
*Burrows, Thomas	
Chase, Philander, Jr.	PE
*Laughton, William	
1818	
Brooks, James	PE
Everett, James	PE
Wilson, Azariah	
1819	
Chase, Moses Bayley	PE
Hambleton, John N.	
1820	
Searle, Addison	PE
1823	
Allison, Burgess	Bap
*Bates, John A.	
Jones, Cave	PE
1824	
Grier, John W.	Presby
McCarty, John	PE
1825	
Addison, John	Bap
Ogilvie, James G.	Presby
1826	
McLaughlin, Edward	Presby
1827	
*Campbell, A. H.	
Hayes, Hervey H.	Cong
*Soter,	
1828	
Fenner, John P.	PE
*May, Charles	
Ridgely, G. W.	PE
Stewart, Charles Samuel	Presby
1829	
Harrison, Timothy J.	Presby
Ryland, William	Meth
1830	
Colton, Walter	Cong
1833	
Jones, George	PE
Wiltbank, James	PE
1834	
Lambert, Thomas R.	PE
1836	
Rockwell, Charles	Cong
1838	
Clark, Peter G.	PE
Elliott, Jared L.	Presby
1839	
*Girard, John F.	
Lewis, Rodman	Meth
Wilmer, Joseph P. B.	PE
1841 (11)	
Alden, Charles H.	PE
Bartow, Theodore B.	PE
Chase, Moses B. 1819	PE
Gillett, Samuel T.	Meth
Harris, Thompson S.	Presby
Jackson, William G.	PE
McKenney, William	Meth
Newell, Chester	PE
Stockbridge, Joseph	Bap
Talbot, Mortimer R.	PE
Taylor, Fitch W.	PE
1842	
Fisk, Photius Kavasales	Unit
1843	
Lathrop, John P.	Cong
Robb, John	Meth
1844	
Frost, Nathaniel	Bap
Grigg, John	PE
Newton, Joel W.	Cong
Swann, George W.	Presby
1845	
Fletcher, N. C.	Univ
Latham, George W.	Bap
1847	
Blake, John	PE
Eaton, Edwin	Bap
Lenhart, John L.	Meth
Stanley, Thomas Coke	Meth (S)
1850	
Bittinger, Edmund C.	Presby
1851	
Dewey, Orville	Unit
Eskridge, Vernon	Meth
1853	
Noble, Mason	Presby
Thomas, Charles W.	Meth (S)

1855	
Given, Robert	Meth
Watson, John L.	PE
1856	
Wood, Henry	Presby
1857	
Davis, Charles A.	Meth (S)
1859	
Junkin, Davis X.	Presby
1860	
Dorrance, George W.	Bap
1861	
Salter, Thomas G.	PE
1862	
Coleman, James A.	Meth
Hitchcock, William A.	PE
1863 (5)	
Hale, Charles R.	PE
McLaren, Donald	Presby
Stewart, William H.	Bap
Wallace, John S.	PE
Winslow, Ezra D.	Meth
1864	
Beugless, John D.	Bap
Henderson, George D.	PE
Hibben, Henry B.	Meth
Smith, George Williamson	PE
1868	
Holway, Wesley O.	Meth
Kane, James Johnson	Bap
1869	
Brittain, M. Cookman	Meth
Cobb, William R.	Meth (S)
Lewis, John Kerfoot	PE
Matthews, John R.	PE
1870	
Crawford, George Artemas	Meth
Rose, Frank B.	Bap
1871	
Rawson, Edward K.	Cong
Van Meter, John B.	Meth
1872	
Tribou, David Howard	Meth (N)
1873	
Clark, Henry Howard	Meth (N)
Hagar, Elijah W.	PE
McAlister, Adam A.	PE
1874	
Brown, James H. H.	PE
Gill, Thomas Augustus	Bap
Hudson, Robert	PE
1875	
Boorom, Sylvester D.	PE
1876	
Hayward, Richard	PE
1880	
*Kramer, Samuel	Meth
1881	
Morrison, William	PE
Royce, Alfred L.	PE
Thompson, Frank	PE
1882	
Hoes, Roswell Randall	Presby (USA)
1885	
Wright, Carroll Quinn	Disc
1888	
Isaacs, Walter Gilbert	Meth (S)
Parks, Charles H.	RC
1889	
McIntire, Joseph Prince	Meth
1890	
Hensley, Alexander C.	PE
1892	
Reaney, William Henry I.	RC
Sherman, Frederick F.	PE
1894	
Edmonson, William Elam	Meth (S)
Goodwin, Montgomery M.	PE
1895	
Chidwick, John P. S.	RC
Frazier, John Brown	Meth (S)
1896	
Jones, Harry W.	PE
1897	
Cassard, William Gilbert	Meth
Freeman, Thaddeus S. K.	Disc
Sykes, Arthur O.	Meth/PE
1898	
Brown, Frederic C.	Unit
Dickins, Curtis Hoyt	Univ/PE
Helms, William T.	Meth
Steele, Robert E.	Presby (US)
1900	
Rennolds, Louis P.	RC
1901	
Brennan, Edward J.	RC
Charlton, Charles M.	Meth (N)
MacGrail, Joseph F.	RC
1902 (5)	
Bayard, George L.	PE
Bellows, Johnson McC.	PE
Fleming, John Frank	Bap (N)
McDonald, Eugene E.	RC
Patrick, Bower Reynolds	Bap (N)
1903	
Gleeson, Matthew Carlin	RC
Stone, Arthur William	PE
1904	
Scott, Evan Walter	Cong
1905	
McGinty, Joseph F.	RC
1907	
Evans, Sydney Key	PE
Stevenson, George E. T.	Bap (N)

1909

MacNair, James Duncan	Meth (N)
Pearce, Hugh M. T.	PE

1910

Brodmann, Edmund A.	RC
Taylor, LeRoy Nelson	Meth (N)

1913

Thompson, Thomas B.	Presby (USA)

1914

Brady, John Joseph	RC

1915 (11)

Alexander, Milton O'H.	Bap (S)
Anderson, William Earl	Disc
Behrens, Oscar	Presby (USA)
Bouffard, Irenne J.	RC
Burke, Eugene Sebastian	RC
Duff, Edward Aloysius	RC
Dumstrey, Herbert	Ref
Ellis, Charles Vinton	Bap (S)
Hayes, Allison John	Meth (N)
Kranz, George Boniface	RC
Workman, Robert DuBois	Presby (USA)

1916 (7)

Elder, William Wytche	Cong
Hastings, Charles Henry	Meth (N)
Lewis, Roy Leslie	Meth (N)
Park, Albert Newton, Jr.	Presby (USA)
Riddle, Truman Post	PE
Seidler, Paul Edward	Luth
Short, Joseph C.	RC

FIRST WORLD WAR

War with Germany was declared 6 April 1917. The Chaplains' Division, Bureau of Navigation, with Chaplain John B. Frazier as the First Chief or Director, was established 5 November.

The abbreviations used after the names indicate the following:

NNV or USNRF—indicates chaplains called to active duty from the National Naval Volunteers or from the United States Naval Reserve Force. The inclusion of the name in chronological order is according to the date when the Reserve chaplain reported for active duty and not according to the date of his commission.

USN—Reserve chaplains who transferred to the Regular Navy.

1917 (55)

Alexander, John Lee	Presby (USA)
Ayers, William B., USNRF	Cong
Beckley, Quitman F.	RC
Bloomhardt, Paul F.	Luth
Brokenshire, John J., NNV, USN	Cong
Casey, Joseph T.	RC
Clausen, Bernard C.	Bap (N)
Darche, Harris A.	RC
Darlington, G. S. B.	PE
Day, James Samuel, NNV, USN	Bap (S)
Dyer, Harold Stras	Meth (S)
Edel, William Wilcox	Meth (N)
Foley, George W.	RC
Gatlin, Henry Grady	Meth (S)
Gay, Arthur Royall	Bap (S)
Goldberg, David	Jewish
Gould, Romeo, NNV	PE
Groth, Emil Hartwig	Luth
Haines, Alfred James	Meth (N)
Hall, Wilford Raymond	Bap (N)
Hammersley, Patrick J., USNRF, USN	RC
Huske, Bartholomew F., NNV, USN	PE
James, Sidney Thomas, NNV	PE
Kerns, George L.	Presby (US)
Larned, Albert Cecil, NNV	PE
Lash, Frank Harry	Disc/PE
Leduc, Paul J. A.	RC
Lenski, Gerald E.	Luth
Leonard, Morris Mills	Bap (N)
Lippincott, Haines H.	Meth (N)
Maguire, William A.	RC
Mark, John Nicol, USNRF	Univ/Unit
Mark, Thomas M., USNRF	Univ/Unit
McClelland, Stewart W.	Presby (USA)
McFadden, Francis Leo	RC
Miller, Robert Edwin	Meth (N)
Mitchell, Perry Louis	Bap (S)
Moore, Charles W., NNV	Meth (N)
Moore, John Warner	Presby (USA)
Murphy, Garrett F. X.	RC
Neff, Josiah Luther	Meth (N)
Niver, Edwin Barnes, NNV, USN	PE
Paugh, Ernest LeRoy, NNV	PE
Petzold, Milton H.	Meth (N)
Putnam, John Henry S.	Ref/PE
Regan, Thomas Francis	RC
Rentz, George Snavely	Presby (USA)
Rountree, Hersey E., USNRF, USN	Christian
Savageau, Edmund E.	RC
Stephens, Bart Donnelly	Meth (S)
Torrance, Arthur F.	Meth (N)
Underwood, Joseph F.	RC
Whitmore, George Merle	Presby (USA)
Williams, William P., NNV, USN	PE
Yates, John Henry	PE

1918 (107)

Ackiss, Ernest Lee	Bap (S)
Albert, Francis Lee	Bap (S)
Averyt, Albert Norris	Meth (S)
Bare, Charles Brenton	Meth (N)

Bate, Francis Howard	Cong
Becker, Carl William	Bap (N)
Bennett, Vincent LeRoy	PE
Blackard, William F.	Meth (S)
Blackshear, Robert H.	Meth (S)
Blain, Cary Randolph, USNRF	Presby (US)
Borders, Karl	Disc
Boynton, Edward C.	Cong
Boynton, Morris Russell	Cong
Brinckerhoff, James H., USNRF	Ref
Burke, Thomas Joseph	RC
Burns, Daniel Robert	RC
Burt, Paul	Meth (N)
Byrne, Edwin Vincent	RC
Callender, Stephen J.	Meth (N)
Carhart, Edmund H., Jr.	PE
Carpenter, Guy O.	Meth (N)
Carruthers, John, F. B.	Presby (USA)
Conroy, John B., USNRF	RC
Costello, Edward P.	RC
Dandeneau, Arthur J.	RC
Davidson, Hugh Roswell	Disc
Davis, Edgar William	Meth (S)
Davis, Richard Joseph	Christian Science
Decker, John William	Bap (S)
Donahue, Philip Edward	RC
Duffy, Edward P., USNRF	RC
Dufrane, Leo John	RC
Durrett, William Yancy	Presby (Cumb)
Earnest, Joseph B., Jr.	Disc
Ellenwood, James Lee	Meth
Ely, John Calvin, Jr.	Presby (USA)
Ericson, Gustav Rudolph	Meth (N)
Fedders, John Frederick	Luth
Ferris, Frank Holliday	Cong
Finn, John Henry	RC
Foster, Earle Ware	Bap (S)
Fry, Henry Jacob	Luth
Hagen, John F.	Presby (USA)
Harris, Leonard Clement	Meth (N)
Harrison, William E., Jr.	Meth (N)
Heckel, Paul Frederick	Luth
Henry, Edward Bedford, USNRF	RC
Hester, James M.	Bap (S)
Hildebrandt, Ludwig	Luth
Hindman, Ralph Blake	Presby (USA)
Hopkins, Walter Alves	Presby (US)
Horton, Douglas	Cong
Hurney, Francis Joseph	RC
Janeway, Frank L., USNRF	Presby (USA)
Jones, Edward A. P., USNRF	Meth (N)
King, Philip Coates	Cong
Kirkpatrick, Thomas L.	Presby (USA)
Krauss, Paul Hartzell	Luth
Laherty, John J., USNRF	RC
Ledwig, Francis J.	RC
Leyfield, Claude Hill	Meth (N)
Lloyd, Starr Hanford	Presby (USA)
Maxwell, Samuel LaVerne	Meth (N)
McCann, Christopher B.	RC
McCarthy, John Anthony	RC
McGrath, Thomas S., USNRF	RC
Meadows, Roscoe McK.	Bap (S)
Merrill, Boynton	Cong
Miller, Leslie	Meth (N)
Miller, Wm. McC.	Presby (USA)
Monaghan, Daniel F.	RC
Moyer, Frank Edward, USNRF, USN	Luth
Neyman, Clinton Andrew	Bap (N)
Nichols, Wm. Albert, USNRF	PE
Norton, Powell Holcroft	Bap (N)
O'Rourke, Simon Anthony	RC
Parker, Albert R., USNRF, USN	PE
Perkins, Joseph Andrew	RC
Peterson, Harry M.	Presby (USA)
Preston, Paul Gordon, USNRF	Disc
Quinn, John M. J.	RC
Reagor, William Paul	Disc
Riddle, Truman Post, 1916	PE
Robinson, Daniel Sommer	Cong
Russell, Robert McWatty	Presby (USA)
Schweitzer, Frederick	Presby (USA)
Shrum, Reuben Welty	Presby (US)/PE
Spotts, Milton A., USNRF	Presby (USA)
Steiner, Walter L., USNRF	Presby (USA)
Stone, Albert Edward	Bap (N)
Sullivan, Mortimer A.	RC
Terrell, William Sole	Bap (N)
Thomas, William N.	Meth (S)
Tibbetts, Norris Lowell	Bap (N)
Tomerlin, Joseph A.	RC
Truitt, Razzie W.	Meth (S)
Twinem, Leonard L.	PE
Van Dyke, Henry, USNRF	Presby (USA)
Vogler, Alfred deG.	Moravian/Presby (USA)
Ward, Chester Franklin	Presby (USA)
Wheeler, Creasy Clement	Bap (S)
Whimsett, Grover C.	UBreth
Wilkie, Henry Frederic	Presby (USA)
Wilson, William H., USNRF	Presby (USA)
Witherspoon, Maurice M.	Presby (USA)
Woodward, Isadore C.	Bap (S)
Wylie, David Roswell	Presby (USA)

1919

Wood, Tipton Lee	Presby/PE

1920 (12)

Benson, Joel Harry	Christian Science
Gottschall, Louis D.	UBreth/PE
Holt, William Therrel	PE
Hyde, John Hutchinson	Cong
Hester, James M., 1918	Bap (S)
Miller, Thornton C.	Disc
Mitchell, Perry L., 1917	Bap (S)
Murdock, Geo. Gregory	RC
Peil, Roman Michael	RC

Putnam, John Henry, 1917	Ref/PE
Tedford, Randolph E.	Bap (N)
Thompson, Walter Lee	Disc
1921	
Brooks, Joseph Howard	Meth (S)
Salisbury, Stanton W.	Presby (USA)
Sanborn, Henry R.	PE
Vogler, Alfred deG. 1918	Moravian/ Presby (USA)
1924 (7)	
Brame, Julius D.	Bap (S)
Criger, Earl Mack	Cong
Eure, Otho A.	Bap (S)
Forsander, John Philip	Bap (N)
Johnson, John Edward	Presby (USA)
McNanamy, Joseph E.	RC
Rafferty, Wm. Henry	Bap (N)
1925 (6)	
Dittmar, Charles A.	UPresby
Glunt, Homer G.	Presby (USA)
Hangen, Emerson G.	Presby (USA)
Jastroch, Andrew C.	RC
Mansfield, Colonel H.	Presby (USA)
Robbins, Edward J.	Meth (N)
1926 (5)	
Burke, Charles Patrick	RC
Drinan, Raymond B.	RC
Koch, Arthur E.	PE
Long, Christopher S.	PE
Markle, Geo. LaClede	Presby (USA)
1927	
Hermann, Herman L.	RC
O'Neill, Emmet	RC
1928	
Doyle, John	PE
Morgan, Victor H.	PE/Cong
1929 (7)	
Doty, Walter Pettit	PE
Douglas, Charles Harold	PE
Harp, Edward Blaine, Jr.	Ref
Hohl, August F.	RC
LeBaron, Earle R.	RC
Pierce, Walter W.	Presby (USA)
Trump, Herbert Ray	Luth

1930 (7)	
Cuthriell, Warren Franklin	Bap (N)
Gorski, Vincent James	RC
Hally, Michael Ambrose	RC
Hugues, John Francis	RC
Pritchett, Lester C.	Cong
Sitler, Carl Moses	Luth
Tunner, John William	RC
1931	
Hamilton, Frank Roosevelt	Cong
Quinn, David Long	PE
1932	
Knox, Thomas Joseph	RC
1933	
Johnson, Lon Portivant	PE
Linaweaver, Paul Glenwood	PE
Marken, Roy Rockwell	Disc
Mechling, Frederick Wm.	RC
1936 (6)	
Andrews, Edgar Clessie	Bap (S)
Bishop, Roy Edward	Meth (S)
Boslet, John Raymond	RC
McManus, Francis Joseph	RC
Robinson, John Frederick	RC
Wuebbens, Everett Peter	Luth
1937 (10)	
Bauer, Edwin Theodore	Meth (N)
Dickman, Paul Wm. Joseph	Luth
Dreith, Joseph Floyd	Luth
Faulk, Roland William	Meth (S)
Hacherl, Charles John	RC
Herrmann, Otto David Fred.	Luth
Redman, Emil Frederick	Luth
Rosso, George Aloysius	RC
Schwyhart, Robert Marion	Bap (N)
Twitchell, Martell Herman	Meth (S)
1938 (6)	
Bennett, Samuel Beecher	Meth (S)
Burke, Francis Andrew	RC
Ferris, Thomas	RC
Howe, Harris Winchester	Bap (N)
Johnson, Thomas Edward	Meth (S)
Volbeda, Frederick	Presby (US)
1939 to 8 September (7)	
Irwin, William Grover	Meth (S)
Maddox, Otis Pendleton, Jr.	Bap (S)
Mahler, Walter Albert	RC
Mannion, Joseph Patrick	RC
Murphy, John Patrick	RC
Schmitt, Aloysius Herman	RC
Wood, Harry Clinton	Presby (USA)

APPENDIX II

A CHART SHOWING RELIGIOUS AFFILIATIONS OF UNITED STATES NAVY[1] CHAPLAINS FROM 1778 TO 8 SEPTEMBER 1939

Including the following six duplications: M. B. Chase, 1819 & 1841; T. P. Riddle, 1916 & 1918; J. H. S. Putnam, 1917 & 1920; A. deG. Vogler, 1918 & 1921; P. L. Mitchell, 1917 & 1920; and J. M. Hester, 1918 & 1920.

	1778–1785	1798–1810	1811–20	1821–30	1831–40	1841–50	1851–60	1861–70	1871–80	1881–90	1891–1900	1901–10	1911–16	1917	1918	1919–29	1930–39	Total
BAPTIST	—	—	—	2	—	4	1	4	1	—	—	3	2	6	14	7	5	49
CATHOLIC	—	—	—	—	—	—	—	—	—	1	3	6	6	12	23	10	17	78
CHRISTIAN	—	—	—	—	—	—	—	—	—	—	—	—	—	1	—	—	—	1
CHR-SCIENCE	—	—	—	—	—	—	—	—	—	—	—	—	—	—	1	1	—	2
CONGREGATIONAL	1	1	—	2	1	2	—	—	1	—	—	1	1	2	8	2	2	24
DISCIPLES	—	—	—	—	—	—	—	—	—	1	1	—	1	1	5	2	1	12
JEWISH	—	—	—	—	—	—	—	—	—	—	—	—	—	1	—	—	—	1
LUTHERAN	—	—	—	—	—	—	—	—	—	—	—	—	1	3	6	1	6	17
METHODIST	—	—	—	1	1	5	4	7	4	2	5	3	3	11	17	2	7	72
MORAVIAN	—	—	—	—	—	—	—	—	—	—	—	—	—	—	1	—	—	1
PRESBYTERIAN	—	—	1	5	1	3	3	1	—	1	1	—	4	6	23	10	2	61
PROTESTANT-EPISCOPAL	—	2	9	4	5	9	1	8	6	4	3	5	1	9	7	8	3	84
REFORMED	—	—	—	—	—	—	—	—	—	—	—	—	1	1	1	2	—	5
UNITED BRETHREN	—	—	—	—	—	—	—	—	—	—	—	—	—	—	1	1	—	2
UNITARIAN	—	—	1	—	—	1	1	—	—	—	1	—	—	—	—	—	—	4
UNIVERSALIST	—	—	—	—	—	1	—	—	—	—	1	—	—	2	—	—	—	4
UNKNOWN	1	24	30	4	1	—	—	—	—	—	—	—	—	—	—	—	—	60
Total	2	27	41	10	9	25	10	20	12	9	15	18	20	55	107	46	43	
Grand total		29	70	88	97	122	132	152	164	173	188	206	226	281	388	434	477	477

[1] Including the Continental Navy.

APPENDIX III

BIBLIOGRAPHY

General Sources

Alden, Carroll S. and Westcott, Allan, *The United States Navy.* New York, 1943.

Bayard, Samuel J., *A Sketch of the Life of Com. Robert Stockton.* New York 1856.

Bell, Frederick J., *Room to Swing a Cat.* New York, 1938.

Benham, Edith and Hall, Anne M., *Ships of the Navy and their Sponsors, 1797–1913.* [priv. print.], 1913.

Benjamin, Park, *The United States Naval Academy.* New York, 1900.

Brooks, Joseph H., *Distinguished Service.* Shanghai, 1933.

Bryant, Arthur, *Samuel Pepys, The Years of Peril.* New York, 1935.

Callan, John F. and Russell, A. W., *Laws of the United States relating to the Navy and Marine Corps.* Baltimore, 1859.

Chase, Philander, *Reminiscences; An Autobiography.* Boston, 1848.

Chidwick, John P. S., *Remember the Maine.* Winchester, Virginia [1935].

Clark, William Bell, *Gallant John Barry.* New York, 1938.

Clowes, William Laird, *The Royal Navy.* London, 1900.

Colton, Walter, *Deck and Port.* New York and Cincinnati, 1850.
Land and Lee. New York, 1851.
Land of Gold; or Three Years in California. New York, 1860.
The Sea and the Sailor. New York, 1851.
Ship and Shore. New York, 1835.

Dana, Richard Henry, *Two Years Before the Mast.* New York, 1907.

Daniels, Josephus, *Our Navy at War.* New York, 1922.

Davis, C. H., *Life of Charles Henry Davis, Rear Admiral 1807–1877.* Boston and New York, 1899.

Dewey, Orville, *Autobiography and Letters of Orville Dewey.* Boston, 1833.

Egan, Maurice F. and Kennedy, John B., *The Knights of Columbus in Peace and War,* 2 Vols. New Haven, 1920.

Encyclopaedia Britannica, Vols. 1–28. Eleventh Edition, New York, 1910.

Evans, Robley D., *A Sailor's Log.* New York, 1901.

Farragut, Loyall, *The Life of David Glasgow Farragut.* New York, 1879.

Franklin, S. R., *Memories of a Rear-Admiral.* New York and London, 1898.

Germain, Dom Aidan Henry, *Catholic Military and Naval Chaplains, 1776–1917.* Washington, 1929.

Goldberg, David, *Sky-Piloting the Great Lakes—A Valedictory.* Great Lakes, 1920.

Goldsborough, C. W., *The United States' Naval Chronicle.* Washington, 1824.

Griffin, Martin I. J., *Commodore John Barry.* Philadelphia, 1903.

Hawks, Francis L., *Narrative of the Expedition of an American Squadron to the China Seas and Japan,* 2 Vols. Washington, 1856.

Harris, Frederick, *Service with the Fighting Men,* 2 Vols. New York, 1922.

Hibben, Henry B., *Navy-Yard, Washington, History from organization, 1799 to present date.* Washington, 1890.

Hollis, Ira N., *The Frigate Constitution.* Boston, 1900.

Jones, George, *Sketches of Naval Life,* 2 Vols. New Haven, 1829.
U. S. Japan Expedition; Observations on the Zodiacal Light (Volume III of Hawk's *Narrative*). Washington, 1856.
Jones, Harry W., *A Chaplain's Experience Ashore and Afloat,* New York, 1901.
Battle of Santiago. New York [1913].
Julian, John, *A Dictionary of Hymnology.* London, 1908.
Knox, Dudley W., (ed.), *Quasi-War Between the United States and France,* 7 Vols. Washington, 1938.
Leech, Samuel, *Thirty Years From Home or A Voice From the Main Deck.* Boston, 1843.
Lewis, C. L., *Matthew Fontaine Maury, The Pathfinder of The Seas.* Annapolis, 1927.
Littleton, William S., *Trumpeters Hand Book and Instructor.* Kansas City, Missouri, 1902.
Lovette, Leland P., *Naval Customs, Traditions, and Usage.* Annapolis, 1936.
Macdonough, Rodney, *Life of Commodore Thomas Macdonough, U. S. Navy.* Boston, 1909.
Maclay, Edgar S., *A History of the United States Navy from 1775 to 1894,* 2 Vols. New York, 1894.
Maguire, William A., *Rig for Church.* New York, 1942.
Mahan, Alfred T., *The Influence of Sea Power upon History, 1660–1783.* Boston, 1943.
Marryat, Frederick, *Mr. Midshipman Easy.* New York, 1866.
Mayo, Claude Banks, *Your Navy.* Los Angeles, 1943.
Melville, Herman, *White Jacket or The World on a Man-of-War.*
Metcalf, Clyde H., *A History of the United States Marine Corps.* New York 1939.
Monk, William Henry, *Hymns, Ancient and Modern.* New York, 1870.
Moss, Lemuel, *Annals of the United States Christian Commission.* Philadelphia, 1868.
Murrell, William Meacham, *Cruise of the Frigate Columbia Around the World.* Boston, 1840.
Nordhoff, Charles, *Man-of-War Life: A Boy's Experiences in the United States Navy.* Philadelphia, 1856.
Parker, William Harwar, *Recollections of a Naval Officer, 1841–1865.* New York, 1883.
Paullin, Charles O., (ed.), *Out-Letters of the Continental Marine Committee and Board of Admiralty, August, 1776—September, 1780.* New York, 1914.
Pratt, Fletcher, *The Navy.* New York, 1941.
Preble, George H., *Our Flag.* Albany, 1872.
Puleston, W. D., *Mahan.* New Haven, 1939.
Reynolds, John N., *Voyage of the United States Frigate Potomac.* New York, 1835.
Rockwell, Charles, *Sketches of Foreign Travel and Life at Sea,* 2 Vols. Boston, 1842.
Roddis, Louis H., *A Short History of Nautical Medicine.* New York, 1941.
Roosevelt, Theodore, *The Naval War of 1812.* New York, 1897.
Sands, Robert C., *Life and Correspondence of John Paul Jones.* New York, 1830.
The Second Division American Expeditionary Force in France. 1917–1919.
Text by O. L. Spaulding and J. W. Wright. New York, 1937.
Sheppard, John H., *The Life of Samuel Tucker.* Boston, 1868.
Snow, Elliot and Gosnell, H. Allen, *On the Decks of "Old Ironsides."* New York, 1932.
Soley, James Russell, *The Boys of 1812.* Boston, 1887.
Historical Sketch of the United States Naval Academy. Washington, 1876.
Sprague, William B., *Annals of the American Pulpit,* 7 Vols. New York, 1857–[69].
Sprout, Harold and Margaret, *The Rise of American Naval Power, 1776–1918.* Princeton, 1942.
Stewart, Charles S., *A Visit to the South Seas in the U. S. Ship Vincennes, during the years 1829 and 1830,* 2 Vols. New York, 1831.
Stewart, Robert Armistead, *The History of Virginia's Navy of the Revolution.* Richmond, 1933.
Taylor, Fitch W., *The Broad Pennant.* New York, 1848.
The Flag Ship, or A Voyage Around the World in the United States Frigate Columbia, 2 Vols. New York, 1840.

Thompson, Robert Means and Wainwright, Richard (editors), *Confidential Correspondence of Gustavus Vasa Fox, 1861–65.* New York, 1919.

United States Wars with the Barbary Powers, Vols. I-III. Washington, 1939–41.

Upton, Emory, *A New System of Infantry Tactics.* New York, 1871; 1874.

Williams, Michael, *American Catholics in the War.* New York, 1921.

Wines, E. C., *Two Years and a Half in the Navy,* 2 Vols. Philadelphia, 1832.

Government and Official Sources

American State Papers, *Naval Affairs.* Washington, Vol. 1, 1834; Vols. 2 and 3, 1860; Vol. 4, 1861.

Bureau of Navigation pamphlet, "Nav. 5-K-Service," 28 April 1933. Washington, 1933.

Bureau of Naval Personnel Manual. Washington, 1942.

Circulars, 1832–63. Issued by the Navy Department, bound and in Navy Department Library.

Compilation of Court-Martial Orders, 1916–37, Vol. 1. Washington, 1940.

The Congressional Medal of Honor, The Distinguished Service Cross, and the Distinguished Service Medal issued by the War Department, since April 6, 1917 up to . . . November 11, 1919. Washington, 1920.

General Orders and Circulars Issued by the Navy Department from 1863 to 1867 Compiled by M. S. Thompson. Washington, 1887.

A General Register of the Navy and Marine Corps of the United States, 1787–1847. Washington, 1848.

The Hymnal Army and Navy. Washington, 1920.

Introduction to U. S. Navy Signal Code. Prepared by Commodore Thornton Jenkins. Washington, 1867.

Journals of the Continental Congress. Washington, 1904 to ——.

Laws Relating to the Navy. Washington, 1922.

List of Officers of the Navy of the United States and of the Marine Corps from 1775–1900. New York, 1901.

Manual for the Government of United States Naval Prisons. Washington, 1923.

Naval Manual for Buglers. Washington, 1940.

Naval Militia Registers, 1907; 1909–1917; 1917.

Navy Department Bulletin, *Information Concerning the U. S. Navy and other Navies.* May 1925.

Navy Department General Orders, Series of 1935 (Washington, 1935), with additional orders issued from time to time.

Navy Registers. Navy Register 1805–6. Philadelphia.

Official Records of the Union and Confederate Navies in the War of Rebellion, Series 1.

Proceedings of the Court of Inquiry appointed to Inquire into the Intended Mutiny on board the United States Brig Somers. 1843

Register of the Commissioned and Warrant Officers of the Navy of the United States; including Officers of the Marine Corps. 1819.

Register of the Commissioned and Warrant Officers of the United States Naval Reserve Force, 1 January 1921. Washington, 1921.

Regulations for the Navy of the Confederate States. Richmond, 1862.

The 1916 Regulations, Origin, History & Laws of the United States Naval Home. 1916.

Reports of the Secretary of the Navy.

Rules and Regulations for the Government of the Naval Academy. Washington, 1850.

Rules of the Navy Department, Regulating the Civil Administration of the Navy of the United States. Washington, 1832.

Ship and Gunnery Drills, 1927.

Signal Books. 1858, 1898, 1908, 1913, 1920.
1861—compiled for the use of the Confederate Navy.

Song and Service Book for Ship and Field. 1941 and 1942.

Tables of Allowances of Equipment, Outfit, Stores, etc. Prepared by C. S. McCauley and G. S. Blake. Washington, 1854.

Senate Documents: 22 Cong., 1 sess., No. 30; 45 Cong., 1 sess., No. 1551; 57 Cong., 2 sess., Nos. 10, 129, 138.

Senate Reports: 45 Cong., 3 sess., No. 626; 46 Cong., 1 sess., No. 209.

House Document: 57 Cong., 2 sess., No. 42.

House Reports: 22 Cong., 1 sess., Vol. II, No. 743, Rep. 124; 31 Cong., 1 sess., Vol. 1, No. 583, Rep. 171; 33 Cong., 1 sess., No. 124; 45 Cong., 2 sess., No. 4477.

Report of House Committee, 25 Cong., 2 sess., No. 996.

Executive Documents, 35 Cong., 2 sess., No. 59.

Congressional Record, 29 January 1879.

Rules for the Regulation of the Navy of the United Colonies; Pay Tables; Articles of Enlistment and Distribution of Prize Money. Philadelphia, 1775.

Naval Regulations Issued by Command of the President of the United States of America, January 25, 1802.

Rules, Regulations, and Instructions, for the Naval Service of the United States: Prepared . . . in Obedience to an Act of Congress, Passed Seventh February, 1815. . . . Washington, 1818.

Rules and Regulations of the Naval Service, 1821: "alterations and additions" to "the rules and regulations for the naval service . . . of 7 February, 1815. . . ."

General Regulations for Navy and Marine Corps. Washington, 1841.

Regulations of Navy, 1863.

Regulations for Government of Navy, 1865; 1869; 1870; 1876; 1893; 1900; 1909.

U. S. Navy Regulations, 1920.

Regulations for the Uniform and Dress of the Navy and Marine Corps of the United States, 1852.

Uniform Regulations of the United States Navy, 1853; 1863; 1869; 1883; 1895; 1897; 1899; 1913; 1917; 1922.

Revised Statutes.

United States Statutes at Large.

Other Printed Sources

American Bible Society, Annual Report. New York, 1821; 1842.

Annual Reports of the YMCA, 1914–1915; 1915–1916. New York.

Annual Survey, YMCA, 1910. New York.

Federal Council of Churches of Christ in America, Biennial Report, 1914; 1917; 1918. New York.

Frazier, John B., *The Navy Chaplain's Manual.* Washington, 1918.

Handbook of the National Catholic War Council. Washington, 1918.

Jewish Welfare Board, First Annual Report. 1919.

Johnson, Lorenzo Dow, *An Address to the Pastors and People of these United States on the Chaplaincy of the General Government Viewed in its Connection with Extending the Redeemer's Name in the World.* Washington, 1857.

Chaplains of the General Government. Washington, 1856.

King, Ernest J., *Our Navy at War.* Washington, 1944.

Minutes of the General Assembly, Presbyterian Church, U.S.A. Philadelphia, 1918.

Riddle, Truman P., *Preparing for Divine Services on Board Ships of the United States Navy.* USS *California,* 1935.

(editor), *Protestant Ritual, Adapted to Service Needs.* Los Angeles, 1935.

(editor), *Religion and Welfare in the United States Navy.* USS *California,* 1936.

Strobridge, G. E., *The Chaplains of the United States Navy, Their Unfair Treatment.* Stamford, Connecticut, circa 1902.

United States Catholic Chaplains in the World War. New York, 1924.

Year Books of the YMCA, 1910–1911, 1911–1912; 1912–1913. New York.

832159—49——18

Manuscript Sources

Chief of Chaplains Office, Bureau of Naval Personnel

Neyman, Clinton A., "A Manual for Chaplains of the United States Navy." Cavite, 1938.

"Some Brief Notes in regard to Chaplains in the United States Navy." Compiled from official sources. 1902.

Witherspoon, Maurice M., "Naval Autobiography."

Also various chaplains' files, certain topical files.

Captain Harold E. Cook, U.S.N. (Ret.) Diary, Chaplain Thomas Lambert.

W. W. Edel Collection

Edel, W. W., "Navy Chaplains from 1775 to 1917."

Hoes Notes.

Tribou Notes.

The Essex Institute, Salem, Massachusetts

Tucker, Samuel, *Journal.*

Harvard University, Cambridge, Massachusetts

Alumni Records, John S. Wallace File.

Library of Congress, Washington, D. C.

Danvers Historical Collection. Vol. 7.

John Paul Jones Collection.

Lippincott, H. H., "The Church Pennant."

The Massachusetts Historical Society, Boston, Massachusetts

Original Balch Commission.

National Archives: Office of Naval Records and Library, Washington, D. C.

Navy Records Secretary's Office.

Naval Records Collection:

Almy, John J., *Naval Discipline and Corporal Punishment.*

Book of Resignation, USS *Vincennes.*

"A Short History of the New York Navy Yard."

Appointments and Resignations.

"Chaplains and Professors," File "O"—1859.

Chaplains' File, "The Chaplain's Mistake."

Church Pennant File.

Circulars and General Orders.

Commandants' Letters.

"Corporal Punishment and Spirit Rations, Reports of Officers, 1850."

Felch File.

General Letter Book.

"Index for General Courts Martial and Courts of Inquiry 1799-1861."

Johnson, P. C., "Journal."

"Lake Champlain and White Hall Stations" (1813-4).

Letters from the Asiatic Squadron for 1877.

Letters from Captains.

Letters from Officers.

Letters to Officers, Ships of War.

Logs of: the *Congress;* the *Constitution;* the *Cumberland;* the *President;* the *Vincennes.*

Miscellaneous Letters.

Muster Rolls of: the *Boston;* the *Constitution;* the *United States.*

Navy Register, 1864.

Original Plans of the Frigate *United States.*

Preble, George H., "History of the Boston Navy Yard in Charlestown, Mass., from 1797 to 1875."

"Report of the Operation of First Provisional Regiment of U. S. Marines in Nicaragua," Register No. 2677.
Signal Book, 1813.
Uniform File.
New York Historical Society, New York
Chauncey Correspondence.
Original Hull Letter Book.
"Medical and Surgical Journal" of the frigate *Java* for 1816.
Navy Collection.
Pitts (Pitty?), James, "Journal."
Porter Correspondence.
The United States Naval Academy, Annapolis.
Chaplains Office, Clark File.
Army and Navy Department, YMCA, New York
Pond, B. C., "History of the Army and Navy Young Men's Christian Associations."

Periodicals

American Catholic Historical Researches.
The Army and Navy Chaplain (formerly *The Army Chaplain*).
Evans, Sydney Key, "Notes by a Navy Chaplain," April 1931.
King, Terence, S. J., "The First Priest in the U. S. Navy," April–May 1942.
Army and Navy Chronicle.
The Baptist.
Neyman, C. A., "My Parish," 15 April 1922.
The Century Church Bulletin.
Christian Advocate.
Edel, W. W., 27 June 1918.
The Christian Century.
Decker, J. W., "The Making of a Pacifist," 6 November 1924.
Schmalz, Alfred, "Blessed by Chaplain Ellis," 6 May 1936.
"The Chaplaincy Question," 16 January 1935.
"Get the Churches out of the Chaplaincy Business," 20 November 1924.
"Why This Continued Espionage?", 17 January 1924.
The Churchman
Stockton, C. H., "Religious Forces in the Naval Service," 20 July 1912.
The Colonization Herald.
The Friend of Temperance and Seamen, Honolulu.
"History of the Bethel Flag," 1 June 1846.
Journal of the Association of Military Surgeons.
Bradley, G. P., "Sketch of History of Medical Corps, Navy," May 1902.
The Leatherneck.
Drury, C. M., "First Marine Burial in Japan," July 1945.
The Lord's Day Leader.
Naval Magazine.
Pacific Northwest Quarterly.
Drury, C. M., "Early American Contacts with the Japanese," October 1945.
Polynesian.
Sailor's Magazine.
Shipmate.
"The Heart of the Academy," June 1942.

Southern Literary Messenger.
The Trident.
Thomas, W. N., "Sea-Going Clergymen," Christmas 1942.
United States Naval Institute Proceedings.
Alden, Carroll S., "Officers and Gentlemen in the Making," October 1935.
Drury, C. M., "Famous Chaplain Teachers of Midshipmen," May 1946.
Edel, W. W., "The Angel of the Church that is at Annapolis," August 1942.
Hunt, Livingston, "A Forgotten Commode—Thomas Macdonough," November 1931.
Kearney, Thomas A., "St. Peter's Chapel, Navy Yard, Mare Island," July 1932.
Luce, S. B., "Commodore Biddle's Visit to Japan in 1846," September 1905.
McCracken, A. R., "The Church Pennant," August 1930.
Morris, Charles, "The Autobiography of Commodore Charles Morris, U.S.N.," January 1880.
Riddle, T. P., "Enlisted Men's Families," January 1939.
Robert, S. C., "Extracurricula," April 1945.
Robert, S. C., "The Naval Academy Chapel," October 1945.
Skerrett, Robert G., "The Baptism of Ships," August 1909.
Sturdy, H. F., "The Founding of the Naval Academy," October 1935.

INDEX OF CHAPLAINS

GENERAL INDEX

 ☆ U. S. GOVERNMENT PRINTING OFFICE: 1949—832159

Made in the USA
Monee, IL
04 November 2026

67385390R00164